# Landscape Architecture as Storytelling

This book introduces students, practitioners, and laypeople to a comfortable approach to learning landscape architectural design free of design jargon and derived from their existing knowledge. A step-by-step process has readers consider their knowledge of language as metaphorically related to basic design and landscape design. Through information delivery and questioning processes, readers build on what they already know, their tacit understanding of language as applied to problem solving and storytelling. Everyone is a storyteller.

Taken one step at a time through a three-tiered analogy of language, basic design, and landscape design, readers learn the makeup and role of such design features as points, lines, planes, volumes and sequential volumetric spaces that make up their worlds. With that, in a sense, new world view, and numerous questions and examples, readers begin to see that they in fact daily read the environments in which they live, work, play, raise families, and grow old. Once they realize how they read their surroundings they are helped to recognize that they can build narratives into their surroundings. At that point the existence of authored landscape narratives finds readers understanding a design process that relies on the designer-as-author, landscape-as-text, and participant, user-as-reader. That process has the reader write a first- or second-person narrative, visually interpret the written narrative into a storyboard, and turn the storyboard into a final design, the physical makeup of which is read by those who participate in it.

**Bob Scarfo**, BLA, MLA, and PhD, is an Emeritus professor with Washington State University. He is a registered landscape architect in Washington and Massachusetts. *Landscape Architecture as Storytelling* is a dream come true, one he looks to share through invited workshops and talks.

# Landscape Architecture as Storytelling

## *Learning Design Through Analogy*

Bob Scarfo

Routledge
Taylor & Francis Group
NEW YORK AND LONDON

Cover image: Front cover watercolor by Ken Spiering, sculptor, artist;
Front cover layout by Robert Parton, designer, illustrator.

First published 2023
by Routledge
605 Third Avenue, New York, NY 10158

and by Routledge
4 Park Square, Milton Park, Abingdon, Oxon, OX14 4RN

*Routledge is an imprint of the Taylor & Francis Group, an informa business*



*Library of Congress Cataloging-in-Publication Data*
Names: Scarfo, Robert A. author.
Title: Landscape architecture as storytelling:
learning design through analogy / Bob Scarfo.
Description: New York, NY: Routledge, 2023. |
Includes bibliographical references and index. |
Identifiers: LCCN 2022016797 (print) | LCCN 2022016798 (ebook) |
ISBN 9781032261850 (hardback) | ISBN 9781032261867 (paperback) |
ISBN 9781003286981 (ebook)
Subjects: LCSH: Landscape design. | Landscape architecture.
Classification: LCC SB472.45 .S229 2023 (print) |
LCC SB472.45 (ebook) | DDC 712—dc23/eng/20220722
LC record available at https://lccn.loc.gov/2022016797
LC ebook record available at https://lccn.loc.gov/2022016798

ISBN: 9781032261850 (hbk)
ISBN: 9781032261867 (pbk)
ISBN: 9781003286981 (ebk)

DOI: 10.4324/9781003286981

Typeset in Adobe Garamond
by codeMantra

*Sylvia and Arthur Scarfo*

*For your pokes and prods; questioning attitudes; boundless patience; laughter; and all you endured preparing me for my journey.*

*Wish you were here to enjoy the results.*

# Contents

# Figures

# Preface

Unlike the old adage that for medicine to be considered good it must taste bad, learning design doesn't have to be a traumatic experience. Yet, ask any number of landscape architectural, architectural, or interior design students, or for that matter faculty, and you'll likely get nods of agreement that their undergraduate, and sometimes graduate, learning experiences were fraught with discomfort, angst, and self-doubt. Contrary to their reactions, learning to be a landscape architect can be both comfortable and fun if we follow the lead of many writers. I have always been pulled to the nature writers and those non-nature writers who include nature, almost as a character, in their writings. The number of fiction and non-fiction writers who realize the connections between language and landscape helps us learn about landscape architecture through language. Elizabeth Kryder-Reid notes "The study of landscapes strikes at the core of how we make sense of our world: how our lives are permeated by our physical environment, perhaps as fundamentally as by the language we speak" (1996, 228).

With the aid of metaphors, similes, and analogies we can move from what we know, through a series of small steps, to an understanding of what we don't know. This way of learning, as an exploration of the unknown, is comfortably experienced through a reliance of one's past experiences. However, there is no escaping the fact that learning something new means delving into the unknown. For most, there will always be moments of discomfort. There's bound to be a sense of uncertainty as you test new skills and knowledge in front of others, particularly a teacher and a roomful of your classmates. But, when we start with something we already know how to do, like reading, writing, and telling stories, learning design occurs through activities and skills most of us have practiced for years.

## MY STORY

As an undergraduate student, my introduction to landscape architectural design came as a shock. I felt I'd been taken to the end of a dock, thrown into the water, and as I flailed about trying desperately to keep my head above water, I was told to swim. I believed, at the time, that I had no prior experience in design problem solving, design process, and design principles. I was wrong. I had been solving design problems throughout my life, but wasn't made to realize that at the time. When I transferred into the landscape architectural design program at the University of Massachusetts, Amherst I brought with me a strong math, science background that had just been capped off with a year of Civil Engineering. Confronted with my first landscape design problem, the only question I knew to ask was, "So, what's the formula?" My landscape architecture instructor's answer brought my faith in the value of formulas face to face with the ambiguous world of design.

*Among the characteristics identifying a profession as a distinctive community: "within the areas of communal action there is a common language, which is understood only partially by outsiders."*
*(Goode, 1957, 194)*

I distinctly remember being awed by the design knowledge and aesthetic wisdom expressed by my instructor. After all, I didn't understand anything he was saying; therefore, I assumed it had to be important and worth my learning. His strongly stated response convinced me I was ignorant of how to produce a new landscape and unaware of the arduous journey about to unfold before me. Even so, I knew I was lucky to be in the Bachelor of Landscape Architecture program. I nodded in agreement to all that I heard in a desperate struggle to appear creative, intelligent, and technically competent.

Invigorated by my instructor's energy, I unrolled my trace paper expecting a design to flow from my pencil. Nothing flowed. Not only was my pencil not responding, I felt I had hit a mental block the size of a wall. I was confronted by the magnitude of the potential held on the surface of that blank piece of paper. I felt that my anxiety must be the state of mind that characterized what design was all about. I began to think design must be a Zen thing. At just the right moment of criss-crossing speculations – ***Slap***! – and suddenly I'd fathom design, and just as quickly set it to paper. I'd be one with the space to be designed and with the people intended to use it. Still, a part of me thought, "So, what's the formula?"

> *Lawyers [read designers] run an entire set of legal [read, design] institutions that are by definition supposed to be responsible and responsive to the public. Those institutions are not in fact accessible to the public, however, in part because the language used in the interaction that maintains them is impenetrable.*
> *(Philips, 1982, 183)*

Thus, as a first-year landscape architecture student I began, and have continued over the years, to search for a formula that would simplify and clarify design as a process and a practice. During 40 years as a teacher and a designer three insights influenced my search. First, in order to "do design" I must understand why people make the spaces that they do for themselves. How do personal and public spaces relate to those who use them? I began to recognize that design is, as John Simonds says, made of experiences and not things (1961, 225). Yet, in developing the kinds of spaces/places that provide those experiences with which people identify and find meaningful, or full of meaning (Norberg-Schultz, 1980), people strive to exhibit aspects of themselves in the form, character, and content of their surroundings. This sense of being integral with your surroundings is embodied in the realizations of Clare Cooper Marcus's *House as a Mirror of Self* (1995), Norberg-Schultz's *The Concept of Dwelling* (1985), J. B. Jackson's *The Ghosts at the Front Door* (1977), and Dennis Wood and Robert Beck's *Home Rules* (1994).

Later, as a teacher, my second insight related to my ability to communicate my understanding of design in ways that go beyond a one-way teacher-student (or client) transmission of knowledge. Students' active participation is paramount in their learning. I don't want to "talk at" students. I want them to be active participants in their own learning experiences, ideally while they contribute to improving their communities. For many of my students, the community was their classroom. Critical to activating students' learning is the ability to organize and deliver engaging exercises that trigger learning experiences.

Over the years my curiosity with design as practice, as teaching, and as learning merged into what I came to call The Analogy, or how to learn design based on how we learned to read and write. The metaphor of landscape as a medium of purposeful expression has served me well as both an educator and a design practitioner. I became more and more engrossed in the idea that the landscape expresses messages. As I pieced together The Analogy I realized that it isn't as much language that binds us but how we use it. As social products our landscapes, like letters, reports, notes, memos, and texts and tweets, evidence the relationships that bind us with each other, with our environments, and with each other's environments. Within the metaphor of reading landscapes my initial impression was that of the landscape as text. But, even then my first impression felt incomplete. What I realized was that I saw the landscape as expressing statements here and there, I had not yet realized that the landscape was more a continuous sequence of developing messages. The statements were part of a larger story, a narrative. Fortunately, the idea of landscape-as-text evolved into a realization of landscape-as-narrative, a progressive sequence of contextually related and developing messages, a story. The real power of a landscape lay in its sequential delivery of information or what we can call storytelling. The landscape-as-narrative realization convinced me of the value of The Analogy as a tool for learning, and as a foundation upon which to build my understanding of how to develop good landscape, architectural, and interior designs.

The third insight driving my exploration of design as process and product was my need to comprehend what constitutes good design. Landscape narrative had to be not just readable but also meaningful to the users as readers. For years I felt good design was an orchestration of our senses in the production of experiences. Good, and appropriate, experiences could not be just any collection of experiences, they had to be particular sequences with which people identified and in which they found meaning and to an extent maintained a sense of personal and social orientation. The form and content of the environments I designed had to be meaningful to people, it meant that they had to be an outgrowth of the people's cultural history and personal biography. My feelings in this matter have since been validated in the works of architects (Bonta, 1979; Greene, 1962; Norberg-Schulz; 1985; 1980), landscape architects (Church, 1983; Halprin and Burns, 1974; Lynch, 1972), planners (Aminzadeh et al., 2016; Rapoport, 1982; 1969), philosophers (Berleant, 2005; Brady, 2005; Light, 2005), sociologists, and social geographers (Barthes, 1982; Relph, 1987; Tuan, 1978; Williams, 1973; Wolff, 1981).

My understanding that design is deeply related to and draws upon people's backgrounds convinced me that most 18-year-olds, or university freshmen of any age, whether small-town born and bred or well-traveled, have all the necessary experiences to become a superior designer. Tapping into my students' backgrounds was the key to their learning to do landscape architectural design well. The Analogy gave me the means to tap into those backgrounds. No matter where they had grown up they had read landscapes: as pedestrians, bicyclists, skiers, and drivers of cars, motorcycles, and trucks. I could use The Analogy to awaken their taken-for-granted experiences and use their new awareness as a starting point to learning to be a designer. The Analogy prompted students to relate their personal experiences to design theory, process, principles, and products. With the help of The Analogy design was no longer learned in a vacuum. Learning design illuminated processes integral to their daily lives. Students became personally involved in what they were learning. Abstract ideas were tied through interpretation to the physical, measurable world. Design was learned through one's reflections on and interpretations of daily life and the routines that made up what they felt as personally meaningful.

Even with only fading memories of elementary English, the language metaphor allowed me, with a great sense of security, to unpack this thing called design. Letters of the alphabet, words, sentences, grammar, syntax, and narrative forms and strategies each provided additional metaphors that further illuminated design elements, rules, principles, and the products to which they gave rise. The parts of The Analogy, its typology, the message content and delivery systems that it illuminates, and the deeper symbolic significance of landscapes that it permits access to helped me understand, translate, interpret, and learn landscape architectural design. Through years of academic design projects and workshops with architecture, landscape architecture, land planning, nursing, and gerontology students, I now know that it can do the same for students and practitioners alike. When we develop designs that touch a person's heart we instill legacies, long-lasting products that leave a place capable of being as good, if not better, than it was. That is what The Analogy offers.

Bob Scarfo
December 1996

## A LATER INSIGHT INTO THE ANALOGY

I went back to school in 1987 to both focus on and get assistance in understanding the idea of landscape-as-text, designer-as-author, and user-as-reader. It was a period of realizations. First, the importance of realizing and admitting to what we are born into. Second, that landscapes are not encountered as individual, freeze-frame plan-view drawings but as eye-level experiences. I came to realize there is more certainty to life than just death and taxes. We are born into a family, a social group, and a language. Well, actually we are born into and grow up with a number of languages, each with its respective signs and symbols. There are sensory languages beyond our basic five, such as proprioception, vestibular, and haptic. However, our unpacking of The Analogy will stick with our five primary senses: visual, auditory, olfactory, tactile, and gustatory. The signs and symbols "read" by these senses individually and in combination are what we rely on for personal safety as we move about landscapes, buildings, and interiors. The various expressions of our sensory vocabularies give content and character to our environments and, in turn, our feelings in the places we move about, e.g., happiness, sorrow, sensuality, mystery, wonder, and more. *Landscape Architecture as Storytelling* explores, interprets, and applies these languages as they express themselves through landscape narratives. Landscapes were to be experienced at eye level as sequential progressions. But, the question then became, "How do I not only teach it but justify it?"

My learning about landscape narratives, the ways in which we read them, and the ways in which we, as designers, can author them as environmental designs, was accomplished in three stages: first, as explored through a three-tiered analogy; second, in the application of The Analogy to the authoring of a landscape narrative; and third, by interpreting the narrative through storyboarding before moving to a final design.

For me, the first stage interrelated a three-tiered analogy made up of language, basic design, and landscape design. I learned the language tier provides a simplified review of what makes up our language as it is written and spoken. Through that review I explored the structural aspects of language: letters of the alphabet and words, how through syntax and grammar they become sentences, then paragraphs, and finally a story or narrative. The second tier metaphorically relates the structural aspects of language to those found in basic design: point, line, plane, and their coming together as an actual or implied three-dimensional volume or sequence of volumes. Why basic design as the second tier? I found it to be a comfortable step in coming to understand connections between language and built environments. Also, because it is one of the fundamental building blocks of all forms of environmental design and, as you will see, the parts that make up language are easily related to those making up basic design. The Analogy*'s* third tier, again easily, and metaphorically, related basic design to language and was comprised of the primary features that make up landscape design: objects in the landscape, linear features, ground, vertical, and overhead planes, and finally the volumetric spaces making up entire landscapes. At this stage in my developing understanding of The Analogy I was asked a question by the late Harry Porter back in 1989 at the University of Virginia: "So, what do we do with The Analogy and its parts?" What do we do with this understanding of the fact that landscape features express, in a sense, statements?

Honestly, when Harry asked me that question, I didn't know. I was too excited about what I was learning at the time, to have gone any further. It took a few years to answer Harry's question. But he'd planted a seed that grew into what I came to learn about design and about my purpose as a professional landscape architect. It was worth the wait. The simple answer to Harry's question, as an interpretation of the three tiers, is that landscape designs express narratives read by their users. We read our environments. Stated another way, both laypeople and professionals author environments. My home has a public face and a private face. Features tell visitors how to safely get to the front door or through the side yard to the back where the deck party is taking place.

So, as I began to realize that people can, and do, author the spaces in which they relax, play, work, study, raise families, and retire, I also began to realize that as a professional designer I could purposefully author narratives that then became my designs. My students helped me a lot at this point. In my efforts to answers students' questions about how to identify what should make up their designs I realized the role of themes (e.g., happiness, comfort, mystery, spirituality) and the qualities of coherency, consistency, and continuity that a thematic approach brings to the design process. The design process was no longer a selection of objects and spaces "played with" until they looked good. Once a theme was identified, the design process became a more logical and sequential series of steps leading to a final design. But, where does a design's theme come from? I found an answer to that in John Simonds realization that we design experiences, not spaces, places, and things (Simonds, 1961, 225).

Since 1989, I frequently found myself thanking Harry Porter for his question. The answer to it here provides an understanding of landscape narrative as developed by a designer-as-author giving rise to a landscape-as-text. That text is not just any text. It has a narrative form applied to landscape design and a theme that guides the selection of landscape features, their materials, the materials' finishes, and their sequencing as they convey a narrative. With some unexpected help, answering Harry's question, "So, what do we do with The Analogy and its parts?" took much less time than my previous efforts to respond to his question.

Fortunately, the twists and turns of my life found me first as a student of and later good friends with the person who would become my mentor, John Conron. John was an English professor at Clark University in Worcester, MA. While working on my PhD in Social Geography, I took every course John taught. Through a close reading of literature, poetry, and paintings John helped me learn more about landscape architecture than anyone else. Through John I learned to see narratives. In that learning process I developed what is the third stage of this book, a way to apply storytelling to the development of landscape designs (or for that matter architectural and interior designs, too) through the authoring of narratives.

Again, in three steps – honest I'm not married to the number three, it just seems to happen that way – we follow an application of The Analogy to the development of the design of a healing garden at a hospital in Spokane, Washington: the authoring of a narrative, the interpretation of that narrative into a storyboard, and the interpretation of the storyboard into a final design. This sequence

| | | | | | | |
|---|---|---|---|---|---|---|
| | **Tier I – Language** | | | | | |
| **Reading And Writing** | **Phonemes Consonants Vowels** | **Letters of the alphabet** | **Sentences** | **Paragraphs** | **Volume(s)** | **Stories** |
| | **Tier II – Basic Design** | | | | | |
| **Basic Design** | **Art Materials and their Qualities** | **Points** | **Lines** | **Planes** | **Volumes** | **Sequence of Volumes** |
| | **Tier III – Landscape Design** | | | | | |
| **Landscape Design** | **Landscape Materials & Qualities** | **Objects in space mass/void** | **Paths edges & boundaries** | **Ground walls overheads** | **Volumetric spaces** | **Sequential volumes** |

Figure 0a.1 Three-tiered analogy interrelating language, basic design, and landscape design.

was reinforced with the outcomes of a number of week-long workshops between 2001 and 2004 with landscape architecture students at Cal Poly Pomona.

Will what you get here be complete, nicely packaged? No. There remain plenty of opportunities for more study, more research. As far as I know only a few design firms employ narration and storyboarding. My hope is that a deeper, more complete understanding of what *Landscape Architecture as Storytelling* can bring to learning landscape architectural design will come from undergraduate senior projects, master's students' theses, and doctoral students' dissertation research. And should you want help defining your questions, identifying how to expand upon The Analogy and the narrative-storyboard- design process, I look forward to hearing from you and working together in a workshop or seminar to build on what you already know. The journey is worth the effort.

Bob Scarfo
October 2019

# Acknowledgments

First and foremost, I am grateful for having met John Conron, an English professor at Clark University in Worcester, Massachusetts. John's passion to illuminate and follow the influences of the picturesque as a landscape genre caused me to learn and see my profession anew. I considered John my mentor and still use his questioning attitude as my guide. John was passionate about the American landscape and delighted in that of all the things that might drop into his lap it would be a landscape architect. He is the lynchpin around which my efforts to understand narrative in its many forms continue to evolve.

Second among those who challenged me is the late Harry Porter. At the conclusion of a talk I gave, he asked one question. "Now that you've given me the three-tiered analogy, what do I do with it?" That's what I heard. What he actually asked was, "What's the moral of the story?" Honestly, at the time, I didn't know. His questioning of what was then a barely developed Analogy was like a pebble in my shoe. Efforts to answer his question prepared me for what John Conron brought into my life.

My gratitude for Mom and Dad is never-ending. They bear so much of the responsibility for who I am, especially in regard to my love of reading. I can honestly say that as a youngster I didn't appreciate Sylvia and Arthur each night when they sat with me for an hour and had me read out loud. I hated reading. Yet they stuck with me. Through only the kinds of complaints a child can produce they provided the kind of love only parents can provide. They put up with me. And they too, like Harry Porter, helped prepare me to recognize, maybe not understand, what I needed to learn, who I needed to draw into my life, and what questions I needed to ask.

And then there is Angelina Alessandroni, Nana, who instilled in me a curiosity that manifest itself as a love of learning for the sake of learning. One day I found Nana sitting on the front lawn with three of my high school buddies. They were talking sex, drugs, and rock and roll. I later asked her why. Her casual response was, "I may not agree with what I'm hearing but I have to know about these things."

Thank you Richard Forster for introducing me to the C. G. Jung Foundation in Toronto and all the magnificent speakers. Learning about Jung's personality types helped me begin to realize the degree to which the depth and character of a landscape design must reflect the depth of the personalities of those who experience it. And thank you Richard for introducing me to the late Marion Woodman. I am grateful for Marion agreeing to take me on as an analysand for the year before leaving Canada to start my PhD. Not only do I look back at that year as the most psychically fulfilling and comfortable but also as the most enriching as to how to improve my teaching and working with public clients.

Thank you Ted Baker, posthumously. Ted was the most personable, huggable curmudgeon I've ever known. His anguish over what drove good, no great, education and what was called for by landscape architects in public practice drove many hours of in-person and on-the-phone discussions. So enriching was Ted's need to live up to his expectations in others.

For Ken Spiering I share the gratitude of a good friend that goes well beyond friendship. What we would come to share over the years started with an introductory bag lunch and stroll in Spokane's Riverfront Park. I was in awe of this well-recognized painter, sculptor. As we walked and talked our enjoyment with learning from each other became obvious. An awareness, and slowly an understanding, of the artist's eye is what I found in Ken. Ken, whose admission surprised me, talked of his understanding of sculptural form but not at the scale of a landscape. Since that day, that varied scale of art, form, character, and message content has filled numerous coffee shop discussions, our creative process with Providence Sacred Heart Hospital's healing garden, and most recently discussions regarding the cover graphic of *Landscape Architecture as Storytelling.* I look forward to more coffee shop discussions wondering where they will take us. Thank you Ken.

I owe a debt of gratitude to those designers through history who produced clearly legible environments that excited and comforted people. I owe the same debt of gratitude to those who didn't. Both provided me with learning moments regarding people's living with the products which designers provide. In generating questions in me about people-landscape interactions I was able to introduce my students to discussions that piqued their curiosity.

To the many many students I interacted with over 40 years, thank you for your excitement, your challenges, and your creativity. Thank you for the questions I never thought to ask. You helped form me not only as a person and a designer but also as an educator. You helped me develop an approach to learning I am now helping to grow in an alternative high school in Spokane, Washington. That approach is summed up by Lisa Mattson, the Principal of On Track Academy, as a school without walls where the community is the classroom.

Much appreciation to Lisa Drost, great friend and informal editor, Deb Finck, intellectual and compassionate explorer, and Cristina Olsen, editor of an early version of *Landscape as Storytelling.* I undoubtedly failed to recognize a number of people who influenced my thinking and practices as an educator and designer. For that I apologize. Please know I wouldn't be the person or designer I am without your input.

And finally, Kate Schell, Senior Editor, Planning, Landscape, and Urban Design at Routledge and my Editorial Assistant (and publishing Guardian Angel) Megha Patel in Planning, Landscape, and Urban Design at Routledge/ Taylor & Francis Group. Thank you for believing in an idea that started taking shape in 1986.

Thank you Terri Scott for, in no uncertain terms in 2019, telling me to get off my butt and finish the book.

# Introduction

Interesting how my life journey had me cross paths with so many thinkers who saw the connections between language and landscape. They were like farmers sowing seeds, ideas, that would germinate, take root, grow, and cause me to learn so much about my profession. As Thoreau notes, this learning doesn't take place over night. "If a man is rich and strong anywhere it must be on his native soil. Here I have been these forty years of learning the language of these fields that I may better express myself" (Thoreau, 1973, xvi).

In learning how to see more clearly, I learned how to express myself more clearly. Loren Eiseley (1957; 1978), probably my favorite environmental observer and teacher, time and again helped me refine what I saw and how I thought about it. Eiseley also helped me realize I didn't have to pigeonhole myself, I didn't have to nail down who or what kind of landscape architect I was.

> *Some men are daylight readers, who pursue the ambiguous wording of clouds or the individual letter shapes of wandering birds. Some, like myself, are librarians of the night, whose ephemeral documents consist of root-inscribed bones or whatever rustles in thickets upon solitary walks.*
>
> *(Eiseley, 1978, 26)*

While Thoreau and Eiseley directed me in seeing details in the world around me, John McPhee made me realize the extent to which landscapes express themselves and how, as I developed designs, I could use their narrative capabilities. "Rocks are records of events that took place at the time they formed. They are books. They have a different vocabulary, a different alphabet, but you learn how to read them" (McPhee, 2000, 164).

Landscape-as-text, designer-as-author, and user-as-reader comprise the fundamental premise of a three-tiered analogy that interrelates language, basic design, and landscape design (see Figure 0b.1). From this premise The Analogy opens us to ways to understand landscape design as processes of production, use, and the sequential, additive experiences of dwelling and place making. The Analogy's approach to learning design through an understanding of the production of environments for others rests on how most of us learned to communicate through speech, reading, writing, and particularly the application of those media to storytelling. By relating to what most of us have as a general knowledge of the building and communication of narratives, the all too often unfamiliar world of design language and thinking is explored here from a familiar and comfortable starting point. The following ten chapters explore reading and writing, basic design, and landscape design as each informs the other and each relates to the social production of space and the responsibilities of those who take on the responsibility of designing such spaces. While The Analogy is directed at landscape architectural design, it is easily applied to interior design, architecture, urban design, and, to an extent, land planning. That broader application of The Analogy to include the other professions is why you'll find I often refer to environmental design rather than just landscape architecture, although you must allow me a few digressions into my own profession and professional ego in my frequent references to landscape architecture.

DOI: 10.4324/9781003286981-1

| | | | | | | |
|---|---|---|---|---|---|---|
| | **Tier I – Language** | | | | | |
| **Reading And Writing** | Phonemes Consonants Vowels | Letters of the alphabet | Sentences | Paragraphs | Volume(s) | Stories |
| | **Tier II – Basic Design** | | | | | |
| **Basic Design** | Art Materials and their Qualities | Points | Lines | Planes | Volumes | Sequence of Volumes |
| | **Tier III – Landscape Design** | | | | | |
| **Landscape Design** | Landscape Materials & Qualities | Objects in space mass/void | Paths edges & boundaries | Ground walls overheads | Volumetric spaces | Sequential volumes |

Figure 0b.1 Three-tiered analogy interrelating language, basic design, and landscape design.

*From the stage to the city is more of a leap in scale than symbolism. Built environment has the capacity to transmit messages which range from the conceptual and verbal to the abstract and inchoate. These intimations do not consist of objective information but rather reflect an internal mental assembly of symbols and images. The message can be 'read' and understood because the mind is versed in the language.*
*(Smith, 1977, 16)*

Chapter 1 introduces the three-tiered analogy as a stepping-stone process to learning landscape architectural design. The Analogy is fundamentally an approach to doing design that sees the designer-as-author, the landscape-as-text, and the participant user-as-reader.

My arrival at what I came to call The Analogy, then the years of peeling back the layers of potential embodied in The Analogy was stimulated by three needs. First, my personal need to know a design process that made sense and could be replicated with equally good results each time. Second, my need to be able to communicate that understanding in an equally sensible and articulate manner. And third, my need to be able to interpret that understanding into learning experiences for my students. Students can be tricky. In my very first studio class, on my very first day of teaching I talked about "space" only to realize three of the students thought I was talking about outer space, not spaces in the landscape. From that moment on, communicating well was of paramount importance to me and likely set the stage for *Landscape Architecture as Storytelling: Learning Design Through Analogy*.

Why was learning design so confusing? My students helped me realize that the confusion rested in instructors and then students mixing the language, logic, and methods of three areas (or domains) of knowledge: the physical world of things and the physical sciences, the social world of people and the social sciences, and the expressive world of arts and aesthetics. Our somewhat sloppy way of thinking and expressing our thoughts is quite common. Start to closely read and listen to the news; often the human body or an ecological issue is discussed in terms of mechanics and machines. The idea that goes with that outlook is that "It can be tinkered with and fixed. Hey, it's a machine."

That confusion, as we come to explore it here, is compounded by the fact that so much of design is based on assumptions. When an instructor and a student (or client) discuss a design, both assume a mutually agreed upon understanding of what is being discussed. Both, to an extent, assume a mutually agreed upon understanding of the terminology being used. And they both assume they see and agree upon what is contained in the designs' images, when so often no such agreement exists. Now imagine that multiplied by a studio full of students each with his or her own assumptions.

The three-tiered analogy presented in Chapter 1 is the foundation upon which the subsequent chapters continue to clarify how design may be expressed in ways that provide for mutually agreed upon understandings. While Chapter 1 provides a background rationale for The Analogy, Chapter 2 introduces a typology along with rules that guide the application of the typology's building blocks in the formulation of landscape texts. Our alphabet, combined in ways that become words, provides the first tier of the three-tiered analogy. Understanding the typology's second and third tiers, of basic design and landscape design respectively (Figure 0b.1), rest on and are derived from the language tier. Language is the medium applied, through rules such as grammar and syntax, to basic design principles and eventually to the authoring of landscape narratives which express themselves on three interrelated levels of interpretation: practical, socio-cultural, and deeply rooted archetypes or visceral meanings held by all humans.

What we will find is that even more interesting than picking apart the three-tiered analogy is the opportunity to apply the typology to the development of a landscape narrative as a particular story set in a particular place and guided by a particular theme. Storytelling, as a form of narrative, is sequential and additive. Words become sentences, sentences become paragraphs, and paragraphs become a story.

The idea of story, or narrative, helps us understand the role of basic design and then landscape design in a landscape participant's emerging experiences. The reader of landscapes like the reader of books gradually identifies a theme and then the character and content of the story or place. They acquire their first impression. As they continue along their route of travel, let's say as they walk into a calming environment of poetic sensations, or into an edgy, mysterious place of anticipation, they pick up on the character and content of those landscape settings by reading their surroundings. So, you begin to see how the sequential character of reading, word after word, is a lot like what you do as you are walking, bicycling, or driving a car or, for that matter, water skiing, rock climbing, or running a marathon. Our activities are primarily linear in nature, moving through space and time. Moment by moment you acquire more and more information. This movement, or what designers usually refer to as "circulation," is key to a designer authoring the dialectic between user and the objects and spaces making up their landscape.

Another benefit to using language as the basis of The Analogy is how its parts (letters of the alphabet give rise to words, words give rise to sentences, and so on) and its structure (a theme responding to the rules of grammar and syntax) can be related to the parts and principles of basic design and landscape architectural design. With this understanding The Analogy introduces the building blocks of the landscape-as-text metaphor.

*Vol-ume, noun*
*1. A book forming a part of a work or series;*
*2. The amount of space that a substance or object occupies, or that is enclosed within a container especially when great.*
***(Oxford Languages https://languages.oup.com/google-dictionary-en/)***

Derived from Tier I, grammar and syntax guide our understanding of the rules that give basic design (Wong, 1972) and landscape design (Simonds, 1961; 2006) their identifiable forms and respective character and content. The building blocks of language are readily seen in the related building blocks of basic design and then in landscape design. Simply put, a word is like a point, a sentence is like a line, and paragraphs are like two-dimensional planes that give rise to books (stories) as three-dimensional volumes. Catch the double meaning?

To guide our understanding of The Analogy's typology, the word-point-object metaphor as part of the language-basic design-landscape design metaphor is introduced in Chapter 2, Typology. Like a point in basic design, an individual spoken or written word in order to express a meaningful message requires a context or setting in which it is located. To move from a collection of individual messages to that continuum we call narrative calls for the planned, sequential organization of the other building blocks to develop a continuity across a site. As designers, once we come to know the probable users of our design, the reasons that bring them to our design, and how they likely expect to interact with the landscape, we can begin to massage the typology's building blocks so they gain a cohesive and contextual character that provides for the sequencing of messages that make sense under a given theme.

> ***In her description of the** Room in Context, **Katherine Benzel notes that while an individual room** "can be seen and 'read' apart from its context…this visual reading makes the interior room appear removed from reality." **She goes on to say that** "we are able to see and read a word alone without its context, but what counts is the meaning a particular word has in the structure of the work through its use in relation to the surrounding words. Surrounding words joined together in patterns of construction support and unfold the meaning of a text". (Benzel, 1998, 15–16)*

Consider coming upon a small, public courtyard. You see it about 50–60 yards up ahead and readily identify it as a public space. The objects defining the approach and entryway, the objects' placement, and their organization tell you this is a place where you can relax if you chose. In that you identified a courtyard, you then consider the sequential expressions of a narrative that speak to your courtyard-related expectations. This is no minor thing. Focusing on a courtyard theme calls into play courtyard-related shapes, spaces, activity settings, and not those of any number of other non-courtyard activity settings: softball, fishing pier, outdoor BBQ.

Now let's say what you identified as a public courtyard is located in a part of Boston, near the base of historic Beacon Hill. Or, maybe in a more modern setting, e.g., Dubai's artificially produced Palm Island, Palm Jumeirah. Draped in their respective social and cultural histories, personal biographies, and aesthetic appreciations the physical, structuring, aspects of each narrative readily bring into play a uniquely distinctive, yet cultur-

Figure 0b.2 1890 Acorn Street, Beacon Hill Boston, MA. Courtesy of Historic New England.

Figure 0b.3 2020 Acorn Street, Beacon Hill, Boston, MA. Item ID: 501710236. Courtesy Sean Pavone.

Figure 0b.4 The walk at JBR, Dubai Marina, Dubai. Courtesy, Fabio Achilli, Photographer (Sept. 3, 2013).

ally coherent character to each environment. As we spend time in one or the other location, the place becomes meaning-full. Through time, we come to identify with and find meaning in the place's features and activities. Through our repetitive use, the character of the place comes to provide us with a sense of being in a place, or being oriented in time and in our respective lives (Norberg-Schulz, 1980; 1985). As we look at transforming spaces into places, objects become so much more than mere things to be moved around. They take on a purpose. One-, two-, and three-dimensional objects convey practical messages: "come here," "go around," or "look at me." Or as adjectives and adverbs they enhance or modify the character of the fundamental messages: "come here slowly," "stay and enjoy the aroma and color," "enjoy going over there and being happy."

Unpacking The Analogy's typology as it interrelates language, basic design, and landscape design begins in Chapter 2 with the letter-of-the-alphabet, basic-design-point, and object-in-the-landscape analogy. This cross-characterization of word-point-landscape object helps us understand the expressive qualities of landscape places, spaces, and things. With that understanding we begin to see how each tier's introduction of guidelines (syntax/grammar, basic design principles, and landscape design concepts) structures individual messages into developmentally coherent sequences that evolve into a complete narrative. In this way design, like storytelling, evolves into a thematically driven sequence of thematically related messages, and not into a haphazard, hodgepodge of expressions.

Chapter 2 gets us thinking about built designs as more than physical environments into which we enter and carry out certain behaviors. Through the typology we begin to see designed and constructed spaces as interactive expressions, as a dialectic between a physical environment made up of messages and its participant readers. In vertically interrelating each portion of The Analogy, making connections between tiers (Figure 0b.5), we begin to explore the parts of each tier as they inform and form each other. A more detailed exploration of the typology starts in Chapter 3 with words as basic design points and objects in the landscape. That beginning is expanded in Chapter 4, Lines; Chapter 5, Planes; and finally Chapter 6, Volumes and Volumetric Spaces.

For an example of a word-point-object analogy (Figure 0b.6), take a blank piece of paper, or computer screen. Look at it. What do your eyes do? They wander, they roam. There is nothing to settle on. Place a single letter of the alphabet or single word anywhere on the empty space and all sorts of wonderful things begin to happen. Suddenly there are aspects of here and there, me and it, a sense of orientation where previously there was none; the designer-as-author's communication with the user-as-reader has begun along with a hint of the narrative's theme and character. As the objects and their attributes become more meaning-full, we begin to recognize that they are as much a part of our purpose for being there as

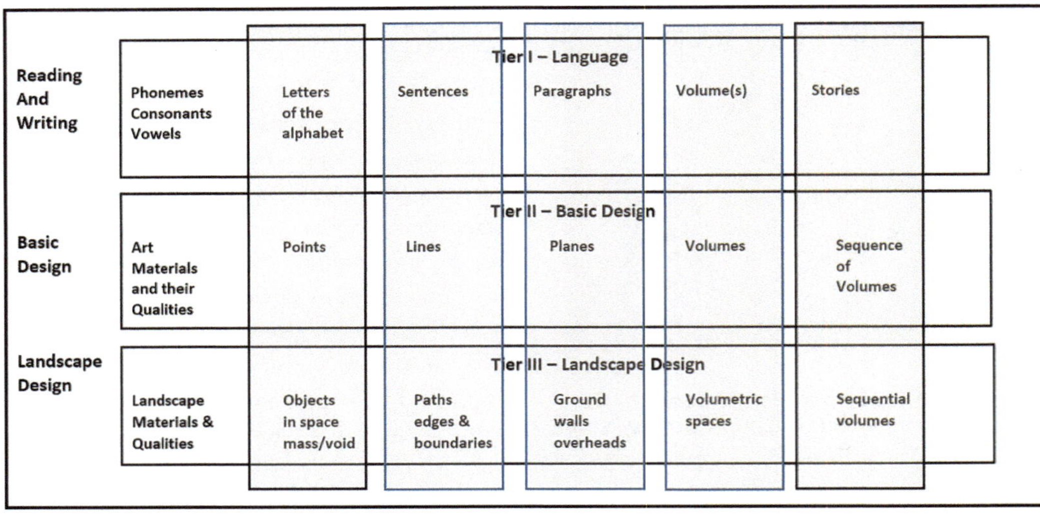


Figure 0b.5 Analogy's vertical relationships linking language, basic design, and landscape design.

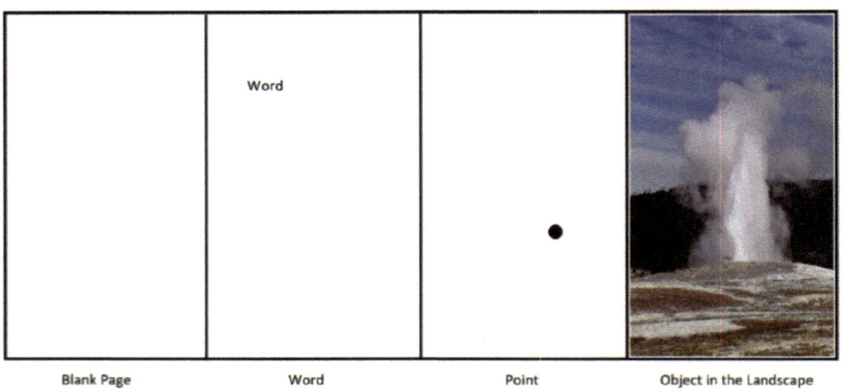


Figure 0b.6 Visualizing the three-tiered analogy.

we are for their existence. Selected relative to a design's intent and well placed across our field of vision, the collection of individual objects begins to meet our expectations. The designer has started off well.

What we typically consider as things in space or points in basic design, I like to think of as gatherers and conveyors of relationships: you-me, here-there, I-Thou, up-down, left-right. When we overlay the progression of a space's messages with rules pertaining to grammar, syntax, and idiom, we are guided in the production of landscape narratives. In a similar fashion the production of two- and three-dimensional design is guided by related rules and principles (Wong, 1972; 1977) as is landscape architectural design (Alexander, 1977; 1979; Church, 1983; Simonds, 1961; Treib, 1991). The fact that we move from point to point throughout any given day calls for those points to be connected with linear features such as paths, hallways, sidewalks, and roads. Chapter 4 delves into these linear features and how they enable a coherent continuity across a narrative, basic design, and landscape.

In that initial act of placing (locating) a word (object) on a blank piece of paper, empty computer screen, or barren site, we initiate and introduce a degree of expectation in the yet to appear construct of letters and words that will follow. The linear construct of a line, a sentence, or path, or walkway is guided by the rules and normative practices associated with the media you are using –

writing, music, mathematics, oil painting – and the context or setting in which they are placed. An exploration of our ability to communicate through various media helps us recognize that our ability to produce landscape narratives involves both implicit and explicit material forms, such as letters of the alphabet that are explicit words or those that are not words but in how they sound imply words: b, be, bee; c, see, sea; t, tee, tea. Becoming a good designer also requires an ability to use pauses. Spaces in landscape design do as much to hold objects together as the objects do to define the spaces. Points of pause along a path or walkway allow us time to see where we are and where we are going – to become reoriented. Points of pause allow us to take a breather before continuing along a line to the next point.

Before getting into the narrative contained in or brought to a landscape through an authoring (design) process we explore a message's coherent development across space: its continuity across messages. You have probably begun to realize that the development of a landscape architectural design is not a willy-nilly process of randomly selecting objects, their materials and the materials' finishes, their placement, and the spaces they define because you like them. Chapter 4 addresses bringing together words to make sentences as coherent strings of message segments or statements. In basic design, as a point moves through space it inscribes a line. In a landscape a line connects points. Speaking or writing a sentence is not a process of random word selection. That would be gibberish. The selection of a word as the subject of a sentence calls into play a selection process that simultaneously identifies and limits the words from which you can select subsequent words. To introduce boating, sailing, or marina design immediately eliminates words and phrases associated with urban plazas and parks, tennis courts, or elementary schools. Similarly, selecting a point or landscape feature of a particular visual character asks that it be followed in a basic design by lines that reflect that character. In a landscape a linear feature typically builds upon the character of the point from which it started or eventually ends. There is a continuity between the points and the lines and their surroundings, those areas to which the linear features begin to give shape.

Planes, the subject of Chapter 5, explore the relationships between sentences as they become paragraphs, lines as they become planes, and linear landscape features making up what we know as ground planes, vertical planes or walls, and overhead planes as ceiling or sky. Metaphorically, in much the same way as lines move through space to inscribe a plane, sentences multiply across a page to become paragraphs, which are also two dimensional. In landscapes such two-dimensional message delivery features do as much to express statements as direct the sequential arrangement of those statements which people read in their surroundings. Much of our understanding of how these features are organized to provide for coherent stories is drawn from a blend of oral history (Rubin, 1995), the social production of space (Giddens, 1984; Lefebvre, 1994), and the deeper meanings and beliefs that bind people as a culture (Campbell, 1974; Marx, 1981; Rapoport, 1982).

A portion of Chapter 5 is given to making clear distinctions too often ignored by designers. Planes are two-dimensional areas. Yet often architects, landscape architects, and interior designers speak of them as being three dimensional. With years of training and more years of practice, designers look at a two-dimensional plan of an interior, an architectural structure, and a landscape and in their mind's eye see three-dimensional environments. All too often the assumption is that their clients have the same ability, which, more often than not, they don't. The confusion is further complicated by two factors. One factor is when an architectural or landscape ground plane or floor plan includes changes in elevation, a problem not an issue in the two-dimensional representations in basic designs and paintings. The other factor is that all too often to understand landscape, architectural, and interior plans designers unconsciously consider movement (Aminzadeh et al., 2016; Bacon, 1974). The fourth dimension of time is introduced but again not often discussed to where it is mutually understood and agreed upon. However, an excellent exploration of time and movement as factors in a designed sequence of views along a pathway is provided by Aminzadeh et al.'s application of scenario writing as the "relationship between time-related experiences while moving through a landscape and the structure of scenario making in cinema and dramatic narratives" (Aminzadeh et al., 2016, 175). Throughout Chapter 5 these implicit and explicit qualities are looked at in literature, basic design and paintings, and environmental design. As the three primary planes – ground planes, vertical planes, and overhead planes – are brought together, the idea of volumes or spatial rooms begins to take shape.

Chapter 6 expands the development of basic design planes in their coming together as three-dimensional volumes. In language our sentences produce paragraphs. As paragraphs evolve from one to the next and the next again they become a book or a volume. Landscape painters use ground planes, vertical planes, and overhead planes to develop visual places or sequences of places as landscape rooms (Conron, 2000, 92–102). In landscape designs such rooms are designed to accommodate activities and direct people's movement from one room to the next. This movement throughout a narrative from place to place brings with it two realizations. First, planes are employed to develop spaces, more precisely volumetric spaces. Second, the volumetric spaces in a planned or authored sequence give us a foreground, middle ground, and background progression. To better understand these relationships, Chapter 6 deals with ground planes, vertical planes, and overhead planes separately and in detail. As they come together we begin to consider the role of movement in people's experiences. Our movement from where we were to where we are and then to where we are going traveling through a planned design sequence is what Cullen calls serial vision (Cullen, 1961, 17). Bacon calls it an unfolding or a sequence of sensations (Bacon, 1974, 53–57). Whether serial vision or an unfolding, we have the culmination of our piecing together of The Analogy's typology and the beginning of its application to the designing of landscapes and authoring of landscape narratives.

As a preface to our delving into landscape as narrative, Chapter 7 opens on a cautionary note. In the areas of art history and philosophy of art there exists a lot of information regarding narrative in general and landscape narrative in particular. However, many of the studies on people reading their environments ignore the existence of landscape all together. Architecture dominates. Fortunately, at a more pedestrian level we find analyses of narrative message content in the ever-changing dynamics of the Mississippi River (Twain, 1984); in homes, neighborhoods, and urban districts (Alexander, 1977; 1979); in the philosophical musings on Paris and the Eiffel Tower (Barthes, 1982); and in the Alps and Grecian landscapes (Durrell, 1969; Lawrence, 2007).

Chapter 7 draws on The Analogy to start with the smallest part of the language metaphor and build up through words to sentences, to paragraphs, and finally to full narratives. How The Analogy's parts are applied to the authoring of a landscape narrative moves from the smallest characteristics of the objects and spaces used up to the larger qualities that enable people to identify what a landscape is expressing. The smaller message mediums are the materials which combine to make landscape objects, the surface finishes given those objects, and the kinds of plants used. From materials' qualities a narrative grows in the story it is telling through the objects selected, their shapes and forms, and the spaces they define. Finally, the sequential, additive nature of the spaces allows us to comprehend the unfolding of volumetric spaces through a site that calls our attention to the narrative's introduction, body, and conclusion.

The sequential aspect of a narrative as it winds its way across a site brings an additive quality to the messages. One message is added to by a second one and so on. What those messages are and how they are perceived though an orchestration of our five senses are explored with the help of such literary pieces as Jewett's *Country of the Pointed* (1956) and Poe's *The Pit and the Pendulum* (1904). There are three levels to the messages conveyed by a landscape narrative. First is the practical, "Can I get there from here? Will I be safe?" Second are the aspects of place and heritage: Colonial New England versus Art Deco Miami versus the Romanesque as found throughout Europe. The third level is that of archetypes or what Carl Jung called our collective unconscious: Hero, Trickster, Guardian.

The component parts of narrative are explored in the latter part of Chapter 7 through a balanced approach between the more philosophical and literary fundamentals of landscape and the everyday aspects of Philadelphia's Mill Creek neighborhood (Spirn, 1998) and Wood and Beck's analysis of the social etiquette two parents work to teach their kids regarding their living room and front porch (1994). In reviewing the meaning context of the Mill Creek neighborhood and a residential living room, we deal with the place making and meaning content of places we have all encountered. In doing so we are able to step back and look at not just what the narrative contents are but how they came about and are perpetuated. Chapter 7 gathers together the component parts of landscape as narrative and brings us to where we are prepared to apply those parts in a three-step process to author/design a landscape narrative.

The narrative-storyboard-design process in Chapters 8 and 9 is introduced by much of what Kevin Lynch outlines as an approach to doing design. Lynch's thinking is clarified as you draft your narrative and discuss it with your peers. Further clarification occurs during discussions over how well the narrative and the storyboard complement each other. Authoring your landscape narrative starts with your understanding of who the probable users of your design are, their values, their reasons for coming to and being in your design, and their expectations. With that, your eventual landscape narrative is built upon people's experiences. The spaces, places, and things that make up your design follow naturally but are secondary. In doing so, Lynch's "three fundamental patterns of location in space and time – the pattern of activity, the pattern of circulation, and the physical forms" (Lynch, 1971, 246, 277) – are given their needed consideration from the beginning of your design process. In that authoring a landscape narrative focuses on people's experiences, not objects and spaces, the design process from its inception avoids Lynch's concern for designers separating out individual "aspects," as he calls them, that could "lead to the neglect of larger relations" and "cause serious difficulties if maintained as a complete design process" (277).

Chapter 8, Authoring a Landscape Narrative, explores the expressive content imbedded in and related to landscape narratives. Supported by the works of designers (Church, 1983; Halprin and Burns, 1974; Lynch, 1972), geographers (Barthes, 1982; Relph, 1987; Tuan, 1978; Williams, 1973; 1983), anthropologists (Rapoport, 1969; 1982), sociologists (Larson, 1977; Wolff, 1981), and critical theorists (Giddens, 1984; Habermas, 1979), three levels of message content are seen to exist in built and natural forms. The first level, that of functional messaging, is the most explicit and immediate to daily life. Functional messages support our ongoing concern for personal safety and comfort: "Can I get there from here?" "How exposed will I be?" "Can I find my way back?" The second level of expression relates to aspects of social culture and heritage that allow us to maintain a sense of orientation in regard to who and where we are. These narratives of period and place, or space and time, are more subtle than those of personal safety and comfort, and yet longer lasting in their binding a people with a place's history with the present and the future. The third level of narrative is the most subtle and yet the most enduring and long lasting. These are the expressions of deep symbolism, deeply rooted values found in what Jung calls the collective unconscious (Stors, 1983). The collective unconscious "is a term coined by psychoanalyst Carl Jung and refers to the unconscious mind shared by all of humanity. It is composed of archetypes, which are simple [yet very powerful] representations of universal figures and relationships" (GoodTheraphy, 2019). "While common to all mankind…each person experiences them his own way" (Stevens, 1982, 39). If the Jungian concept is a little too unconventional for some of you, then consider the deep emotional convictions inspired by national and religious symbols, or cultural beliefs such as "la familia" the family or clan, nationalism, or religion. Chapter 8 explores these levels of communication in terms of the kinds of messages they express: function; socio-cultural heritage; and visceral, heart-pounding archetypes in the unconscious.

Combining statements into coherent message sequences moves the concept of landscape-as-text into the realm of landscape-as-narrative. Statements are singular expressions. "Stop." "Stop now." "Have a seat." "Look to your left." These individual statements lack context, purpose, anticipated outcome. They are situated and likely stationary in that as individual expressions they are tied to particular objects and spaces. But this is not how we exist and live in our environments. Our movement through an environment is not a collection of disconnected, separate moments. We do not beam in and beam out of locations along a walkway or road. On the contrary, our reading of a narrative is a developmental continuum made up of transitions across space and time, whether that space and time is a page or painting or a landscape.

With the help of a case study, you are introduced to students' responses to a request to design a healing garden for Providence Sacred Heart Hospital in Spokane, Washington, then asked to author a short narrative of your own. Once reviewed with friends or work cohorts, the students then interpreted their narratives into storyboards, which itself was then critiqued by peers and interpreted into a final design. Through this study you will see that the process was dialectical, each phase improved through interaction with the others as the narrative was reflected off the storyboard and further refined when the storyboard was reflected off the developing final design.

We prepare landscape designs for people not objects. Specifically, we are told by John Simonds, we design for people's experiences. To that I add that we design for people's anticipated experiences. In Chapter 8, with the help of three authors, fiction writer Colleen Mariah Rae (1996), poet Ted Kooser (2005), and novelist and non-fiction writer Anne Lamott (1994), we are provided somewhat differing approaches to writing. Their own experiences as authors, and not always good experiences, help us become less anxious about writing a first- or second-person narrative. The narrative you will be asked to write can be for a healing garden in a hospital complex or a project of your choosing. The idea is to experience first-hand what you are reading about: authoring a short narrative, turning it into a storyboard, and then interpreting your storyboard into a landscape architectural design.

I ask you to write a first- or second-person narrative because they express people's experiences, the happiness of finding what they anticipated, of being surprised, and sometime not finding what they anticipated and of being disappointed. You are not asked to write a third-person narrative because third-person narratives tend to be descriptive of places and settings. Descriptive narratives like much of what is typically practiced as landscape architectural design leave people's experiences to be assumed. A designer can describe a person's walk to a seat and the person sitting down. With that the designer and the client are left to assume the person's experiences. The existence of a mutually agreed upon understanding of the person's experiences, their sequential experiences, remains assumed. The narrative-storyboard-design process helps avoid that from happening.

Chapter 8 speaks to minimizing designer-client assumptions by exploring two benefits that come with writing a first- or second-person narrative text. First, the linear sequential nature of a narrative easily echoes the linear sequential nature of someone walking, driving, cross country skiing, bicycling, pushing a carriage toward, into, and through a designed landscape. Second, we are allowed to author and then review the additive sequential nature of people's senses gathering information from near and far. At a given point visual sights may dominate while sounds and aromas remain in the background yet supportive of the primary messages. A few steps further and the primary visual scene becomes secondary to what are now dominant sounds and aromas. Narrating what your landscape participant users experience through their senses is much like an orchestration. The basses come forward while the wind instruments remain in the background. Later the strings take stage front supported by a background of basses, winds, and percussion.

I imagine many designers look at writers in much the same way many writers look at designers. "How do they do that?" Hard work. Stick-to-itiveness. Perseverance. Fiction writer Rae (1996) and poet Kooser (2005) help us understand writing as a mostly fun creative endeavor. The key approach to writing, the one I most enjoy sharing with others, is non-fiction writer Lamott's "Shitty

First Drafts" (1994). While Rae and Kooser deal with the repetitive nature of writing, the time commitment, and the refining of sentences and flow of thoughts, Lamott speaks to what those of us afraid of writing easily understand: just get it down on paper. It may be a mess, but no one will see it.

Most interesting is that being a designer is much like being a writer. Chapter 8 helps you become less anxious about writing through examples of other's early drafts of narratives. Given even the simplest narrative drafts, much like the stick figures famous film directors used to sketch up simple storyboards, reworking your narrative through discussions with cohorts brings them more and more to life. Shitty first drafts eventually become wonderful landscape designs.

A major benefit to producing a storyboard, as found in Chapter 9, is how closely related it is to a written narrative. A storyboard is the visualization of a written narrative and as such provides a smooth progression from landscape users' experiences to a designed landscape-as-narrative. Both text and storyboard are linear, sequential mediums that help replicate the movement of participants in a landscape. What you find is that reflecting one off the other, especially in discussions with staff, classmates, or friends, quickly illuminates gaps in the flow of your landscape participant's experiences and the environment through which he or she is moving.

Refining the characteristics of objects and spaces in your storyboard that come to make up your design brings up the question of "How do I know a particular combination of objects, materials, and finishes will in fact contribute to someone having the experience(s) being called for?" Chapter 9 provides an exercise, an Emotional Matrix, that as you fill it out answers that question. Working through the Emotional Matrix carries you step by step to those sensual criteria that will most likely allow you to identify the objects that will make up your final design. Filling out the matrix is an exercise in self-reflection. You come to realize, almost simultaneously realize, how much you know but take for granted regarding the environments making up your daily routines, and the similarities across built environments that contribute to the experiences you and others encounter.

As an intermediate presentation to your client(s), the narrative-storyboard-final design process is a powerful tool. Its power is in the fact that it is personal. First- and second-person narratives captured in storyboard visuals are quickly and easily identified with by clients and potential built design's participants. Assumptions regarding what your final design contributes to as the participant-users' experiences are minimized. There are three other benefits to this process. First, your presentations are through the eyes and sensual experiences of those who will come to enjoy your design. Second, you easily avoid describing your design through the language of geometry and design jargon, an approach that relies on assumptions often never shared with clients. And third is that since your whole design process is, in a sense, derived from people experiencing your design there is no reason to use such statements as "I decided." "I wanted." "I put this here." Such statements are common in many architectural, landscape architectural, and interior design presentations. Clients and public attendees' natural responses to such statements are questions commonly directed at you: "Why did *you* do this or that?" The narrative-storyboard-design approach gets questions, but they are focused on the design participants' experiences, with which your audience often readily identifies.

Ethics in Landscape Architectural Design Practice, the focus of Chapter 10, may at first seem a strange concluding chapter to a book on doing designs that give shape, form, and character to the lifeworlds of others. Actually, Chapter 10's introduction of the ethical responsibility of landscape architectural designers for those for whom they design is a logical conclusion to *Landscape Architecture as Storytelling*. Almost nothing exists in the education, certification, and professionalization of landscape architects that speaks to their responsibilities to those left to live with what they design. The landscape users as participants in the design are mentioned but rarely in terms of their sequential experiences. Chapter 10 argues for an understanding of design as an outgrowth of the personal and social contexts in which designs are placed. A summary of the socialization of design professionals illuminates the growing separation of designers from their clientele, the greater public, and the places in which that public resides. The contemporary existence of this separation is further supported by the education and works of planners, landscape architects, and architects (Scarfo, 1988; 1989).

Chapter 10 explores designers' ethical responsibility to meet the expectations that comprise people's daily routines as those activities help sustain a place and its resources. A close look into what constitutes design practice highlights a discontinuity between doing design for the sake of doing design and doing design for the sake of its expressive content as narratives. This final look at design deals with professional ethics in terms of designers' responsibilities to people and their live, work, play, and learning environments. Chapter 10 asks the question, if designs do not at least meet people's experiential expectations, or exceed them, are we being ethically responsible as the designers of other's worlds?

Our responsibilities to those we serve are grounded in the reading and writing of landscape narratives, of spatial texts. That approach brings designers "down to earth," to the eye-level sequential interpretative realm of their proposed designs' eventual users. By narrating people's eye-level experiences, the designer avoids getting personally removed from a proposed design's activity areas and people's experiences as happens when they use geometric, technical, and financial jargon. An eye-level interpretation of landscapes helps avoid a trap many land planners fall into regarding land use. For land planners, yellow on all land-use maps denotes single-family residential usage. But the eye-level sense of character and of place varies if the "yellow" land use is applied to a 200-year-old village or a new community. Language, and later language-as-narrative, helps us understand and apply that knowledge of a place's eye-level character to landscape, architectural, and interior designs, to the cultural surroundings and subjectivities of everyday life.

With this kind of thinking a look into the content of the American Society of Landscape Architects' Code of Professional Ethics brings out the fact that little attention is given to those individuals left to live with our designs. A further review of the American Society of Landscape Architects' Constitution and Bylaws brings to light a similar finding (Vernon, 1987). Throughout most of the profession's rules of conduct, academic accreditation, and professional licensure the responsibility of professional landscape architects to people is an underlying assumption. The existence of such assumptions is considered in Chapter 10.

While the idea of landscape-as-narrative is not new to landscape architects (Alon-Mozes, 2006; Potteiger and Purinton, 1998; Spirn, 1984; 1998), or for that matter architects (Benzel, 1998; Smith, 1977), interior designers (Benzel, 1998), and geographers (Light and Smith, 2005), the application of a design's users' experiences as a narrative to the teaching, learning, and practicing of design is. An excellent collection of papers on landscape as narrative, published between 1995 and 1998, is found in *Landscape Review* published out of Lincoln University in Canterbury, New Zealand. The papers predominantly deal with landscape messages as singular statements in separate locations throughout a site and not as narratives across a site (Swaffield, 1995; 1996). Benzel (1998) and Bacon (1974), while not explicitly recognizing the narrative content of spaces, do recognize that life in, through, and between spaces is a continuum or, as social geographers would say, a state of becoming. From a design perspective, I tend to think of it as an ongoing state of transition; step by step we constantly transition from one moment to the next, from one experience to the next. Benzel's expanded understanding of interior design recognizes the interrelatedness of contexts along a continuum. She recognizes that a room is only understood when we know it "as inseparable from its environment and, once removed, cannot be seen either in the round or as a pattern within the whole" (1998, 15). In basing his *Design of Cities* (1974) on paths and pathway systems, Edmond Bacon provides one of the better introductions to people reading their environments as a continuum. For Bacon, the fourth dimension's interrelationship of space and time is critical to understanding a place. While each author's perspective is in itself a valuable addition to the premise of designer-as-author, and user-as-reader, we still need to understand the greater cohesiveness and wholeness brought to people's experiences of a place through narrative. An "authored" or designed continuum linking, to borrow from Benzel (1998), the city to the room through the neighborhood, landscape, and buildings, is what *Landscape Architecture* as *Storytelling* introduces to the practice of design based on the landscape-as-text metaphor, or better landscape-as-narrative.

Over the years I noticed that students who use the landscape-as-text and landscape-as-narrative metaphors seem to more readily recognize the value of interrelationships between their landscape architectural and elective classes. Using a landscape narrative approach to developing a design provides a steppingstone in the use of metaphors to consider new knowledge, not as separate bits of information, but as contributing to an integrative and interdependent whole. As students gain more knowledge from classes other than design their world becomes broader, their narratives become clearer as does the content of the messages making up their environmental narratives. They become increasingly aware that the selection of specific materials and finishes needed to communicate particular messages requires not only knowledge of history, sociology, and art but also knowledge of proper construction techniques, growing mediums, spatial forms, exposure, and more. Site engineering, construction techniques, and plant identification and ecology all contribute to their understanding of landscape design vocabulary. This is what constitutes the admixture of art, science, and sociology that is landscape architecture.

As students acquire an understanding of design vocabulary and narrative form and content, the idea of "doing design" becomes more than just selecting objects because they look good at the moment or feel right. Knowing what objects "say" calls for a knowledge of the objects' basic attributes and how those attributes relate to a landscape users' cultural and historical backgrounds. With an awareness of the meanings associated with various design features, the designer-as-author moves beyond being a youngster who repeats words ad nauseam and toward an author who labors over the development of each word's selection and meanings. People can speak words that express intent, but not write poetry. Designers can call for particular design features, but the organization and character of the resulting spaces may speak only to the designer (Sorvig, 1996, 3). In such instances when a project is completed the users quickly alter the signs and signals in an effort to make the messages in the text more readable to themselves, more identifiable with who they are. Again, here, consider the time you spend after visitors leave your dinner party rearranging your apartment or home, putting things back "into place." When landscape users do not identify with a design, especially one that is part of their daily lives, they consciously or unconsciously work to change it to something that does speak to them, that reflects who they feel they are.

One other thing to be aware of as you read *Landscape Architecture as Storytelling* is that narrative is used throughout the book in two ways. First, narrative pertains to the narrative the designer-as-author writes as a short first- or second-person narrative story which is then turned into a storyboard and then interpreted as a final design. It is in the storyboard and the final design that the second use of narrative, that of landscape-as-narrative, becomes evident in the site design process and final built product.

Finally, I have two hopes, first that designers' acceptance and application of the narrative-storyboard-design approach generates sufficient discussion and research as to cause its foundation in people's landscape experiences to be brought into the American Society of Landscape Architects' Code of Professional Conduct. My second hope is that The Analogy is accepted into accredited landscape architecture design programs, and further explored in undergraduates' senior projects, master theses, and doctoral dissertations.

# 1

# Three-Tiered Analogy

In 1859, Samuel Langhorne Clemens, at 23 years old, earned his steamboat pilot's license. While a terrific achievement, I'm more in awe of how Mark Twain wrote about his relationship with the river. Like Loren Eiseley, Twain brought life to the expressive character of the Missouri and Mississippi Rivers. He was not single minded. He intermixed the science of hydrology with the engineering mechanics of steam engines and the social curiosity of those who would become his readers in ways that only he could express how clearly the river spoke to him. Language is everything, if you can read it. Clemens could read it.

*The face of the water, in time, became a wonderful book – a book that was a dead language to the uneducated passenger, but which told its mind to me without reserve, delivering its most cherished secrets as clearly as if it uttered them with a voice. And it was not a book to be read once and thrown aside, for it had a new story to tell every day.*

*(Twain, 1984, 94)*

Figure 1.1 Bowl and Pitcher, Riverside State Park, Spokane, WA. Photo by author.

## RATIONALE BEHIND THE ANALOGY

I struggled as a landscape architecture student to transition from problem solving in the formula-driven worlds of science and mathematics to the subjective and seemingly ambiguous problem-solving world of design. In that struggle I found some clarity in three interrelated realms of knowledge: physical sciences, social sciences, and aesthetic expressions, or we could say science, society, and the arts. My desire to understand how, or even if, the three realms of knowledge are interrelated in the socialization of designers led me to realize that most of what is learned as design perpetuates two flaws in what many of us exercise as design process (Scarfo, 1989). First, design students tend to define and solve for social and aesthetic problems using technology. In doing so they focus on "how" a design is carried out or "how" people physically interact with a place independent of "why" it is designed the way it is and "why" it evokes the particular experiences it does. Second, when they do argue for social-psychological relevancy and aesthetic goodness they all too frequently do so from a position of personal opinion. Rarely do they define social contexts or explore local expression, a people's place in history and their sense of orientation within their heritage. The acceptance of opinion over facts is a learned form of discourse (Scarfo, 1989). The distinction between the form

DOI: 10.4324/9781003286981-2

| | | | | | | |
|---|---|---|---|---|---|---|
| | **Tier I – Language** | | | | | |
| **Reading And Writing** | Phonemes Consonants Vowels | Letters of the alphabet | Sentences | Paragraphs | Volume(s) | Stories |
| | **Tier II – Basic Design** | | | | | |
| **Basic Design** | Art Materials and their Qualities | Points | Lines | Planes | Volumes | Sequence of Volumes |
| | **Tier III – Landscape Design** | | | | | |
| **Landscape Design** | Landscape Materials & Qualities | Objects in space mass/void | Paths edges & boundaries | Ground walls overheads | Volumetric spaces | Sequential volumes |

Figure 1.2 Three-tiered analogy interrelating language, basic design, and landscape design. Drawn by author.

and content of discourse, or how we say what we say, is explored more fully in Chapter 7.

Ignorance of the three realms of knowledge and their respective problem-solving strategies and production skills is not confined to designers but is found in most modern professions (Boyer and Mitgang, 1996; Forester, 1985; Larson, 1977). Many designers, like much of Western society, have yet to recognize that social problems cannot be solved with technical solutions. Neither can technical problems be solved with only personal appreciations or political strategies (read, without facts). With an introduction to and application of The Analogy design students come to inescapably recognize their proposed designs' social and psychological contexts, their humanity. This recognition of what we could call the poetry of life occurs because design is approached through the eventual users' view of the design, their reasons for being there, and their anticipated experiences.

The authoring of a design narrative, written as a first- or second-person narrative, causes a designer to be grounded in their design's period, place, and people. Writing a narrative on behalf of a person living out the poetry of daily life excludes the use of abstract design jargon such as axis, focal point, and circulation system. By writing a narrative that expresses people's experiences the separation from daily life found in abstract design language is replaced by a sequencing, or choreography, of human expectations, behaviors, and participation. The appropriate role of the three realms of knowledge – a designers' efforts to determine "what" is to be expressed (the science and technology), "why" it is to be expressed (the sociology), and "how" it can be expressed (the arts) – as applied to the authoring of a landscape narrative is explained in more detail in Chapters 7 and 8.

Development of The Analogy grew out of my concern with what design students in general, and landscape architectural design students in particular, come to believe constitutes good design. Listen closely to what students say in their presentations, what they feel is important enough to be communicated about the goodness of their designs. Between September 1986 and Spring 1987, I followed, interviewed, and recorded a cross-section of first- through fourth-year

*What are the sources of our environmental information? What happens when we substitute one kind of evidence for another? How do we supplant inadequate or erroneous views and stereotypes? .... [and what*

*happens when] second-hand data are superseded by first-hand experience, and the impact of this replacement process on exploratory decisions?*
***(Lowenthal & Bowden, 1976, 7)***

University of Toronto landscape architecture students. That work was supplemented with a more limited study of University of Guelph and Rutgers University landscape architecture undergraduate students and faculty (Scarfo, 1989). That research provided over 100 hours of audio and video recordings. When reviewed, the oral and visual student-student, student-faculty, and faculty-faculty interactions showed that all to too frequently the students' language and reasoning was that of geometry and technology, how this or that was produced, what the final design product looked like in two-dimensional, plan-view geometric forms, and how those forms related to each other in two dimensions (Scarfo, 1989). These observations are further substantiated in Chapter 10 through a review of aspects of Donald Schon's *Reflective Practitioner* (1983). After that research, numerous workshops with designers and planners showed me that the phrase "land use character" often refers to the two-dimensional character of land use areas seen on a map or functional (activity) areas on a site plan. Yet, the phrase is seemingly accepted as implicitly referring to the eye-level character of a place, when it doesn't.

Designers' predominant attention to objects in plan view, as elevations, and even Sketch-up models eliminates attention being given to people's subjective experiences, their appreciations, feelings, and emotions. In doing so, why a design has the eye-level character it does, and why the spaces flow into each other as they do remain assumptions.

*The place of an object or instrument, even if sometimes precisely assigned [i.e. objectively through agreement with others] does not derive from the nature of the object itself, but it is through me that place…is realized*
***(Sartre, 1966, 370–407).***

Throughout The Analogy I argue for landscape architectural design to be recognized as a merger of the three domains of knowledge. Design is as much the efficient and cost-effective construction of details as it is the fabrication of places that respond to social norms and evoke lasting emotional responses in their reflection of personal memories (Hough, 1990; Marcus, 1995; Potteiger and Purinton, 1998; Schama, 1995). Our understanding of environmental design must go beyond Henri LeFabvre's recognition that space is a social product (1979) and therefore a human construct and into what those social products say and how they are read. Landscapes, buildings, and interiors are each a human concept, and a changing one at that. Winston Churchill had it so right, "We shape our buildings and afterwards our buildings shape us" (Roberts, 2018, 8). The dialectic of evolving changes makes it clear that over time people form their landscape as it forms them (Nash, 1989; Schama, 1995). Our appreciations of interiors are formed and informed by what is functionally, psychologically, and aesthetically expressed (Beecher and Stowe, 1987; Girouard, 1978; Rybczynski, 1986; Watts, 1975, 320–346). As agents of change, as designers, we are integral players in the ecology of a place. While our role as the designers of others' lifeworlds may, at times, be seen as a bit perverse, wasteful, or self-destructive, we are still an inescapable part of the emerging content of a design. In considering landscape as a form of narrative, The Analogy pushes us to consciously consider not only a coherent story across a landscape but also people's experiences as related to the sequential experiences that link personal spaces with larger public places.

The rationale driving The Analogy is the development of a person-centered design approach to landscape architectural design grounded in John Simonds' epiphany that we design "…not places,

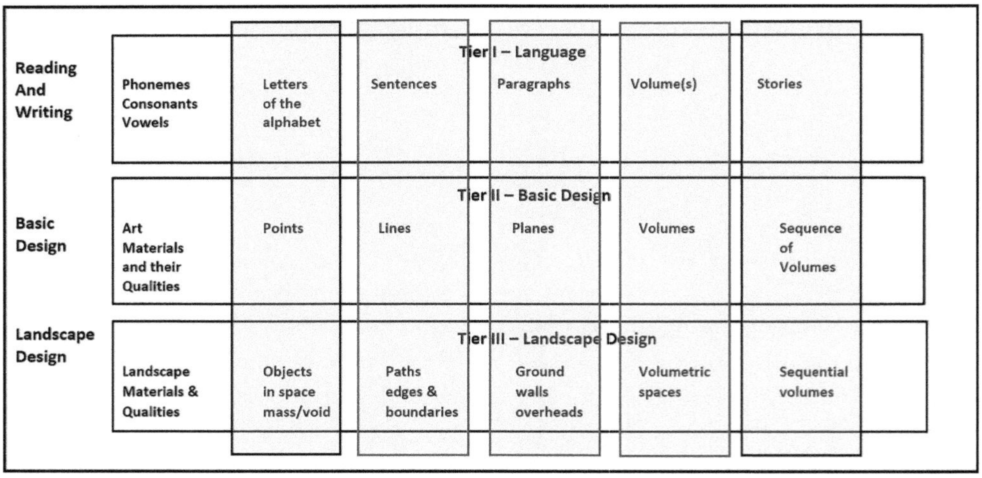


Figure 1.3 Analogy's vertical relationships linking language, basic design, and landscape design. Drawn by author.

spaces, or things [but] *experiences* (my italics). The places, spaces, and things take their form from the planned experience" (1961, 225). In a phrase, The Analogy is an experiential approach to design. The Analogy puts people's expected experiences at the beginning of the design process and moves the benefits derived from geometric and basic design languages toward the end of the process where they are most beneficial. Within The Analogy, I consider Tier 1 to be Reading & Writing, or Language. Tier 2 is that of Basic Design, and Tier 3 is Landscape Architectural Design. As we will see, it is very easy to interrelate each tier's constituent parts and in doing so gain a deeper understanding of landscape architectural design.

## TIER 1: LANGUAGE AND COMMUNICATION

An exploration of a landscape's message-delivery capabilities begins with the first of the three tiers that comprise The Analogy: language. A close look into what constitutes a text's basic parts, the ABCs, as it were, is provided here along with the role of the organizing principles of grammar and syntax or the rules that hold the parts together. That there is an environmental syntax or basic rules used to produce well-formed sentences, or for that matter environments, is laid out by Smith relative to

> *...a kind of syntax for the design of cities. The term 'syntax' is used to convey the idea of 'rules of communication'. There is a constant transaction between the mind and the environment which is predominantly a matter of the brain interpreting environmental signals. Something as complex as urbanism communicates numerous 'messages' simultaneously. It is able to communicate because there seems to be a significant level of collective agreement about the 'meaning' of shapes and spaces both on the cognitive level of information schemas and the 'deeper' levels of instinct and symbolism.*
>
> *(Smith, 1977, 14)*

The building blocks that comprise The Analogy's basic typology are, in later chapters, used as metaphors for their counterparts in Tier 2: Basic Design and Tier 3: Landscape Design. Simply stated, the letters of the alphabet are like points, points in a landscape by which people orient themselves, in which people gather, or intersections where people's routes of travel intersect.

A courtyard is a good example of a point. People gather in it; or, we could look at it from the perspective that it draws people to it. Another example of a point would be a fountain set in the courtyard. People gather around it or use it to orient themselves. Expanding the language tier, sentences are like lines, paragraphs like planes, and volumes are, well, like the three-dimensional volumes we call books. Used as metaphors, grammar and syntax help us to understand design principles, or what directs the organization of basic design elements (Ching, 1978; Wong, 1972; 1977). Even a minimal understanding of grammar and syntax can be used to inform our understanding of design themes and the principles that make those themes communicable and coherent. It is likely that we can all distinguish between a mystery and a comedy. Both are built upon the same alphabet, and many of the same words, and yet with a little reflection we begin to identify what makes them different. We can then bring that understanding into a design that calls for a sense of mystery, a bit of humor, a moment of reflection or meditation, or out-and-out happiness.

> *Grammar: System that guides the combining of words into larger units to convey ideas and information*
>
> **(Guth, 1961, 5)**

> *Syntax: Specific rules that order word usage; topical contexts: historical, geographical, and cultural. They call into play characteristic categories of viewpoints and word usage, e.g. history is time and geography is space; all spatially-related activities such as sailing and dance have their own particular nouns, verbs, and modifiers (adjectives and adverbs).*

> *Idiom (usage): A word often combines with other words in an expression that is not molded on common grammatical patterns but that nevertheless becomes the habitual way of conveying a certain idea…ex. Bear in mind; change one's mind; take one's mind off…*
>
> **(Guth, 1961, 80)**

In recognizing that we read the arrangement, placement, orientation, and sequencing of "things" (as objects or masses and as spaces or voids), we come to recognize that there exist rules by which they are organized. For this reason, it is safe to say that the landscape-as-language approach to environmental design helps in the application of Alexander's *A Pattern Language: Towns, Building, and Construction* (1977). In merging *A Pattern Language* with The Analogy, we come a step closer to that formula I sought as an undergraduate design student. With the landscape-as-language metaphor design decisions rely on the more comprehensive unfolding of a story than on the singular pleasure of how a group of objects is organized. In addition, an application of The Analogy shifts our focus from objects to the relationships between the objects in the spaces delineated by the objects (Bacon, 1974; Sitte, 1945) or the places where people live out their experiential lives.

What is interesting here is the similarity between grammar, syntax, and idiom and the

I don't pronounce my R's. I call a liquor store a packie, a garbage can a barrel, a water fountain a bubblah, sprinkles are jimmies, a basement is a cellar, a living room is a Pahhlah, a dresser is a Beaureu, dinner is Suppah, and a milkshake is a frappe. To me any place west of Rte. 128 is the Boonies and Cape Cod is a piece of heaven here on earth. I ride the T. I know bettah than to swim in the Charles, but I do love that Dirty Watah…I'm a Bostonian, and I'm WICKED proud of it! Repost if you're proud to be from Boston.

Figure 1.4 Talk like a Bostonian. Anonymous Facebook post 2019.

three realms of knowledge noted earlier as the hard sciences, social sciences, and the expressive arts. The hard sciences provide the same building blocks used in the development of a landscape architectural design no matter where the design is located. Water runs downhill all the world around. In a similar manner, the English language and its 26-letter alphabet give rise to a relatively finite collection of words used by mothers and fathers, teenagers and technicians, politicians and policemen. The words each social group uses are organized in particular ways (syntax) that identify and distinguish one from the other. These topical contexts are further refined, or individualized, into local and regional (idiomatic) dialects of the particular language usage that identifies one profession or trade from another (medical vs. design, law vs. religion, and plumbing vs. carpentry). Carried to a humorous extreme (seen to the left) is the bit of Boston Pride written by an anonymous, I'm guessing, Bostonian.

Vocabularies become personalized by professions, individuals, and regional cultures. Alternative word usage provides for a local expression that becomes identified as the idiomatic expressions of a place set within its greater regional context. In basic design these structuring systems, or what we call design principles, make shapes and forms comprehensible and meaningful to the artists and the public. An extreme example of an idiomatic expression of art is found in folk art. So, as a spectrum of expressiveness of art the generic forms, or basic art forms, are refined through cultural interpretations (east/west; European/Middle Eastern) (Newton, 1971; Rapoport, 1969, 1982) and then further refined through the idiomatic expressions of more regional and local interpretations. The spectrum of design principles employed in landscape, architectural, and interior design practices rests upon psychology, sociology, anthropology, and art, and is used to guide the authoring and communication of local and regionally appropriate narratives.

We are each born into our language. We all come to know how to communicate (body language, sign language, oral and written language). As we grow older, our knowledge of communication becomes a tacit or taken for granted understanding of our language. With little to no

Figure 1.5.1 Cultural depictions of carved bird figures. Courtesy Bryce at Inuit Gifts.

Figure 1.5.2 Cultural depictions of carved bird figures. Courtesy Eleni, Handmade with Love, Greece.

Figure 1.5.3 Cultural depictions of carved bird figures. Courtesy Gabreil Vandervort, ancient resource: http://www.ancientresource.com/.

thought we daily apply the rules of grammar and syntax learned in grade school. Through our everyday conversations we combine words to formulate sentences, sentences multiply to produce paragraphs, and paragraphs build into complete stories. We rarely question our knowledge of our language at a conscious level, yet we readily string words together as we interact with bankers and bakers, lawyers and lovers, and parents and children. Each time we speak or write we reinforce our language (the words we use, the ways in which we use them, and their meanings), our integral associations with our society and culture, and in turn our social and cultural values toward others and toward the land (Hall, 1969; Knapp, 1986; Rapoport, 1982).

The same associations of tacit knowledge and its being reinforced through use exist between us and our environments, be it an interior, a landscape, or a city. If this wasn't the case, then Benzel's interrelating a room to a region would be impossible (1998). As with our use of language, the associations that bind us through our environments with our social and cultural realms remain tacit. The purpose of what follows here is to transform your tacit knowledge into a collection of useful tools that contribute to a more comprehensive understanding of both basic design and landscape architectural design. Put another way, designers cannot take for granted how they associate with various landscape settings, let alone the details that make them up.

## TIER 2: BASIC DESIGN

Basic design, an early stage in the landscape architectural design process, is based upon fundamental shapes and forms brought together in ways that make sense. The best way to "bring together" basic design features – point, line, plane, volume and series of volumes – is through an application of one or more design principles such as unity, hierarchy, and gradation (Wong, 1972, 1977). Of course, once you grasp the "rules" you are in a better position to knowingly bend or ignore them. Such basic shapes work to provide a narrative's fundamental messages. As basic shapes they have yet to be refined through landscape design needs such as material selection, finish, color, and maybe sounds and aromas but they still get a basic message across.

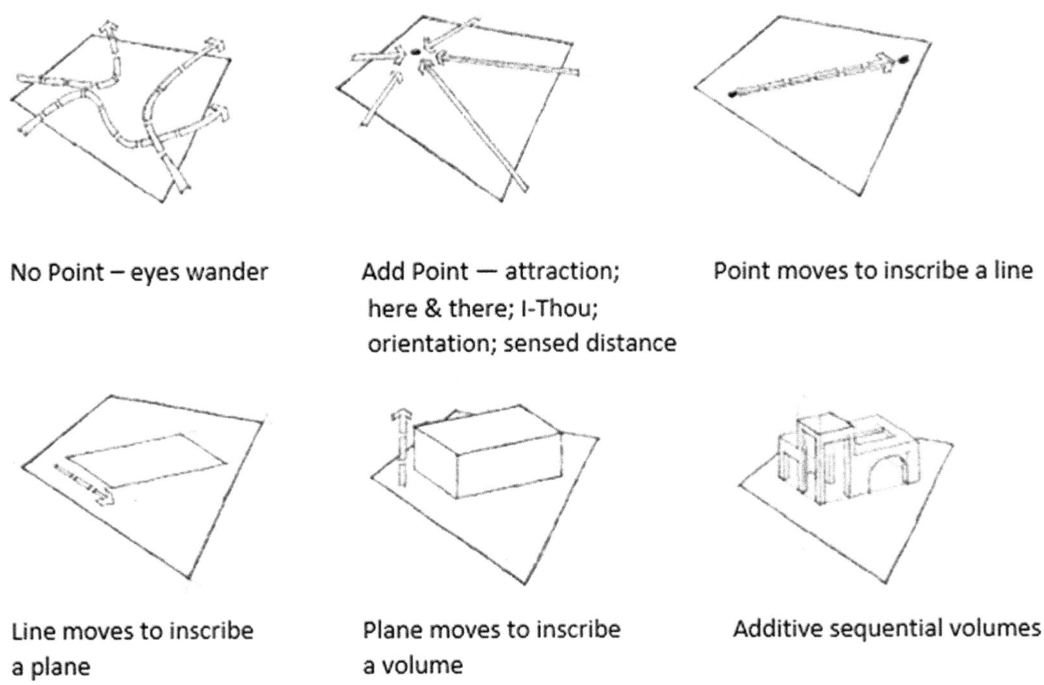


Figure 1.6 Basic elements of design: point, line, plane, volume. Drawn by author.

*...spatial perspective matters. Don't look down on the city from the skyscrapers, says Lewis Mumford, Jane Jacobs, and de Certeau; interpret from the sidewalk.*
*(Eckstein & Throgmorton, 2003, 29)*

Understanding the role of basic design is critical to the landscape narrative design process. An unpacking of the parts and principles of basic design, the basic messages expressed by the shapes making up every landscape design are fundamental to good design. The process of interpreting a written narrative into a storyboard is made easier as you determine the messages to be expressed. This will become all the more clear when you fill out the Emotion Matrix in Chapter 9. By extension, the beauty of adapting a storyboard into a final design is that as first- and second-person narratives it holds the author/designer's feet to the ground. The narrative content of a text, come basic design, come landscape design, occurs at eye level and develops a person's on-the-ground experiences. What is surprising about this approach is that the identification of design features really isn't all that difficult. As you'll come to realize, writing about people participating in your designed space brings into view the design features that come to make up the design. Your written narrative's characters will help you see what comes to make up your design's landscape narrative.

As a reminder, to avoid confusion, two forms of applications of narrative are used throughout The Analogy. One form of narrative is what you will write as a way of illuminating your design's users' experiences and expectations. The other form is what the users of your built design will read in their surroundings as they move toward, into, and through your design.

Basic design, as both process and product, is the foundation of the various professions comprising the fine and applied arts. As a bridge in the three-tiered analogy, basic design connects (or spans the gap between) language and environmental design. Basic design strengthens the idea that landscape architecture is an art form. For that matter, we can consider each of the environmental design professions an art form that interprets and applies the sciences and technologies in the planned development and construction of their products. As such, this realization keeps alive my thought that there exists a formulaic approach to not just "doing design" but doing great design. Basic design gives us the ability to manipulate math and science into aesthetically pleasing habitable environments of appropriate social, cultural, and historic significance. By merging *the* world (the physical domain of objective knowledge) with *our* world (the social domain of subjective knowledge) and our personal and collective expressions (the emotional domain of aesthetic knowledge), the objective qualities of a space give way to the expressive qualities of place. This metamorphosis of the objective into the subjective is integral to landscape architecture, architecture, interior design, and planning in its various applications, including ecological planning and design. Or, as Randy Hester refers to it, "the soul of the place" (Hester, 2006, 118–127).

*Scarfo: Within minutes of being given a site and the request to design a small, vest pocket park, I overheard students discussing bench materials and, no joke, the diameter of bolts to be used in a bench all without even knowing if a bench would even be included in their design.*

Seeing basic design situated in The Analogy, as shown in Figure 1.2 between language and landscape architectural design, helps newcomers to the practice of design avoid the tendency to pick nifty landscape stuff out of thin air and sprinkle it about their designs. As a bridge linking language and landscape design, basic design allows us to focus on the relationships that are part and parcel of basic design elements, e.g., points gather energy and draw attention to themselves (Ching, 1978, 20; Dondis, 1973, 40). And yet, the basic design elements of point, line, plane, and volume are abstractions. As an outgrowth of a design's experiential narrative, points, lines, planes, and volumes are molded to simultaneously fit with the users' expectations and their surroundings. Why a designer would want to use a particular basic design feature, or composition of features, quickly becomes apparent when considering the greater context, the body of the narrative text, or story. Within the language tier of The Analogy, dealing with a word instead of a point, a sentence instead of a linear feature, or a composition of paragraphs instead of a series of volumes allows a design student freedom to deal with basic forms and put off struggling with the minutia of what's to be communicated by the design later in the design process.

More detailed design decisions come later in the narrative-storyboard-design process. Some detailed decisions will be called for as you turn your narrative into a storyboard, but the majority of the detailed decisions will appear when the storyboard is turned into the final design. With this progression from broader to increasingly detailed decision-making The Analogy's design process becomes less complicated, less messy. The same holds true when a landscape design is first considered in terms of its basic design elements. Rather than the myriad of problems to be solved regarding plant selections, sun angles, construction techniques, and who's doing what and where, the designer is able to concentrate on the spatial structure of a proposed built environment. The design's spatial structure is guided by a narrative that expresses the users' expectations. Then that basic design is massaged in response to increasingly more detailed requirements of the design's setting: level, sloping, rural, urban, densely or sparsely populated.

## TIER 3: LANDSCAPE DESIGN

The third tier of The Analogy focuses on landscape architectural design, the narrative's four-dimensional built form. However, as noted earlier, The Analogy is easily applied to the work of architects, interior designers, and, to a certain extent, land planners.

By combining our innate understanding of narrative form with basic design, we quickly come to recognize the contextual bases on which we make design decisions. Excellent examples linking life and urban environments – basic design and landscape – are found in Kevin Lynch's *Image of the City* (Lynch, 1960). His five-year study focused on urban residents' use of mental maps to better identify with and find meaning in their surroundings. Lynch helps us understand that learning landscape architectural design best occurs in conjunction with people's everyday situations and is not the discrete, personal expression of a designer. A narrative, any story that presents users' perceptions and experiences, immediately introduces one or more contexts or themes. With that, the selection of things, characteristic spaces, and materials to be used becomes narrowed. And that's a good thing.

Take a moment and reflect on a simple design project with which you've been involved. Maybe building a bookcase. Consider all the contexts that came into play: living room, bedroom, study, hallway to kitchen. Once you recognize the contexts in which the bookcase will be set, the choices of what can and cannot be used in your project are reduced. Design components are seen for what they are, contextual and context-driven, relevant to the social and physical surroundings in which they are located (Chatman, 1980; Schama, 1995). Whether a side yard or urban public garden or plaza, a place-specific narrative, a story derived from and set in a particular location, cannot escape its social, historical, geographical, and ecological contexts. To introduce a design with little or no relationship to its contexts would have it seem "out of place" (Hough, 1990).

Put another way, a narrative may express the satisfaction of arriving at a destination, having seen it from a distance and finally arriving through a series of discoveries. Such a narrative occurs in thousands of places, millions of times a day: bus stop, front door of child's school, coffee shop on the way to work. When designing a sequence of discoveries, moments of reorientation, and the satisfaction associated with arrival at a destination, the selection of basic forms that make up this progression reduces the number of ways it can be expressed. This is not to say that your opportunities to challenge your creativity are reduced. They aren't. As your design's context(s) become a factor, the design features you can chose from are becoming narrowed. A design's location in a particular place – Seattle vs. Los Angles or Boston vs. Miami, or a particular neighborhood in one of those cities – reduces the selection of material resources that can be used, and the answers to design decisions to be arrived at. This thinking is why the post-modern corrugated metal house set in a historic neighborhood says the designer and homeowner are either ignorant or disdainful of the house's socio-cultural setting and of the neighbors and the neighborhood's sense of place.

> *That is the great distinction between the sexes. Men see objects, women see the relationships between the objects.*
> *(Fowles, 1965, 373)*

> *They [men] are interested in "objects" and "things" rather than people and feelings" (16)... the Venusians [women] are more concerned with living together in harmony, community, and loving cooperation. Relationships are more important than work and technology.*
> *(Gray, 1992, 18)*

> *Because men tend to be good with mechanical and spatial activities, they get emotionally attached to building, fixing, and chasing things (11)... Life is much more of a process for women than it is for men (12).*
> *(Farrel and Farrel, 2001, 11–12)*

In other words, working within a given setting, drawing on the surrounding color, material, and spatial palettes, environmental designs are less likely to be considered "out of place." They fit. They look like they belong. They are one more part of a much larger whole. Whether designers work with (in harmony) or against (in contrast) a given context, the important fact is that they come to do so consciously, respectfully. Design decisions are helped and in a sense guided by their larger social and geographical settings. As such, information drawn from a project's contexts contributes to the justification and legitimization of design decisions. Recognizing contexts, and responsibly working with them, quickly involves students in expressing "why" they did this or that rather than the "how." The "how" of design, the technological and economic wranglings giving form to design decisions, remains more appropriately a response to the design's production (construction and site engineering) than a major factor used to justify a design's form and content.

## MESSAGES, MEANINGS, AND NARRATIVES

Looking back, I admit that my early exploration of The Analogy remained focused more on things, individual messages ("walk here," "sit there," or "enter here"), the physical objects of language and design, than on the relationships or dialectical interactions that both formed and informed those things and the people who lived with them. Thanks to the late Harry Porter, at the University of Virginia, I came to realize that as texts, landscapes went well beyond communicating discrete one-liners or singular statements. Landscapes are four dimensional. They are not snapshots or still-life images. Working up a design is like scripting and shooting a movie (Aminzadeh et al. 2016). Landscapes embody and reflect the ongoing development of people's morals, values, ideologies, and personal and collective aesthetics (Duncan, 1990; Ley, 1983). Landscapes are the manifestation of a continuum of interrelationships across spaces. Landscapes are narratives.

In 1989, at the conclusion of my presentation of the three-tiered analogy, Harry Porter's simple question, "What's the moral of the story?" took me years to realize that he had simultaneously provided me with a question and its answer. At the time I thought he meant "What is the moral of my presentation?" and said I was confused by the question. He broadened his question to something like, "Using the Analogy to design a landscape, what's the moral of the story or design that the users of the design walk away with?" I didn't know. As I said, at that time I was too focused on the individual *things* that made up the typology. I was aware there was a message content but had yet to consider the messages as comprising a more complete narrative and the implications of narratives as design development tools.

*I came upon my first Military Orchid... I fell on my knees before it in a way that all botanists will know. I identified... I measured, I photographed, worked out where I was on the map, for future reference... Yet, five minutes later...I suffered a strange feeling. I realized I had not actually seen the three plants...I had managed to set the experience in a kind of present past, a having looked, even as I was temporally and physically still looking...I knew I had just fallen, in the stupidest possible way, into an ancient trap. It is not necessarily too little knowledge that causes ignorance; possessing too much, or wanting to gain too much, can produce the same result.*

***(Fowles, 1979, There are no page numbers; look for: Oak January, Derbyshire)***

Back then, my understanding of a landscape's expressive potential was static, still confined to the delivery of single messages. As scary as this now sounds to me, back then I was still of the mind that the environment was more or less a snapshot of an area. That is the way I had been taught, to see and deal with the landscape as a two-dimensional site plan. I later realized that my view of environments and people was derived from my science background and early design education. Much of it objectified people. In many instances, people were little more than assumed to exist and somehow cultural environments are, as if by magic, miraculously altered over time. Like John Fowles' admission to having seen/studied his first Military Orchid but not experienced it, I wasn't yet seeing people as agents of change – as actively involved in place making. Maybe that is what distinguishes the formulas of science from the authoring of a narrative. First- and second-person narratives are explicitly subjective, or at least implied by the personification of landscapes and their features (Berry, 1964; Eiseley, 1957; Jackson, 1977). If I knew then what I know now my response to Harry's question would have been "Everyone reads landscapes. Landscapes are narratives. The Analogy can be applied to authoring landscape designs and as a result provide both practical and emotionally appropriate experiences to those who enjoy our work." Or, something on that idea.

As with most design education, my training, up to when I began to explore the language-narrative metaphor, had been primarily visually based. I was trained to see the designed environment more like a still-life photo than a cinematic dynamic of unfolding experiences. I was yet to realize that landscapes are more like films than snapshots, that people's character is found in a site's evolving social context, that a place's local character is interrelated with its historical context, and that all of our senses play a role in our identifying, reading, and interpreting what we encounter as we move through time and across space.

*...the cultural landscape is fashioned from a natural landscape by a culture group. Culture is the agent, the natural area is the medium, the cultural landscape is the result. Under the influence of a given culture, itself changing through time, the landscape undergoes development, passing through phases, and probably reaching ultimately the end of its cycle of development. With the introduction of a different – that is, an alien – culture, a rejuvenation of the cultural landscape sets in, or a new landscape is superimposed on the remnants of an older one.*

*(Sauer, 1925, 46)*

Years later, with my realization of film, music, and paintings as narrative forms, and my immersion in historical, cultural, and social geography, The Analogy's typology acquired its current three-part foundation: social, spatial, and temporal. John Conron, Geraldine Pratt, and Marvin D'Lugo's courses at Clark University in Worcester, MA, helped me learn that texts are more than linear; they are additive processes of communication. Understanding that a landscape is more than a text, that it is a narrative, brought with it the realization that people interact with their surrounding environments more so than I had ever thought: step by step, moment by moment. There is an ongoing dialectic. As people interact with their surroundings, they acquire information through their senses in ways that blend statements into messages and messages into an even greater choreography.

Landscape narratives are simultaneously repositories for and mirrors of people's values, their appreciations of life, and in some respects their lack of appreciation (Marcus, 1995; 1999). Landscapes not only provide the settings for stories, they are also the stories, whether expository, poetic, purely functional, or spiritually moving in the most sublime sense. A place's displayed material culture is like a physical manifestation of people's oral histories and local histories. Oral history sustains people's deeply held meanings and values over long periods of time. Yet, while the vehicles of expression may change the basic story and the walk-away messages remain the same (Campbell, 1949; Chatman, 1980; Rubin, 1995).

Since I began learning landscape architecture my attention was drawn to cultural landscapes. The physical manifestation of social interactions, the palimpsests of people's overlain experiences building upon artifacts, and their bundles of memories socially given to and reflected off their surroundings. This realization introduced a new dimension to both my skill and my responsibility as a designer-author of landscape texts. Context was not just what I found at and around a given site, context was a place's past as evident in the present and moving into the future (Schama, 1995; Yamin and Metheny, 1996). As a narrative, the landscape text both forms and informs its users within the continuum of their everyday lives as set within the broader biographies and local and regional histories and heritage. Landscape narratives communicate three levels or domains of information: at the surface, the functional and practical; slightly deeper, the socio-cultural aspects of history and heritage; and at the deepest level, archetypes as found in our collective unconscious. Our deepest beliefs that sustain our humanity.

Landscapes, designed landscapes in particular, need to communicate what is noted in the definition of landscape architecture and the other design professions. Designers contribute to environments that serve the public's health, safety, and welfare by designing places that serve useful, aesthetic, safe, and enjoyable purposes. The fact is that as design professionals, we work in the service of others. That service, by definition, involves our identification with and satisfying of other's needs within the contexts of local ecology, social setting, and history. An excellent starting point to recognize those needs is Maslow's hierarchy of human needs (1970). And with those we start with the basic human needs of sustenance, shelter, and safety.

*Exposition...to provide necessary information concerning characters and events existing before the action proper to the story begins. Its emphasis is strongly explanatory... traditionally done in the summary mode.*
*(Chatman, 1980, 67)*

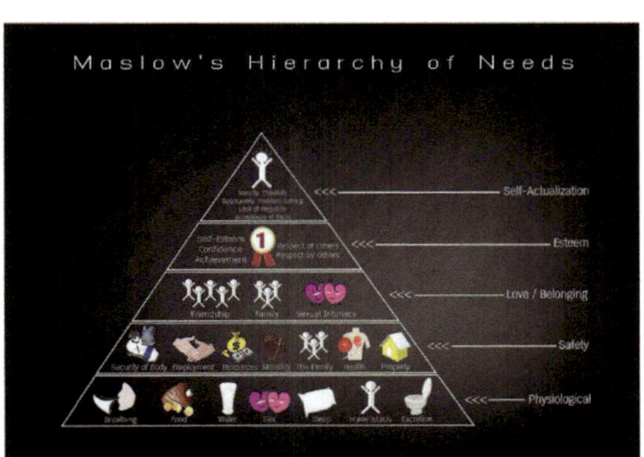


Figure 1.7 Maslow's hierarchy with five levels of needs in human motivation. Shutterstock ID: 554319703. Contributor: Iamnee.

Still niggling away at me, Harry Porter's question prompted me to begin to distinguish between the message and meaning capabilities expressed within a designed environment. As a text, a landscape, its contents and their composition, and the interrelated organization of both, is the grammatical guide through which messages are sequentially expressed (Swaffield, 1995, 1996) and become a narrative. Those messages may be delivered in an expository form, expressing information necessary to people's health, safety, and welfare, such as "walk here," "sit, relax, and refresh," "watch out for this hazard," or "this space will allow you safe passage." Other message types are poetic and prosaic delivering information as more abstract (calling into play contemplative reflection) or pragmatic (local vernacular) forms, respectively. The other part of the distinction is the recognition that the medium does communicate a certain understanding of the designer, the period in which the design was done, and the history and social norms of the designer's setting. This is a gross simplification, so for now I'll note that the critical theorists, and the various deconstructionist schools of thought, recognize that the medium does provide cues and clues as to the author's values, attitudes, and politics (Giddens, 1984). Another way to consider this is that as designers of others' environments we need to be aware of the values and attitudes we bring with us – and sometimes put them aside to enable us to adopt those of the users.

*More than any other botanist of his time, Thoreau moved past the mere naming of trees – the nouns of the forest – to track the verbs: the birds, rodents, and insects that pollinate flowers or disperse seeds, and all the other agents that shape the forest's structure.*
*(Nabham, 1993, xvi)*

In Chapter 2, the typological units are explored as they contribute to the meanings and messages related to three overlapping message domains: functional, socio-cultural, and deeply emotional, or what Carl Jung identifies as the collective unconscious (Storr, 1983). Functional messages are in a sense expressed as the nouns and verbs of design: objects and actions. Actions as motion bring (or guide) people to the basic messages of everyday activities such as entrance, exit; come over here, go there; "Yep, I can get there from here." This is design in its stripped-down basic state, basic messages yet to be modified/enhanced with environmental adverbs and adjectives.

Based on Maslow's hierarchy, I believe landscape users, in an effort to satisfy their basic needs, look for, read, and interpret the functional domain's messages first. They read the pragmatic messages such as "come here," "go that way," and "stop." Functional messages provide for people's immediate safety and welfare, something I became acutely aware of when I suddenly found myself one winter sporting a pair of crutches. I immediately read my surroundings through a new lens. Suddenly I read the placement, alignment, and organization of, within, and between objects on the land at a functional level for the purpose of maintaining my crutch-dependent welfare. My "perspective" had changed, not in terms of height above the ground but in terms of my stability and space required (now with crutches splaying outward to the right and left) for me to move about safely on one foot instead of two.

A person in a wheelchair reads a landscape's lines, planes, and volumes differently than a child, a bicyclist, a parent pushing a stroller, or a 6′4″ tall pedestrian. People's primary concern is with their ability to make it over, around, into, through, or past a landscape feature, whether they are carrying groceries from the car to their house or finding their way from a parking lot to a park's interpretive center. With a designer's recognition that designed environments need to address a users' practical comfort, or what I consider mundane, everyday aesthetic experiences, design begins to deal with the shapes and forms of a landscape's foreground, middle ground, and background as a cohesive continuum. The functional domain of landscape messages is the foundation for statements that express culture and place, the landscape narrative's second domain.

Aspects of place (space) and history (time) are easily introduced into functional environments designed for functional efficiency. With the addition of adjective and adverb modifiers like color, texture, a material's finishes, a space's basic functional messages acquire the more contextual aspects of period, place, and their emotional content and context. A retail parking lot in a Southwestern town may have the same dimensions as one found in Old Town Alexandria, Virginia, but in Old Town the southwestern materials, colors, and textures would readily be felt to be out of place in the Colonial Virginia setting. Being able to find your way around a place is one thing, being in the right state of mind to properly read and interpret what's there is another. A person's state of mind does as much to structure expectations, what is anticipated in terms of a particular landscape's vocabulary, as does the functional context in which a person executes his or her daily routines. Being lost is when you do not find the shapes and forms you expect as cues and markers to where you thought you were going, even while the spaces function well.

The third domain of environmental messages and meanings looks behind landscape as medium and message, beyond the surficial readings of use, and into the deeper collective values people bring to and reinforce in landscape spaces and landscape character. Understanding the existence and inclusion of deeper meanings is helped through a review of semiotics, or the study of signs, symbols, and symbolism, as embodied in designed and natural environments. The ways in which functional messages are delineated speaks volumes in terms of social and cultural values, a people's attitudes toward others and toward the land. Robert Frost said fences make good neighbors, yet the Berlin Wall divided a people who had lived a different definition of neighbor prior to 1961. Streets may promote commerce between places, but as gangland turf boundaries they have just as often worked to separate people only 60′ apart and contributed to the death of neighborhoods.

Consider the Vietnam Veterans Memorial: directional walkways, you can see the beginning and end at the same time. One wall points to and visually connects you with the Washington Monument,

Figure 1.8 Vietnam Veterans Memorial: visitors reflected in names of those who served and died. Photo by author.

the other in the direction of the Lincoln Memorial. Functionally an easy read. But then the names, the 58,390 men and women's names (as of 2019), in the highly polished black granite in which you see your image reflected as you walk deeper and deeper into the ground. I'll let you consider the messages, the deeper sense of a people and a time in their history. But functionally, "Yep, I can walk the 250' from this end to that end."

Look at the photographs of the Browne's Addition neighborhood closely. At the functional message-delivery level the landscape narrative is pretty straightforward. Sidewalks, steps, walkways to front doors, handrail points up toward doors, and roads for vehicles. Some sidewalks are provided with cooling shade during the heat of the summer. No apparent obstacles or tripping hazards. All the surfaces, lines, and objects are pretty straight-forward invitations to come and visit. You're welcome. Some front doors are even preceded by covered porches with chairs and walkways that connect the street

Figure 1.9 Browne's addition neighborhood, Spokane, WA. Photo by author.

Figure 1.10 Browne's addition neighborhood, Spokane, WA. Photo by author.

to the porch, stoop, and front door area. Some don't. With some houses (Figures 1.13 and 1.14) you can't even see if there are front doors, a statement made less welcoming by the fact there are no walkways from street or sidewalk to the front door. Maybe cars and people only come and go through the garage. At the deepest message-delivery level the houses with strong visual connections to the street (transportation, commerce, public services) call to you in a welcoming way and say that the inhabitants of the houses are not only connected to the greater community but also actively interrelated with it. They express community. Without those statements what do you read? Are you openly welcomed; or welcomed at all? Is the sidewalk-less, walkway-less, front door-less neighborhood just a collection of individuals? Then think about the sidewalks, if there are any. Set back from the street, as opposed to adjacent to the roadway, winters are made much easier when snow isn't plowed from the street onto the sidewalks. What does this say about a city that makes it a hazard to walk in the wintertime? Daniel Kemmis (1990; 1995) provides a wonderful analysis of community, not through walkable streets and interconnected homes but a marketplace in Missoula, Montana. What he shares is, however, easily modified to fit into these photographs in ways that help further express the landscape narratives seen here.

I've not done a detailed analysis, but believe I can date a neighborhood by the presence or absence of sidewalks set along the curb or set back with a green strip between sidewalk and street, walkways connecting street to houses' front doors, readily visible front doors with covered stoops or

Figure 1.11 Browne's addition neighborhood, Spokane, WA. Photo by author.

better yet porches with swings and chairs. The presence, absence, and configuration of those landscape expressions also say something about the prevalent social norms of the time. There is a social-psychology at play here. Given the structures were built with city approval (and paid permits), consider what they say about those who bought them and live in them?

When I became aware of the existence of landscapes' symbolic content I began listening for how it was recognized in designers' works, planners' debates over growth management, and ecologists' views on sustainability. Twenty-five years ago I found little recognition of the deeper meanings local residents held for the sites on which designers and planners worked. I hear more today from designers who live and work close to their design projects, less and less as the distance between home, work, and project site increases.

More frequently than found in design literature, the spirit, the culture, the humanity of a place is read in the works of the "nature writers," or those I refer to as "place writers":

Figure 1.12 Four photos, Manito/Cannon Hill neighborhood, in Spokane, WA's South Hill. Photos by author.

Figure 1.13 Veradale neighborhood, Spokane, WA. Photo by author.

Figure 1.14 Comstock neighborhood, Spokane Valley, WA. Photo by author.

Figure 1.15 Manito/Cannon Hill neighborhood, Spokane, WA. Photo by author.

Emerson (1971), Thoreau (1973; 1993), Muir (1988; 1997), Gary Snyder (1996), Tom Horton (1987; 2002), and Annie Dillard (1990), and any number of writings by Wendell Berry (1964; 1968; 2013), Loren Eiseley (1957; 1978), and Barry Lopez (2019) to name a few. I know we can argue about this, but generally speaking, while many earth artists (Goldsworthy, 1990; 2015; Lippard, 1983; Smithson, 1996) are assuredly in touch with a place's spirit(s) I don't deal with them here because earth art may be ephemeral, visited and experienced but rarely lived in.

*Mythic forms...may be regarded as either pointing past themselves to mysteries or universal import, or as functions merely of local ethnic or even personal idiosyncrasies.*
*(Campbell, 1974, 11)*

The line place writers walk between fiction and non-fiction has a lot to offer environmental designers' understanding of what contributes to a sense of place. Meanings, values, heritage, romanticism, these are the invisible forces of both consistency and change that rarely appear as part of a designer's site inventory or GIS database. The invisible forces that bind people and place are what is read between the lines. One of the few landscape architects I am aware who recognized that interiors, architecture, and landscapes are read by their users was Andrew Jackson Downing (1969). As for me, in working with the physically impaired and people of other cultures, I am again and again made to realize people's relationships with their environments are the basis for how well they maintain the character of their surroundings: e.g., personal objects may be used again and again yet seem to remain in their places. I believe that if designers work with those human dynamics, such as sense of identity and meaningfulness, then a design as it includes its users will, in a sense, take care of itself. Work against the grain, ignore people's deeper values that form and inform their daily routines, and their sense of identity, and a design quickly disintegrates into what the users identify with which isn't the product they got.

Harry Porter's question of how The Analogy could be applied to design stayed with me for years. Looking back, I realize that as I worked to answer his question I became aware of the social, cultural, and personal meanings and values that overlay and are embedded in the physical aspects of an environment. What I now understand as a landscape's potential to communicate messages, to simultaneously build and bind places and people, took on a dynamic and interactive character. Whether landscape or architectural, personal or cultural, interior or exterior, space is a social product (Lefebvre, 1976; 1979; 1994; Lund, 2012). Seeing built environments as social products was the first of my realizations of the interplay between people and their physical surroundings and the language and abstractions designers use to record and recreate those surroundings. Lefebvre's statement allowed me to comprehend John Simonds' realization that as designers, we "plan experiences. The places, spaces, and things take their form from the planned experience" (Simonds, 1961, 225). I also came to realize that people's experiences are gained through a dialectic form of social interaction (Giddens, 1984; Gottdiener, 1985) between them and their places, spaces, and things. The structure embedded in the idea that as designers we author narratives was becoming set in my mind. So to was the idea that planning or designing experiences was the cornerstone on which authoring landscape narratives would rest.

Working with the three domains of landscape narrative's expressions (immediate practical, less immediate socio-cultural, deep collective unconscious) easily moves our exploration of the meaning filled content of people's landscapes into the social ecology of dwelling in place (Ardalan and Bakhtiar, 1973; Bacon, 1974; Norberg-Schulz, 1985; Rowles and Bernard, 2013; Sitte, 1945) or the routines or continually evolving relationships people have with the spaces they inhabit. As a continuum, the ongoing ebb and flow that is daily life helps move our realization of individual messages to their being sequentially linked into narratives.

As a cultural medium, whether vernacular or professionally designed, the serial nature of landscape, architectural, and interior spaces involves a continuum of unfolding messages that become a story to the user/reader of that narrative. The unfolding is generally an outcome of a person's movement from a foreground vantage point, on through the middle and then into the background which in turn becomes the foreground, over and over. In reading about narrative I came to understand that the form, function, and structure of oral history contribute a further clarification to landscape as narrative, and to the introduction of narrative into design process.

The particularly interactive-interpretative processes employed by the users of designed environments are clarified as we explore oral history (Rubin, 1995) and the narrative structure employed in fiction and film (Chatman, 1980). Bringing an understanding of story and discourse into the lived-in world of design process and designed environments is further helped with Giddens's structuration and communication theories (Giddens, 1979; 1984) and the analyses of spatial readings by such geographers as Duncan (1990), Duncan and Ley (1993), Barnes and Duncan (1992), and Wood and Beck (1994). These authors, and more, recognize that people spaces are read and that there are multiple, but not exhaustive, potential interpretations a reader may make of a given text. They also note that by recognizing and working to develop a design's theme boundaries are introduced (Rubin's semantic context) within which design (narrative) decisions are made.

Stated differently, a design's intended theme provides designers with focus. The design of a dog park likely would not include messages associated with an off-road vehicle park. Accepting to design an outdoor wedding venue would....well you decide what you would find there: some things could be included, some shouldn't. Message boundaries derived from a place's theme provide the designer-author with narrative constraints. The act of doing design, of dealing with seemingly infinite decisions to be made, is, in reality, not as hopelessly open ended as many students tend to believe. A park with a river running through it provides a wholly different set of activities than a park without a stream or river. As a result, one provides for a somewhat different narrative that the other cannot.

As clarified here, creativity is not a license to do whatever a designer pleases. There is a bounded, developmental character to the authoring of a story. Within a given theme, let's say inclusive play area, or public transit stop, or a personal getaway, what is to be communicated, how it is to be communicated, and the spatial and temporal sequences used to express each narrative gain their organization from the "recursive character of social life" (Giddens, 1979, 41). The repetition of those messages associated with particular shapes, forms, and sequences builds our landscape vocabulary, and in turn our landscape grammar. In other words, people cannot escape who they are, their personal biography, or their cultural history. In understanding this, Giddens sees people's regularly repetitive activities as reinforcing, or as giving structure to and being structured by their heritage, where they grew up. The idea of developing a part of a landscape narrative by what spatially and temporally precedes it is found in such concepts as Rubin's "constraint" (1995), Chatman's "reading out" (1980), and Rapoport's "cues" (1982) all of which are used to outline ways to apply narration to design process. These approaches are dealt with further in Chapters 7, 8, and 9: Narrative, Authoring a Landscape Narrative, and Storyboarding, respectively.

The Analogy's progression of linking basic design elements to deep-rooted social and cultural meanings identifies and combines the observable world as filtered through people's memories and those memories' contribution to a place's local character. This blend of qualitative and quantitative information enables the "doing of design" to act as a bridge that joins basic design elements with people's everyday appreciations and comfort. As masses and voids, basic design elements are the fundamentals of a landscape's character. The selection, articulation, and arrangement of those features embodies and reflects people's investment in their surroundings, a commitment to dwelling, a continuous movement toward comfort, whether in a residential property, neighborhood, community, or region. Being on the land over time, people's acts of "becoming" are the human dimensions of both social and natural ecology. People are both natural resources and stewards of the natural resources. Once arranged to be comfortable and meaningful, as articulated by Norberg-Schulz (1985) and Rowles and Bernard (2013), the user group(s) tend to care for and sustain their immediate environments (sustain their sense of identity). We could say they steward their surroundings. This idea is seen in how people maintain the arrangement of objects on their desks. Move a stapler or cup of coffee on someone's desk and within minutes they will be moved back "into place," meeting the individual's personal sense of space/place. People are the bearers of social constructs that daily reaffirm their personal, social, and cultural sense of identity, of being in place.

Contrary to my first encounter with design, I now realize design is not a mystery. Design is at the intersection of the dynamics that bind the lives of the people we serve and the land they inhabit. The results of that intersection are lasting, even when the designs aren't. Learning design is a sweet not sour experience. This is not to say that there aren't many, many difficult, pertinent questions in need of research and that there aren't individuals who seem to have been "born" with a "natural" talent for design and for teaching. Design, though, as process and product has bases that can be readily identified, taught, learned, and communicated clearly to others. This is the goal of *Storytelling for Landscape Architecture: Learning Design Through Analogy*.

# 2

# A Typology

Christopher Alexander (1979) and Raymond Williams (1983) both speak to the languages that bind us to each other, to our cultures, and to the physical worlds we inhabit. Alexander speaks to the architecture of the built world about which we move and socialize. Williams helps us better understand our world of words that make up our politics, religion, social groups, and individual selves. They impress me because they are of such different camps and yet share many of the same thoughts.

*The [pattern] language, and the processes that stem from it, merely release the fundamental order which is native to us. They do not teach us, they only remind us of what we know already, and of what we shall discover time and time again, when we give up our ideas and opinions, and do exactly what emerges from ourselves.*
*(Alexander, 1979, xv)*

The idea of sharing in Alexander's use of the collective "we" is reinforced when Williams says,

*Within our common language, in a particular country [social group or profession] we can be conscious of social differences, or of differences of age, but in the main we use the same words for most everyday things and activities though with some variations of rhythm and accent and tone.*
*(Williams, 1983, 11)*

*If that analogy [narrative as landscape] is any good, then the way to create computer narratives is to define the features of the landscape to be explored, and let those define the path. Before I go any further (so to speak) I want to explain how I came to believe that narrative really is landscape.*
*(Hughes, 1997)*

The Analogy's typological basis to learning how to develop landscape architectural designs based on how we learned to read and write bolsters our understanding of the landscape as a repository and medium of expression. Messages expressed in the shapes and forms of our built worlds communicate functions, aspects of culture and heritage, and the values associated with our deepest collective beliefs. The typology asks us to identify and organize the fundamental elements of expression within each tier of The Analogy: language, basic design, and landscape architectural design. Once identified, the basic components of reading and writing, basic design, and landscape design are more easily reviewed for their relationships with

DOI: 10.4324/9781003286981-3

each other and the relationships they enable us to have with others and with the physical world. With this understanding, we can study basic landscape components, as expressive signs and symbols, with regard to the greater wholes we construct through their use. However, before we unpack The Analogy's three tiers we need to become aware of the social dynamics in which they are both products and processes.

> *As Richard Lyon puts it: Noise is as much psychology as physics. Car buyers, Lyon points out, are apt to spurn a car if its doors shut with a tinny clank. "Sound is information".*
> *(Wolkomir, 1996, 56)*

Reading and writing, basic design within the arts, and landscapes, structures, and interior spaces are simultaneously social products and social processes, each reacting to the other. They exist through our collective need to use signs and symbols that gather, record, hold, and communicate messages and meanings. The visual and audible (oral/aural) signs we accept as words enable us to interact with each other. As speech acts, spoken thoughts enable us to recognize, record, verify truths, and reuse agreed-upon signs as a means of sharing information, retaining what we have learned from our experiences, and generating new knowledge. In turn, our evolving knowledge of the landscape's signs and symbols contributes to our safe use and appreciation of the environments we inhabit. Our ability to remember, interpret, and understand what is communicated by our environments also occurs through our other senses, e.g., the memories, behaviors, and values associated with the smell of a bakery versus a herb garden, the tactile sensation of a cold, exposed aggregate bench seat versus that of a Chippendale end table, and the taste in the air of salt-water spray instead of the rancid smell of refineries or paper mills. Each interaction reinforces value-laden meanings, reinstates personal and social values, and reaffirms people's mutually agreed upon use of the signs and symbols as media of communication.

> *It was that she [Ethel Merman] filled the theater....from the back of the house to back of the off stage area, ceiling to floor, row, by row...On film...she nailed down her spot, feet firmly planted in what would become a familiar stance in which she bent her elbows, held her arms straight out in front of her, and rocked them backwards, elbows going up, then back down, and then arms forward again. Merman rocked.*
> *(Basinger, 2019, 190)*

The expressive characteristics of design fundamentals such as points, lines, planes, and volumes rely on our routinely (re)creating combinations of signs with which we identify, in which we find meaning, and through which we maintain a sense of position, or orientation, in our personal lives and heritage (Norberg-Schulz, 1985; 1980). The identifiable expressions of meaning-full places, memorable places, are anything but a hodgepodge collection of landscape features, building components, and interior furnishings and finishes (Conron, 2000). Rarely can a random selection of furnishings be gathered together to make an interior or exterior space meaningful to people (Alexander, 1977; 1979). Randomness is not part of a good design process. The selection and organization of meaningful design features are outgrowths of people-place-activity contexts. Designed environments identified as meaningful by their users exhibit the difference between a designer knowing a language and being able to eloquently "articulate" it. Individual visual or

Figure 2.1 Wooden structures in Provincetown, Massachusetts. Photo by author.

Figure 2.2 Brick buildings in Back Bay, Boston, MA. Shutterstock Item ID: 7210093. Contributor: Jorge Salcedo.

audio signs, scents, and tactile sensations, identified with particular places and times, people and events, are what geographers refer to as bundles (Hagerstrand, 1970). A bundle might be all the shapes, materials, objects, and odors or aromas associated with a lakeside versus ocean-side marina, tennis courts versus horseshoe pits, or a public skating venue versus an amusement park. Within Tier 1 of The Analogy, that of Reading and Writing, the collection of referential items associated with particular activities, occupations, or areas of study is more often than not governed by specialized languages, shapes, and forms associated with those activities, functions, and topics. The specialized vocabularies of sailing or racquetball, cooking or woodworking, express the nuances of certain situations in the same way a person's particular body motions display a "command or reserve of expressive techniques; repertoire: *a dancer's vocabulary of movement*" (Morris, 1970, 1434). When you think about them, the specialized words and phrases associated with particular settings and activities also refer to identifiable characteristic spaces; the makeup of spaces associated with and supportive of particular activities is their syntax. Consider a people-friendly urban park. You'd expect the associated activity spaces to accommodate sitting, some for small groups and some for larger groups, sun and shade, framed views, maybe small tables or other surfaces for food, or chess or checkers, a separation from nearby roads, and so on.

Even with changes in our landscape-as-narrative vocabulary over time, one aspect remains as much the landscape as a medium of expression. Through that medium we seek three types of information. As we move about this or that landscape, we most immediately seek functional information. Two facts support this conclusion: Maslow's hierarchy of needs (1970) and people's production and use of fundamental forms throughout history. Within the visual vocabulary of inhabited spaces, the fundamental functional messages have remained much the same over time: "Yes, I can get there from here." "I will be safe going that way." "Stop here." "Look around, see where you are relative to where you want to go." The linear nature of a path expresses direction and movement, motion and time. The two-dimensional qualities of an area denote gathering, orientation, and transition. The original gathering areas for carriages at Versailles and The Residence in Wertzberg, Germany are today gathering areas for cars. Receding or advancing, ascending or descending, flat surfaces remain steps the world around. Their associated acts of ascension and descension in turn express both practical and deeply moving meanings. Shapes, forms, and spatial sequences of expansion and compression express messages of movement associated with personal safety and the availability and access to food and shelter. Ching's *Architecture: Space, Form, and Order* (1978) provides a great introduction to spatial forms and sequential experiences.

Figure 2.3 Chelan Crest, Washington State. Photo by author.

Figure 2.4 Centennial Trail, Spokane, WA. Photo by author.

The sequential flow of functional messages across a landscape, coupled with our collective associative memories of those messages, allows us daily to move about in relative personal safety. We identify the differences between lines and planes that distinguish gates from walls and fences, the textures and colors that delineate walkways from planting beds, and edges and dimensions of areas that distinguish bike paths from roads and automobile intersections. Imagine the accidents that would occur without our ability to read these collectively agreed upon distinctions associated with the lines and planes of roads and paths.

Messages of culture and heritage, while more powerful and longer lasting than functional messages come second to questions dealing with personal safety, welfare, and access to food and shelter.

Figure 2.5 Sidewalk mosaic gives neighborhood a sense of identity. Oberklee, Czechoslovakia. Photo by author.

Functional messages come and go. Once safely past a point, let's say a street corner, we are concerned with what is up ahead (overhead and underfoot) and then beyond that much more so than what is behind us. Messages associated with our social milieu, culture, and heritage are deeply engrained in our psyche. They are a part of us, of who we are, and as such are carried within us as we move about.

Our use of the visual signs in landscapes, buildings, and interiors is learned through a subtle socialization process. As we grow up, the organization of environmental signs develops into rules of behavior (Wood and Beck, 1994). A stop sign is a wonderful example. As a hexagonal piece of metal painted red with white letters spelling "STOP," it is sign, plain and simple. It is a statement, an

Figure 2.6 Stop sign. Drawn by author.

international statement. As a symbol it expresses certain socially dictated parameters that are reinforced each time we heed the "order" to stop, knowing that if we don't and get caught we will be punished. This is where the narrative form of The Analogy begins, with the distinction noted by Barthes: "it is 'what it says' that makes a statement a functional unit, not the manner in which it is said" (Sontag, 1994, 262).

Within an environment's second layer of information we read cultural, social, and personal aspects of time and place. In this realm we daily experience the messages that reinforce who we are: American, Canadian, Creole, Japanese; student; baker, executive, wilderness guide. Like

Traditional Russian Matryoshka

Figure 2.7 Matryoshka Russian (Nesting) Dolls. Courtesy, The Russian Store. https://www.therussianstore.com/blog/the-history-of-nesting-dolls/. Last viewed February 1, 2022.

Russian Nesting Dolls our personal histories, our biographies, are nested within our family history, set within local, regional, and national histories. Realizing these bases of how we locate, or orient, ourselves within our society provides for meanings within the messages that express our personal and unique collective senses of being and belonging in a place and a time.

Landscape features, their materials, their finishes, and their particular sequential organizations of space are a few of the media that express our personalities, display our familial biographies, and exhibit the social and cultural histories of which we are a part. Our routine acts of dwelling establish functional messages set into the contexts of a place and time (Norberg-Schulz, 1980; Wood and Beck, 1994). The manner in which place-related information is imparted supports the interplay of Christian Norberg-Schulz's concepts of orientation and of dwelling (1980; 1985) and is wonderfully explored in Yamin and Metheny's *Landscape Archaeology* (1996). It is not difficult to find paths, courtyards, patios, and public spaces across cultures and countries that accommodate similar uses and user groups, e.g., outdoor markets (Ardalan and Bakhtiar, 1973). Even with the same dimensions the details delineating two paths, one in Ming dynasty China and the other in Victorian England, would communicate walking but with quite different distinguishing characteristics of each setting's particular time and place. A wonderfully detailed review of such landscape features across history is found in Norman Newton's *Design on the Land: The Development of Landscape Architecture* (1971). A more concise review of similar landscape features set in different cultural settings is found in William Mann's *Landscape Architecture: An Illustrated History in Timelines, Site Plans, and Biography* (1993).

The second layer of landscape-as-narrative also tends to express (or call for) a place's emotional content (Al-Husain et al. 2013; Light and Smith. 2005) as drawn from our personal and collective memories. Then even deeper, there are our collective beliefs, values, and allegiances.

Within the third layer of message delivery (our collective unconscious) are embedded the deepest, most pervasive, essentials regarding the realities of life, nationalism, and afterlife. As the fundamental beliefs of our relationships with each other and with nature, "symbols (e.g. words and mental representations) get their meanings via correspondences to things in the external world...[as such] symbols that correspond to the external world are internal representations of external reality" (Lakoff, 1987, xii–xiii). Herein lie the basic fundamentals which color the ways we see the environment and interpret its features into the messages that form and inform our relationships and actions. Two wonderful summaries of this relationship with the world are found in the Wendy Faris's *Labyrinths of Language* (1988), and in a collection of works by Roland Bathes, found in *The Barthes Reader* (Sontag, 1994). The existence of related literary works is echoed in the studies of cultural landscapes by such authors as Zube and Zube (1977), Rapoport (1982), Potteiger and Purinton (1998), and Spirn (1998).

An intriguing realization regarding the three layers of communication is that the layer of deep meanings is both the most subtle and yet the most enduring, e.g., iconic architecture embodying culture, nationalism, and people's heritage: cross-cultural archetypes. Functional messages are the most blatant, immediate, and short lived in our mind as we move about a landscape. But, due to the immediacy of our concern for our personal safety as we move along a sidewalk or path from place to place earlier functional messages are quickly replaced by subsequent, immediate, upcoming functional messages. The messages making up a functional narrative are quite transitory. The secondary messages of the socio-cultural realm take time to be identified and interpreted (Potteiger and Purinton, 1998). Over time our initial recognition of a door as a door evolves into recognitions of a New England Colonial door versus a Spanish American door versus a 7th-century cathedral door. Consider visiting a foreign country for the first time, or even a new part of your own country, after a while you begin to pick up bits of the environmental vocabulary with which you were not originally familiar. But, keep in mind, you did recognize the basics and differentiate between doorway and the wall in which it was set.

## THE DUALITY OF COMMUNICATION

Our ability to read narratives in built environments involves an interplay of two aspects of communication: the structural aspects typically associated with linguistics and the interactive aspects of dialogue that are more the study of social theorists. Structuralism looks at human culture, what people do, think, perceive, and feel, as being organized and given habitual form, by larger overarching systems or structures (Johnston et al., 1981, 329–330; also Wikipedia, 2016). The repetitive nature of our interactions with particular spaces in given contexts structures how we come to read, and interpret landscape messages, and texts. Among the social theories that best support this argument is Anthony Giddens's structuration theory (Giddens, 1984). Our interaction with environments that we come to read is through a dialectic. Each time we identify and interpret, or read, a message "correctly" and act accordingly we reinforce that message and the signs and symbols from which it emerged. A pathway says "walk this way." That dialectic, the simultaneous back-and-forth messages, can also call into play our need to make changes to the walkway if we don't quite understand what the designer had intended. This is the back and forth that makes a discussion between us and the environment we are traversing. The repetition of such a dialectic is the living development of a social structure that says and is agreed upon by those who carry it out, in this case "Walk this Way," and I am not referring to a Monty Python skit.

Whether in oral or written forms of communication or landscapes, messages are expressed in and through their parts. The sequential nature of the parts that come together over space and time as a story requires the developmental continuity provided by an appropriate and related (to the story's theme) structure. An example is the mythic hero's story which is found in cultures around the world (Campbell, 1949; Vogler, 1992). The parts are what we will deal with shortly as the typology, first as parts of language and then as the related basic design elements of point, line, plane, volume, and sequence of volumes. Then we explore how the parts combine to provide the continuum of messages we call landscape narrative. The character of those building blocks as found in different places, and how they are physically organized guide our interactions with our environments. In considerably more detail, Anthony Giddens's structuration theory argues that the speech act, or our interaction with our surroundings both form and inform our sense of place and who we are is not a sequential act but a simultaneous one, a dialectic. As a dialectic it happens as part of a larger continuum as you cross a plaza, approach your favorite taco truck, and walk around to a friend's house to arrive at a backyard barbecue.

Each time we use a walkway our actions reinforce the linearity of the feature as it reinforces our understanding of and ability to move in a linear fashion. Our repetitious behavior also reinforces the social courtesy of "staying on the walkway" and off the grass. The fact that someone built the walkway displays a certain authority; to maintain it shows agreement with and support of that authority. To ignore a walkway's request to stay on it and diagonally cut across a lawn is to disagree with the function, meaning, and authority associated with the walkway and the authority behind its construction. To violate the social rules surrounding the proper use of a walkway is to at least show contempt for accepted social practices if not out-and-out disregard for local authority. Yet when enough people speak, their authority gets expressed, e.g., University of Michigan's "Daig" over many years became an acceptable way to "cut across" the open green.

We know the gathering nature of areas and the ascending and descending nature of diagonal features such as stairways and ramps. Our knowledge of such things is tacit, taken for granted. We know how to use language to interact with one another. Yet, if questioned, few of us can readily explain how we do it. Our tacit knowledge of the social rules of life allows us to move about in the contexts of our daily routines without being able to readily, and explicitly, talk about why we do what we do. Dennis Wood and Robert Beck (1994) do a great job of illuminating the socio-cultural roots underlying the social expectations in the organization of spaces and objects making up an American family's living room.

Figure 2.8 Aerial photo of University of Michigan "Diag." Courtesy Michigan Photography, University of Michigan.

Paraphrasing Giddens, regarding the use of landscapes, we daily reproduce the messages, meanings, and values associated with the elements of language through their use. "The repetitiveness of activities which are undertaken in like manner day after day is the material grounding of what I call the recursive nature of social life" (Giddens, 1984, xxiii). Habermas sees this recursive nature of social life as part of his communication theory, specifically the taken-for-granted rules which he calls validity claims. Whenever we are about to speak to someone we do so with certain expectations, or validity claims. Validity claims are presuppositions latent in our communicative interactions with others aimed at achieving at least a mutual understanding if not consensus. We expect that what is said is comprehensible (it is understood), that it is sincerely and honestly stated (it is truthful), that it is proper within the social rules structuring the situation (its normative; appropriate), and that it is true in fact (its propositional content is true). Satisfying the four validity claims constitutes Habermas's "ideal speech situation" (Habermas, 1979, 62–65). Now, consider

Figure 2.9 University of Michigan "Diag." Courtesy Michigan Photography, University of Michigan.

applying the validity claims to landscape architectural design. If our built designs, no matter how attractive, are not comprehensible, sincere, proper to their social setting, and not misleading, is that a good design? Is it an ethical design?

*An image of an ideal environment [a blend of the typical objective standards of residential design and residents' aspirations and values, or social space] theoretically subsumes aspirations about territoriality, accessibility to desired destinations, and ideal site character. Activity orbits (the spatial expression of social reference systems) contribute to and mold this image. Individuals establish affective relationships with particular urban places, routes, nodes through the spatial activity patterns generated by interaction with their social systems. Their expectations about residential area character are also influenced by the norms and values transmitted within this reference system.*

*(Buttimer, 1980, 30)*

In his structuration theory, Giddens highlights the fact that the repetitive nature of our daily actions and uses is situated; we tend to carry out the same activities in the same locations day after day. Our daily actions, occurring in particular and interrelated locations, bring to light the serial nature of our expectations and actions across space and time. "Every individual is at once positioned in the flow of day-to-day life; in the life-span which is the duration of his or her existence…Finally, each person is positioned, in a 'multiple' way, within social relations conferred by specific social identities" (Giddens, 1984, xxiv). Building on the works of Goffman (1963) and Garfinkel (1963), among others, Giddens sees the situated nature of routine actions not only related to locales and settings but also expanding over space and time to involve a "regionalization of encounters," or what can be termed a regional identity (touring the fall colors of New England). Repetitive social activities contributing to "place" are also explored as activity orbits and their spatial profiles by Buttimer (1980). The repetitive nature by which we apply our practical knowledge, our ability to reflect as a means of monitoring our ongoing flow of life, and other social and psychological activities are more fully explored later when we deal with the second and third realms of message delivery. Here, we begin dealing with the structural and structuring aspects of The Analogy.

## BASIC ATTRIBUTES OF THE TYPOLOGY

*What is really happening… which may be very conscious or may be felt only as a certain strangeness and unease, is a process quite central in the development of language when, in certain words, tones and rhythms, meanings are offered, felt for, tested, confirmed, asserted, qualified, changed. In some situations this is a very slow process indeed…in other situations the process can be rapid.*

*(Williams, 1983, 12)*

Each tier of The Analogy, reading and writing, basic design, and landscape design, is introduced below in terms of the fundamental signs identified by our senses. The basic attributes of the fundamental signs are those associated with vision and hearing, or what we write and read and what we say and hear in the communication of messages. This is not to say that even more information isn't communicated through non-words. Psychologists say that non-words such as sighs, gasps, yawns, and the like say a lot. We have all experienced the punctuation of a conversation with a sigh, a yawn, or a gasp. What was your reaction? What did it tell you about the other person? A *Time Magazine* article noted that computer analysis of micro-movements in people's facial expressions may soon replace polygraph tests as a means of detecting whether someone is telling the truth or lying (Golden/ La Jolla, 1999). We also read people's culturally learned behaviors: stance, posture, and gesture. Each of these message delivery systems is based on the identification of lines as edges, the shapes of the areas the edges delineate, the volumes the surface areas form, and the sequence of the volumes to which they give rise. Beyond the visual, our other senses also read the attributes of the objects and spaces that make up our designs. The movie and advertising industries employ this understanding of such aspects as shape, form, color, and sequence to say much more to our egos and vanity than any written text could possibly express (Ewen, 1976; Lynes, 1954). And yet, how many movies of people have you seen in cold winter scenes and their breath is not appearing as vapor?

Figure 2.10 Signaling a cold season: warm breadth vaporizing in cold air. Shutterstock ID: 717801571. Contributor: Robsonphoto.

Texts, cinema, the visual narratives in paintings, dance, opera, and the like carry us from message to message, impression to impression, and idea to idea. At a simplistic level, our interaction with words, motions, and sounds in the recognition of patterns that draw us into artists' and authors' stories is a form of wayfinding and route following. "Wayfinding initially takes place when a route is not previously known even though an origin and/or destination are known" (Golledge, 1999, 21). Our initial encounter with an environment occurs on the functional level of satisfying the need to get somewhere, safely and on time. Subsequent trips along the same route are what is called route following. When leaving on a previously traveled journey we experience an anticipation of particular time frames between points, curves, rates of speed, direction, and the like: expectations to be met. Route following, in subsequent trips, allows people to read more of the environment beyond the objective aspects of finding their way. People build an understanding of a place, they draw from or impart meaning upon, and reinforce the social and cultural contexts of which the route is a part (Buttimer, 1980). The nuances realized along a path traveled several times are similar to seeing the same movie, or reading the same book, a second, third, or fourth time. Previously meaningless or missed words, gestures, motions, objects, and sounds become evident and add to a deepening appreciation of the story in a book, play, movie, or orchestration.

As we consider the data each of our senses identifies, we begin to recognize that different types of information are interpreted through more than one sense. One good example is seen in the 1980s Maxell magazine advertisement. The power of the sounds recorded on Memorex tapes (*Is it Live or is it Memorex?*) is communicated visually with the listener's hair and necktie being blown straight back away from the speakers and, in the television version of the ad, with his chair being pushed away from the speakers. Look to ads, to films, to gourmet meals, and to music and you quickly find ways the advertisement industry's masters of persuasion play with our senses' reading of the "This is for you; You need to buy this" messages (Ewen, 1976; Lynes, 1954). Recognizing how various signals may be employed to communicate and reinforce a given message, as in the Memorex ad, enlarges the palette from which designers can author landscape narratives. Word search "Blown Away Man," then ask yourself, in the old 1980s Memorex advertisement photo, how do you know the power of the sound waves is blowing the seated young man backward?

As I argue later, while much of landscape architectural design remains grounded in the visual world, it is the supplemental orchestration of our other senses that turns a good design into a superior design. Our awareness of a point in a landscape is enhanced or diminished with supplementary audio, olfactory, and tactile input associated with that point. An intersection between two roads is made more memorable by the presence of nearby olfactory landmarks like the aroma of fresh bread or fragrant flowering shrubs. The linear nature of a pathway is echoed in the sound of the material under foot and the rhythmic occurrence of scented herbs along the route. A common example of the use of supplemental audio messages is found with the rumble strip running parallel to the painted outer edges of highway lanes.

*phoneme: one of the set of the smallest unit of speech that distinguish one utterance or word from another in a given language (Morris, 1970, 985). They are the smallest units that can form or change meanings, e.g. sit and sat.*

*Vocabulary: a shared body of words and meanings in our most general discussions...of the practices and institutions which we group as culture and society. (Williams, 1983, 15)*

The Analogy's expressive foundation begins with the visual units we know as the 26 letters that comprise our alphabet. The visual signs' aural-audio counterparts are phonemes, or the smallest units of sound capable of conveying a distinction such as the *f* in fat and the *c* in cat. From their socially accepted combinations come the syllables which have been combined and recombined through history to give us the sounds and symbols we use today considered the phonological building blocks of words (https://en.m.wikipedia.org/wiki/Syllable).

If I may make a metaphorical leap, the phonemes in basic design are the qualities that characterize each particular media and in turn bring a particular character to a design's expressive character: chalk's texture; paint's viscosity or sheen; paper's tooth; graphite's hardness, ink's density; motion's direction, rate, and rhythm; and sound's timbre. The qualities of a particular media, when used to make a visual point, or dot, initiate the production of a message. It initiates the character of a statement, as chosen by the author, and sets up the expectations of what is yet to come that will reinforce the theme. Consider the resulting qualities of a 3B (very soft) lead pencil versus those of a very hard 3H lead pencil, who uses them, and the products associated with the people, and their messages. Yes, I realize few people use pencils anymore but you should try a couple different densities. A point made with a hard graphite, a soft charcoal, or oil paint isn't yet a message. Still, such points can be considered statements, or hints of statements, yet to be enlarged upon. Taken in combination with the other appropriately selected design features they come together to build not just a statement but a statement of particular character. Then, in combination with other statements, the production builds into messages and then a narrative (Rubin, 1995; Sontag, 1994, 257–258).

*...no [structural] level on its own can produce meaning. A unit belonging to a particular level only takes on meaning if it can*

*be integrated in a higher level; a phoneme, though perfectly discernable, means nothing in itself; it participates in meaning only when integrated in a word, and the word itself must be in turn integrated in a sentence.*
*(Sontag, 1994, 257–258)*

Line Shapes as Line Metaphors:

| | |
|---|---|
| Borderline personality disorder | Toe the line |
| Bottom line | Crossing the line |
| Lifeline | Take a hardline stance |
| Come into line | Further down the line |

Syllables, as a combining of consonants and vowels, are the divisions into which we break words. Each division must have a vowel. Keep in mind that the pauses between syllables do as much to provide recognizable distinctions as does the contrast between the pitch and frequency of the sounds, or the hue, size, and shape of the letters in the text. Someone, and I cannot remember who, once said that it is the pauses between the notes that make music.

When gathered in a particular order, visual and audio signals furnish particular ideas. Remember the STOP sign from earlier? Working to understand the dynamics behind our interpretation of the signals as units of meaning, both in their production and in their reception, is an exercise well spent in coming to understand the message content of designed environments. The basic attributes of materials we use as design vocabulary are the structural and chemical characteristics that make them up. It is one thing to know what the crystalline structure of a rock will allow in terms of shape, character, and finishes. It is another to know that rocks left loose, as a dry laid stone wall, act in a fluid manner heaving and settling with frost and hydraulic expansion and contraction. Such walls communicate not only very different messages than stone fixed in concrete, but also the very construction of the message, in the form of a dry-laid wall, which is dramatically different from that of one requiring mortar and a foundation. Different degrees of craftsmanship, different amounts of time needed to construct a product, contribute to differing, deep meanings. In addition, the materials themselves can denote regions, e.g., wood planking, brick, adobe, stucco.

*A terrace...* ***unless there is sufficient*** *room to slope the earth at a stable angle, retaining walls will be needed to hold it back at the edges. These become much more difficult and expensive to construct as they get taller. The problems with...the cost of shifting earth usually impose very strict limitations on the widths of the terraces. The steeper the slope the narrower they get.*

*...simple earthen retaining walls can't be made very tall, so they fit closely to the original form of the land...Different natural landforms provide opportunities for a variety of terraced compositions.*
*(Moore et al., 1988, 27)*

Along a similar line of thought, the cellular structure of wood calls for its selection or placement as planks to anticipate cupping and checking, how they may be treated with oils and stains, and how they may best be used to take advantage of their load-bearing capacities. The elegance expressed in a grand entryway using fine inlaid woods is easily destroyed by poor finishes and poor craftsmanship – a poor understanding of the

Figure 2.11 Warped footbridge planking plank. Photo by author.

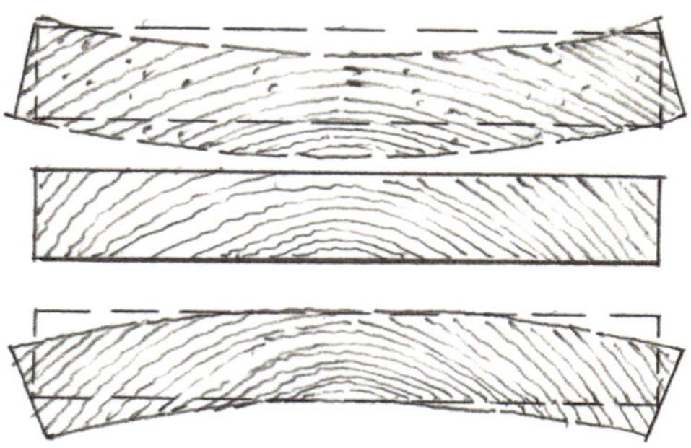

Figure 2.12 Cellular structure of wood. Drawn by author.

basic attributes of wood in differing contexts and climates. Good design is as much good site engineering as good narratives are an intimate knowledge of language and its meaning-full content.

Meanings associated with sounds, the silence between the sounds, and the order in which the sounds occur go beyond those that make up our spoken language. The most subtle, and sometimes not so subtle, forms of message delivery are important to the buying public. Auto, truck, and motorcycle manufactures not only know which phonemes the public most immediately accept, but they are also registering the rights to use particular words so other companies will not use them in the future. Imagine the Ford Bronco having been named the Bambi or Hummingbird. Say the three words: Bronco, Bambi, Hummingbird. Listen to the emphatic sound of the "B" in Bronco over the other two words.

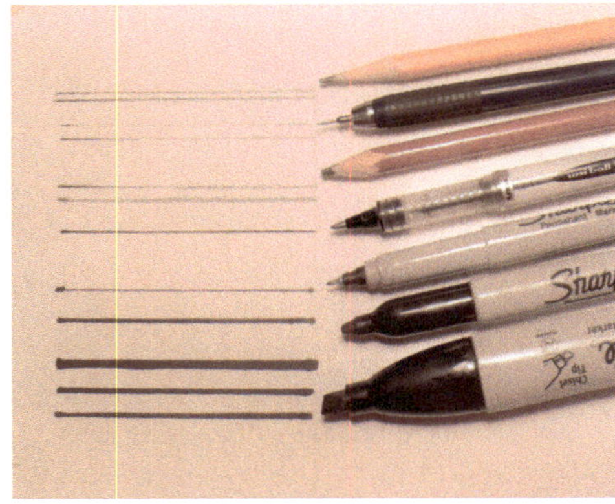

Figure 2.13 Different media produce varied visual messages: engineered precision to rustic character.

*An excellent introduction to how language shapes the way we think is Lera Boroditsky's TED Talk. (https://www.ted.com/talks/lera_boroditsky_how_language_shapes_the_way_we_think?rid=ZtfIml0pxUI3&utm_source=recommendation&utm_medium=email&utm_campaign=explore&utm_term=watchNow)*

As the simplest basic unit of the visual arts, a point quantitatively distinguishes itself from other points not only by its shape, color, and texture but also by its media's qualities. Points may be in the shape of a star, a square, an oval, and more. Each calls attention to itself, yet, in combination, a point's particular shape's edging, hue, texture, and tone calls attention to a distinctively expressed character. Points defined with visually softened edges express traits different than those with well-defined, crisp edges. The body character of a dancer's pause in a Fosse production is easily distinguished from that seen in a classical ballet. Our ability to associate (or project) personality traits on to shapes and forms is the basis of

Norton Juster's love story entitled, *The Dot and the Line* (1963). Similarly, seeing a bright spot set in a darkened forest is special when that accent is a dogwood coming into Spring bloom, as opposed to a spark growing into a fire. The visual accent of the bright bloom of the dogwood, surrounded by darkness shouts, "It is spring, the birth of a new season." Its accent in a darkened forest whispers amour.

A phonetic attribute of a point may also be in the point's seeming absence. Consider an oasis versus a clearing in the woods – mass vs. void. Audio points can be as much the notes or words as the pauses in conversation and music. Pauses in sentences are usually used to emphasize importance. Placed before the response, "Yes, I love you," rather than after the "Yes" speaks volumes. A momentary absence of sound due to a change in texture under foot or air temperature as you walk along a path modifies the primary experience. Visual, audio, tactile, and olfactory accents, focal points, inflections, and highlights may be masses or they may be voids. What is not there often communicates as much information as what is there. The absence of what is expected sets a place apart from other places of similar design function.

*That depends where you want to get to.*

*How do I get to Broadway?....I want to get to the center of things.*
*(Dos Passos, 1953, 4)*

*Quinn was struck by the way Stillman had skirted around the edge of the territory, not once venturing into the center.*
*(Auster, 1990, 82)*

Phonemes bring to language what the fundamental attributes of plants, soils, air, light, and water bring to landscape design, building materials bring to architecture, and lighting and finishes to interior design. Landscape architects select plants not only for their shape, not just for whether they are coniferous or deciduous, but also for whether or not their leaves are more green than yellow, more glossy than matte, more serrate than palmate, or of greater or lesser leaf density. We recognize, at least unconsciously, that this level of a material's fundamental attributes introduces or supplements the character of a message and comes to support the thematic character associated with an evolving narrative. Consider a sugar maple in full leaf, in full fall color, or without any leaves at all. How the phonemes combine are what we know as syllables, the basic units of the words we use to communicate: dic-tion-ar-y, land-scape, ar-chi-tec-ture, in-te-ri-or de-sign. Behind those basic units is the tone in which they are expressed: tone of voice or font's visual character.

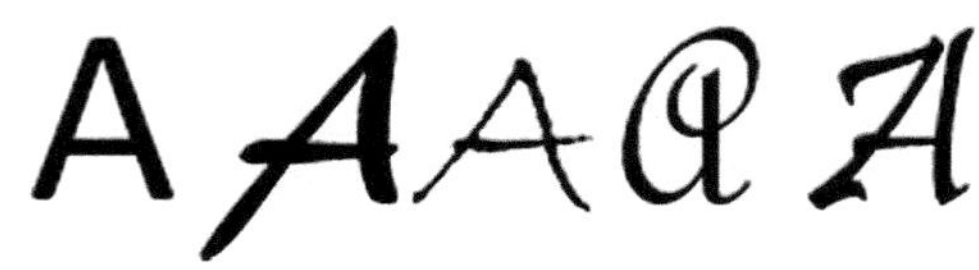

Figure 2.14 Expressive font characteristics.

The phonemes and syllables of landscape design call our attention to the attributes that give trees, shrubs, perennials, and annuals the overall identity we seek to employ in specific design and activity settings. An interesting aspect of plants as the primary medium of landscape architecture is that their attributes change over time as they mature. In a sense, the profession's medium is both an asset and a liability. As a plant matures its leaf arrangement, surfaces, and shapes, branching structure and patterns, leaf density and flexibility in the crown, and structural aspects vary. The beech tree is a good example. Starting out with upright limbs, the beech tree matures into a form with massive drooping lower branches often resting on the ground. An older beech tree's canopy comes to hide from sight what could be readily seen years earlier. As many plants mature they increasingly define spaces, often foreshortening views, and at times eliminating distant views all together. A private residence was deprived of a

Figure 2.15 Mature Beech Tree, Manito Park, Spokane, WA. Photo by author.

view over downtown Baltimore and its harbor as the neighbor's pine trees matured. In Washington, DC a historic estate, built in the 1800s, north of the Mall originally had a view over the city, the Mall, and the Capitol. A forested area downhill from the estate grew and eventually prevented any view of the Capital area. Historians and preservationists argued for years over whether to leave the trees (preservation) or cut them down and return the view (restoration). And to add to the character of a message imagine the difference in a single tree in a clearing or backyard when the tree is a conical pine versus a rounded or oval canopied deciduous tree. What might it express in each season?

Rather than just letters of the alphabet, I consider a word comparable to a point in basic design and a singular object in a landscape design, e.g., light pole, water fountain along a walkway, kiosk. The placement of a word, one word, on a blank piece of paper or computer screen initiates a cascade of events. Words are the basic design points of our language. Single words provide for our simplest statements. Used alone a word punctuates the quiet, typically stops a flow of thoughts, or calls attention to the speaker. In a landscape space, a lone object can arrest our movement, grab our attention, provide for a sense of here and there, and help orient us. Single words carry the phoneme's initiation of a statement/message a step further. The *si* sound in "silence" in combination with the shortness of the word brings a subtle emphasis to the message that the *qwa* sound in "quiet" doesn't.

In language, the evolution of a thought from a word to a series of words makes that linear feature we call a sentence and involves us in applying grammar and syntax: rules of message organization and context setting. In a similar fashion, in basic design, a point inscribes a line as it moves through space. In basic design we gather together design elements in response, or contradiction, to basic design principles (Wong, 1972; 1977). The number of principles vary across the art and design professions. Some designers say there are seven design principles, some say ten, some list even more. A few of the most frequently identified design principles include contrast, balance, emphasis, proportion, hierarchy, and repetition. Our selection and organization of the objects and spaces that make up landscapes, architecture, and interiors are guided by design principles. Some guiding design principles are practical in support of a place's best and safest use. They guide a certain clarity of expression in landscape messages so users' safety and welfare are protected. Some are aesthetically based, while others are derived from a culture's historically established art theory and perceptual psychology (Arnheim, 1969; Kandinsky, 1979; Langer, 1967).

The selection and organization of expressive spaces and objects depend as much on a syllable, the unit of spoken language consisting of a vowel alone or with one or more consonants, as it does on the phonemes. In other words, in landscape design, the basic units contributing to the greater message do so based upon their sequencing and how they are supplemented and augmented, how landscape features are refined, and modified. While the phonemes may be considered the crystalline structure of granite or basalt, or the chemical composition of paint, the syllables are the more immediately sensed characteristics of each material: granite doesn't look like basalt or shale. Viscosity and texture help us differentiate between watercolors and oils as each expresses a different visual character and message quality. As the immediately discernible material qualities of things, syllables allow us to identify objects and spaces for what they are: cedar wood bench, exposed aggregate concrete bench, plastic molded bench. Knowing and applying particular material qualities to a design allows us to impart it with the attributes of particular character – to supplement or diminish the particular theme of a place. In writing or printing, the basic visual unit is a squiggle on a page: a letter or group of letters of the alphabet representing a syllable. This starting point to the construction of messages indicates that there are media that can be combined in ways that are identifiable in both the shape and the qualities expressing that shape. William Carlos Williams' poem "Rain" (Williams, 1965, 745–748) is an excellent example of shape and subject as one. Once the shapes are identified they are then interpreted as meaningful or meaningless. Please understand that I use "interpret" for the sake of simplicity. Our actual acts of interpretation are socially based and occur only during or after we have first recognized a greater context and then identified a particular image, object, or pattern set in its context (Evernden, 1992; Wolff, 1981).

The tranquility of ponds and reflectivity of puddles is seen in rain that falls on level ground. Rain that falls in the hills or mountains "expresses" a forceful vitality and energy in the moving linear forms of creeks, streams, and rivers. Eastern Maryland's mostly flat coastal Eastern Shore is characterized by ponds and when it rains puddles. The western part of Maryland, they would say mountainous side of the state, is characterized by flowing water, streams, creeks, and falls.

A further building of a narrative's initial message qualities is provided by sentences. Like lines, sentences as linear units impart considerably more meaning than their basic units: the single word or point. As sentences multiply, their messages expand and open a developmental (progressive) logic to what is being communicated. Both their physical character (straight, curved, zigzag, vertical) and their message content are linear. Sentences are additive in their leading the reader/user to the author's predetermined location and event, or along a sequence of possible conclusions. The progressive development of messages calls into play a recognition of social conventions similar to those found in "playing a game with agreed upon pieces and rules and procedures" (Moore at al., 1988, 23; also Wittgenstein, 1958, 1960). While there may be a number of messages to be communicated, once begun the available potential content and form of what is to be delivered is increasingly narrowed (see Chatman's "contingency," 1980, 45–48). An author uses this logic of increasingly reduced choices to lead the

"So my life is spent
to keep out
love
with which
she rains upon

the world
of spring

drips

so spreads

the words
far apart to let in

her love

And running in between

the drops

the rain
is a kind physician

the rain
of her thoughts over

the ocean
every

where

walking with
invisible swift feet
over

the helpless
waves--"

Figure 2.16 Excerpt of lines 36–60 from William Carlos Williams poem, *Rain* (1965, 745–748), displays his effort to visualize rain.

reader toward or away from particular conclusions (Rubin's "cues," 1995). In matters of public safety and security the unfolding messages must be blatant and be what the general public would expect in the message sequence and emerging anticipated conclusion. "Do this." Do that." "Yes, you'll get from here to there safely." In matters of humor, a playfulness is set up between what is expected and what is found. This same interplay, done with greater seriousness, can give way to a sense of mystery or drama.

> *Once a label (professor, housewife, Black, chair, automobile, laxative) has been applied, people are much less likely to notice the actual qualities or attributes of that which is labeled.*
> *(Adams, 1976, 81)*

> *We shall not cease from exploration/ And the end of all our exploring/ Will be to arrive where we started/ And know the place for the first time.*
> *(Eliot, 1963, 208)*

Being familiar with the fundamental characteristics of various media used in a design (troweled vs. brushed concrete vs. gravel) heightens a designer's ability to orchestrate music for people's eyes and compose beauty for people's ears. Coming to know the potential nuances of basic design and landscape design is helped by James Adams's *Conceptual Blockbusting* (1976). Adams identifies numerous conceptual blocks that limit how we see and think about the world. He identifies the blocks as perceptual, cultural, environmental, emotional, intellectual, and expressive. One way he suggests breaking through these blocks he calls attribute listing (80–82). His introduction of attribute listing helps us break our tendency to settle for believing we know something just because we've accepted its label. He introduces us to ways to look beyond the immediate and below the surficial. Adams' exercise of attribute listing takes to heart Gregory Bateson's adage that "the map is not the territory." By asking us to list a particular object's attributes we come to know it better for what it is or, to paraphrase a cereal commercial, "to come to know it again for the very first time."

An exploration of the perceptions made by each of our senses carries our understanding of messaging beyond the visual realm of perception typically associated with design. Bricks are literally the building blocks of many things in the landscape. No one could guess the number of times bricks have been specified in projects throughout history. Yet, I have never heard designers discuss bricks as Adams asks us to: "list some of the attributes of a brick. Some of them are: weight, color, rectangularity (sharp edges, flat faces), porosity, strength, roughness, the capacity to store and conduct heat, poor capacity to conduct electricity, [and] hardness" (Adams, 1976, 81). He then asks us to consider the potential uses of bricks based upon each attribute. Yes, some of the answers he shares are strange, but a bunch of the answers do redirect our thinking to uses in the landscape that are both appropriate and unique. If we take this approach with various landscape materials and plants, then we acquire not only an understanding of the materials themselves but also the rarely considered major and minor message-delivery capabilities of those materials.

Exploration of The Analogy's typological units began here with the point and its contribution to line(s). In the upcoming chapters a little more detail given regarding points and lines sets the stage for a progression through planes, volumes, and sequences of volumes. Each unit is looked at (1) as related to language, (2) as related to landscape design, and (3) as building blocks to a landscape narrative. At this point you have been asked to consider the idea that landscape features express information, that information combines across space and through time to express a text, and set in the greater contexts of history and culture the texts can be authored and read as narratives.

# 3

# Points

The Analogy's typology helps us understand the expressive capabilities of the primary medium of our profession, the landscape. In an effort to reinforce that foundation we start an in-depth review here of the fundamental component parts that make up our design palette.

Our exploration of point, as both a design element and a unit of narrative, is based on two premises. First, points are read by each of our five senses, and second, points, as well as the other design elements of line, plane, and volume, are conceptual as well as actual entities, e.g., Lynch's "landmarks" (1960, 78–83).

"A friend in Seattle said of Mount Rainier, 'It is my reference point, my compass bearing. Even on a cloudy day, I know it is there. Portlanders feel the same way about Mount Hood" (Ewert, 1999, 7). Actual points, as physically unique features, stand out from their surroundings. This contrast can be seen in a point's relative size – typically small within its surrounding referential area or its shape relative to its setting, such as the Washington Monument. I say it that way because of the difference between a dot on a page and

Figure 3.1 Washington Monument, Washington, DC. Photo by author.

the three-dimensional Washington Monument on the Mall in the District of Columbia. Roland Barthes provides a fascinating analysis of another landmark, the Eiffel Tower:

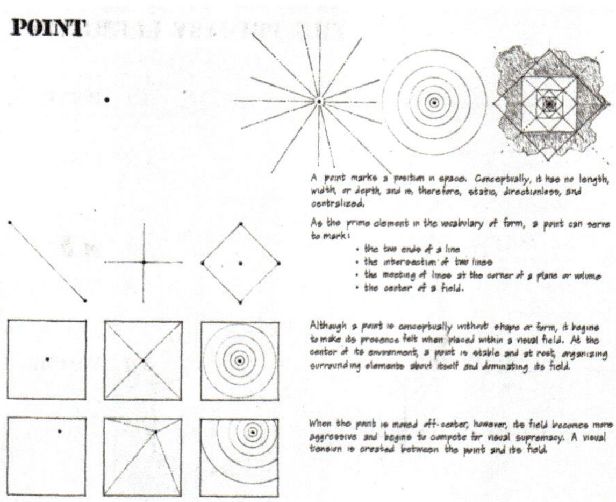


Figure 3.2 Dynamics associated with a point (Ching, 1978, 20). Courtesy John Wiley & Sons.

DOI: 10.4324/9781003286981-4

Figures 3.3.1–3.3.3 We take up points in space: limited personal space, midrange social space, and lots of public space.

*...all this night, too, [the Eiffel Tower] will be there, connecting me above Paris to each of my friends that I know are seeing it: with it we all comprise a shifting figure of which it is the steady center...the only blind point of the total optical system of which it is the center and Paris the circumference.*
*(Barthes, 1982, 236–237)*

Both landmarks can be considered points within their respective settings. Both "standout" from their surroundings. As a result, both draw attention to themselves or we could say "gather our attention to them." They are reference points, moments around which people may orient themselves. "The [Eiffel] Tower attracts meaning, the way a lightning rod attracts thunderbolts..." (Barthes, 1979, 5).

Contrast is a point's primary identifying feature: a momentary lone note heard over a harmony, a yellow garnish set atop a bowl of gazpacho, a splash of sherbet to cleanse the palette between courses, the sudden feeling of gravel at a special location in a garden, or the smell of freshly baked breads at a particular location in a city. These landmark moments often become memories. In the future they trigger memories, and the marker points along a journey.

Points do not occur in isolation. Whether on the written page, in a two-dimensional design, or as part of a landscape or building, points are points due to their relation to their surroundings. Single letters of the alphabet or Christopher Alexander's elementary patterns occur in relation to their contexts. "Each pattern can exist in the world, only to the extent that it is supported by other patterns" (Alexander, 1977, xiii). Contexts may be little more than the edge of a page or the orientation of the area in which the point is located. While a point may remain physically the same within itself, set within a variety of different contexts its implications and meanings can vary. The same dot produced by a pencil may be a decimal point separating dollars and cents or a period at the end of a sentence or it may be an intellectual concept, as in the animated film *The Point*. Points may be personified expressions of worldviews such as those of the characters found in *Flatland* (Abbott, 1991) or personal relationships between *The Dot and the Line* (Juster, 1963).

*Once upon a time there was a sensible straight line who was hopelessly in love with a dot. "You're the beginning and the end, the hub, the core and the quintessence," he told her tenderly, but the frivolous dot wasn't a bit interested.*
*(Juster, 1963)*

Figure 3.4 Moongate Garden, Smithsonian, Washington, DC. Photo by author.

Circular masses and circular voids also provide a spectrum of meanings and uses: a horizontal traffic circle or rotary, a vertical moon gate or bank safe's door, a spherical rotunda or dome. The embedded nature of a point, its strictly physical quality, is inescapable. Chameleon-like, any given point can maintain its identity as a point while its purpose is adapted to and by its surroundings. This is not to say they change, in and of themselves. A moon gate remains static yet shifts in early morning versus mid-day or diffused evening light. It remains a contrast to its shifting surroundings through the day and into the evening. The variety of uses and meanings are what the landscape users bring with them, the activities that they are looking to find and accommodate in meeting their expectations.

Once you decide to go to a particular place you conjure up both conscious and subconscious, context-driven expectations of what you will encounter along the way and when you arrive. The landscape reader-participant brings an aspect of context with him- or herself. Sitting in your favorite living room comfy chair is one context; the context you bring with you lies in the fact that the book you're about to read is a romance novel. Knowing that colors how you read each word. Considering context, the designer-as-author can anticipate and use the participant's expectations for coming to a particular place and give structure to what is seen, heard, felt, and smelled. One's movement through such places is helped by occasional points of pause and reorientation. The users' reasons for going to a particular kind of space greatly contribute to the way the orchestration of spaces and things in that space are read. By understanding the reader-user(s) the designer can better identify their expectations and choreograph their directions, movements, and pauses.

A point's ability to communicate information in the form of visuals, sounds, scents, textures, and tastes, whether in written or spoken language, a piece of art or a landscape, calls into play five criteria. They are the fundamentals of (1) space and (2) time, (3) the interplay of repetition and (4) memory as played out in our oral traditions (Rubin, 1995), and (5) the socio-cultural rules we live by (Wood and Beck, 1994). These criteria introduce the designer-author to the taken-for-granted contexts of many people's personal routines as daily and weekly cycles or, as Buttimer refers to them, activity space profiles (Buttimer, 1980, 29–40).

For now, we will consider the power of points as organizers of space and time. Alexander recognizes the sequential nature of working through the patterns making up his *Pattern Language* (1977), "moving always from the ones which create structures, to the ones which embellish those structures, and then to those which embellish the embellishments..." (Alexander, 1977, xviii). Alexander's recognition of the sequential ordering of patterns is both an application and an exploration of the idea of context. He realizes the connectivity of individual structures to structural systems, the larger systems of which they

Figure 3.5 Georgetown neighborhood, Washington, DC. Photo by author.

Figure 3.6 Philadelphia mixed-use neighborhood similar to Georgetown and others: Elfreth's Alley, Philadelphia, PA; Beacon Hill, Boston, MA; French Quarter, New Orleans, LA. Photo by author.

are a part, and the continuity across the details that hold them all together as identifiable entities. The shops snuggled in among the townhouses of Georgetown, in the District of Columbia, fit in their respective neighborhoods. As contrasting points dotted about the densely packed urban residential area, the shops provide residents with sustenance and news, desserts, and coffee. They also provide local people with the kinds of landmarks that allow them to easily give visitors directions to other parts of town. While their architectural lines and facades clearly denote them as other than residences, they are clearly an integral part of the greater fabric of the neighborhood. As a pedestrian neighborhood, the connection being made between individual structures and their larger structural systems is one of fit.

The shops dotted about Georgetown's neighborhoods provide centers from which radiate traffic. They provide orientation points that give an area its distinctive patterns of landscape and life, more so than if the neighborhood was a homogeneous suburban land use area of single-family residences.

## THE POWER OF THE POINT

*Thus we look upon the geometric point [a period ending a sentence] as the ultimate and most singular union of silence and speech… In the flow of speech, the point symbolizes interruption…at the same time it forms a bridge from one existence to another.*
*(Kandinsky, 1979, 25)*

*Gardens, like buildings, can engage a site in several different ways. One way is for the garden to stake a claim over the site, just as a perfectly formed, symmetrical Greek temple takes a spiritual control of the whole landscape that it surveys. Circles of trees on hilltops, found in the English landscape since King Arthur and Merlin, similarly appropriate their surroundings.*
*(Moore et al., 1988, 22)*

As a contrast to their surroundings, points are energy gatherers. They have power. They allow us to orient ourselves as they introduce a sense of here and there. They accent or punctuate a progression or pathway. What is unique about this power is that as a mathematical entity points do not actually exist; they have no length or breadth. Mathematically and philosophically, they do not occupy space – that other intangible medium of design. Ignoring the mathematical definition, designers work as though points in the landscape, in architecture, and in interiors do have length, breadth, and depth; as though they do exist. Points are places to be gathered in, stopped on, turned around on, seen as a beginning or end, or passed by. As spaces, points get a lot of a designer's attention, much to the dismay, I'm sure, of philosophers (Moore et al., 1988, 26). Carried to its humorously logical conclusion, the power of a point as an existing, nonexistent energy gatherer is found in Italo Calvino's (1968) intriguing exploration of the Big Bang. He introduces that point of inception as that

> *moment when all the universe's matter was concentrated in a single point, before it began to expand in space.*
>
> *Naturally, we were all there... where else could we have been? Nobody knew then that there could be space. Or time either: what use did we have for time, packed in there like sardines?*
>
> *I say 'packed like sardines,' using a literary image: in reality there wasn't even space to pack us into. Every point of each of us coincided with every point of each of the others in a single point, which was where we all were. In fact, we didn't even bother one another, except for personality differences, because when space doesn't exist, having somebody unpleasant like Mr. Pber t Pber d underfoot all the time is the most irritating thing.*
>
> *(Calvino, 1968, 43)*

Points, landmarks, and exclamations mark and demarcate locations in space and time. As orientation points they denote location, places of transition along a journey, whether that journey involves walking or driving, a five-course meal, a symphony or aria, or that moment everyone gets terrified in a ghost story heard around a campfire. Each of these event points is a moment of differentiation in someone's daily life.

Other specific points also help us measure our progress along spatial and temporal continuums: birthdays, rest areas or unique landscapes, a high vista or a narrow gorge. While the campfire, waterfall, or clock tower are themselves points located in space along the duration of an errand, a day, a week, year, or lifetime, people's journeys through each day typically comprise smaller, less monumental, points: coffee shop, corner bus stop, florist, vest-pocket park, cleaners. That special meal may be a high point in your social calendar, while the serving of Prairie Oysters or Baked Alaska may be the bright spot in the meal. Macro- and micro-points are found in the landscape as well. Walden Pond is a special location, along the path around the pond the site of Thoreau's hut is an equally special spot. At both geographic scales, that of the pond and its surrounding ecology and that of the hut's site, as points they exhibit gathering power. Additionally, people intellectually gather at Thoreau's pond and hut as they read *Walden*. People congregate or pivot about points while remaining oriented: a small bowl with trickling water at a juncture of two paths or place of reflection, an ancient Beech tree and its shaded hiding places, a gateway or stairway connecting two gardens.

Figure 3.7 A campfire as a gathering point. Shutterstock ID: 59945146. Contributor: Alexander Ishchenko.

Roland Barthes captures the essence of a point, in this case a vertical point in the form of the Eiffel Tower, and its ability to contribute to our sense of orientation in both space and time. In his musing on the panoramic vision afforded a tourist by the elevation of the Tower's observation deck, Barthes notes:

> *....the bird's-eye view, which each visitor to the Tower can assume in an instant for his own, gives us the world to read and not only to perceive; this is why it corresponds to a new sensibility of vision... the bird's-eye view...permits us to transcend sensation and to see things in their [greater] structure...in Paris spread out beneath him, he spontaneously distinguishes separate – because known – points – and yet does not stop linking them, perceiving them within a great functional space; in short, he separates and groups; Paris offers itself to him as an object virtually prepared...*
> *(Barthes, 1982, 242–43)*

In much the same way in which Barthes communicates his realizations, in a sequential additive manner afforded by written text, so too he says that the meaningful realizations of a tourist are sequentially built during a visit to the Tower. Among the tourists' experiences, Barthes argues, is the realization of a landmark, as a point, in its form as a vertical feature and as one with an inside and outside. The Tower gathers to it people, a meaning of place, a sense of Parisian identity, and a dynamic of here/there and then/now that gives people a sense of orientation.

> *...being enclosed is therefore a function of the rite [of coming to know Paris]. Here, too, the Tower is a paradoxical object: one cannot be shut up within it since what defines the Tower is its longilineal form and its open structure: How can you be enclosed within emptiness, how can you visit a line?*
> *(Barthes, 1982, 248)*

The gathering power of points in the landscape is not new. Throughout history the tree gathered to it philosophers, religious leaders, and the myths of many cultures. Buddha rested under a tree to gain both physical and spiritual nourishment. Many peoples have seen the tree as an axis mundi that connects the three worlds, with its roots in the underworld, its trunk as part of the earth as we are, and its branches in the heavens. Basil Johnson speaks to Ojibwa ceremonies and a central post connecting mother earth and father sky (1982, 116–117).

Lewis and Clark found the Great Plains and Plateau Indians using the lodgepole pine to support their teepees and sweat lodges. The poles supported the shelters of home and the place of spiritual renewal. As the main support of the usually circular teepee or Sweat Lodge, their circling around it emphasized the axis mundi around which all of life revolved. The use of many of the lodgepole's byproducts likely added to the conceived power of this pole to renew, purify, and sustain life. Totem poles are another form of center point around which the spaces of life gathered and were seemingly protected while the spirits oversaw a tribe's well-being. Recently in popular media we were introduced to The Tree of Souls around which the Na'vi gathered in the film *Avatar*.

Nabokov and Easton's *Native American Architecture* (1989) provides wonderful examples of what I refer to as explicit (actual) and implicit (inferred) features that connected, located, and oriented Native American peoples in their actual

Figure 3.8 Eiffel Tower, Paris, France. Shutterstock ID 256252072. Contributor: S-F.

and spirit worlds. Most, if not all, Native American tribes had some form of central ceremonial feature in their wigwams, lodges, tipis, and homes. Some were explicit, actual features such as that found in "over twenty Plains tribes built temporary Sun Dance lodges" in which "the center pieces were a forked cottonwood tree" (168). During Haida winter ceremonies a pole that extended through the smoke hole was planted in the house. "This 'cosmic tree' was climbed by the shaman to symbolize his access to the sky world, with its spirits and powers" (38). The Hadasta earth lodge had "a four post and beam structure that provided internal support for the rafters at the central smoke hole" (127). Additional implied features visually connected the fire pit to central smoke hole or, working in the other direction, a shaft of sunlight would connect the sky world and inhabitants' structure's interior. Other somewhat implied vertical central features stood a few feet tall and aligned with the smoke hole, and by association connected the underworld and sky/cosmos. The Mandan's village center was designated by a tight circular column of "woven willow rods symbolizing the high-water mark of the Great Flood" (140).

Early Native Americans regularly experienced the power of gathering places. The locations of many waterfalls throughout the country were the annual spring gathering points of Native Americans. In early spring, fish, returning upstream to spawn, were slowed, or stopped at the base of many waterfalls. Stories say that even warring nations set aside their differences and lived in relative harmony as they gathered each spring to fish salmon on the Spokane River and alewife on the Connecticut and Charles Rivers of Massachusetts. In Turners Falls, Massachusetts, the annual spring feasting occurred over so many years that the soil below the falls is said to be saturated in fish oil and bones to a depth of 40′. The power of these annual springtime places and their related gathering events provided physiological nourishment, spiritual renewal, and the turning point from long winter to plentiful summer. The points were temporary, yet annually repeated, moments along cyclical journeys.

Maryland's State Tree in Wye Mills, MD was more than 100 feet high and had a branch spread of 165 feet. The marker in front of the Wye Oak said:

> *The Wye Oak. The largest White Oak in the United States. Estimated to be 400 years old (1940). The Wye Oak was believed to be over 460 years old at the time of its destruction during a severe thunderstorm on June 6, 2002.*
>
> *(Wikipedia https://en.wikipedia.org/wiki/Wye_Oak)*

Points as objects are explicit. They are easily recognized. They denote such generic meanings as initiation, transition, and terminus of ideas,

Figure 3.9 Points within points; Duncan Garden, Manito Park, Spokane, WA. Photo by author.

movement in a particular direction, and relations with a place. As spaces, points are, in and of themselves, implicit. They connote many of the same messages found in their associated object forms, but those messages come more through the activities that they allow to happen in them than the objects that surround them. A bench denotes a place of rest and the opportunity for relaxation, reorientation, or entertainment in people watching. Set in the linear space of a walkway, rather than in an area to the side of it, the message of "rest" and the experience of renewal are diminished. Explicit landmarks such as tall sculptural structures, dominant trees, clock towers, and buildings generally call for adjacent and surrounding gathering spaces. The association of vertical point with horizontal area is the knowledge we take for granted as we move toward such objects. Village church spires, seen from a distance and preceding your arrival at the village, foretold of a public space, typically a market area. Together they suggested both spiritual and nutritional sustenance. The Eiffel Tower and the One World Trade Center work as large-scale orientation points in the same way the Wye Oak did as a historical landmark drawing people to itself.

*One of the most universal of human instincts is to raise a landmark from the surface of the earth. This gives a center to a fragment of the world... Landmarks may serve to announce a destination, to create the promise of arrival in a place of human habitation...Instead of marking a center, a landmark may also be used to define a boundary.*
*(Moore et al., 1988, 30)*

Explicit signals call us to gather in their associated spaces: market towers and church steeples. The cathedrals of Chartres and Mont Ste. Michael are each easily seen from 8 to 10 miles distance. Each stands well above the surrounding farm fields from which peasants came for protection, knowledge, spiritual renewal and to pay their tithe. Places of congregation were,

Figure 3.10 Chartres Cathedral, France. Shutterstock ID: 1985033804 France. Contributor: Didier Foto.

Figure 3.11 Mont Saint-Michel, Normandy. Shutterstock ID: 1471944950. Contributor, Ilya Shimanskiy.

and still are, places of shelter, nourishment, and growth through social interaction. Vertical points are intriguing in their association with the fundamental aspects of Maslow's hierarchy (1970): shelter, food, and social interaction. These are aspects of the functional messages associated with points. The narrative aspect of vertical points in regard to deeper social and symbolic meanings is explored below. For now, it is important to begin to recognize the functional messages associated with vertical points.

Often a point as an object works in tandem with a point as a space, or void: the market square and church steeple. One announces the other. Again, we are reminded that points have the power to gather people's attention, if not the people themselves, and their energy. So, by broadcasting its presence, the vertical allows the other to be utilized by those gathering in or by it. A church steeple at the end of a marketplace is as much the marketplace as is the space accommodating the vendors, the words and gestures used in bartering, and the aromas and scents of the goods being hawked. A public square or marketplace becomes associated with the vertical commercial or religious structure. The vertical object announcing the horizontal space and the horizontal space are interdependent. At the edge of a garden, or the approach to an estate, a gate is as much the arrival point as a gateway is a stage on which the experience of arrival is felt. In fact, we often lose sight of the fact that what isn't there (gateway, doorway) is frequently as important as what is there (gate, door). When we refer to a gate, is it the gate or the gateway to which we refer? Is it the space or the objects defining the space that are important? The same questions can be applied to a window (Norberg-Schulz, 1980, 8); is it the wall or the opening in the wall that affords us views connecting interior and exterior?

*The Utility of Not Being Thirty spokes unite around the nave; From their not-being (loss of their identity)Arises the utility of the wheel. Mold clay into a vessel; From its not-being (in the vessel's hollow) Arises the utility of the vessel. Cut out doors and windows in the house (wall), From their not-being (empty space) arises the utility of the house. Therefore by the existence of things we profit And by the non-existence of things we are served*

*(Lao Tzu, 1948, 87)*

In 1889, Camillo Sitte observed the interplay of masses and voids making up public spaces. In his pivotal work *The Art of Building Cities: City Building According to Its Artistic Fundamentals*, he identified and substantiated numerous urban design principles from his personal observations of ancient European urban spaces. He determined the "ancient," as opposed to the modern, rationale for locating fountains and statues in public spaces was to place them where people didn't walk. While such objects signaled gathering places, "this explains why fountains and monuments are not located at places of intense traffic use, nor at the center of public places, nor on the axis of a monumental portal, but by preference to the side" (1945, 14).

*If the building of a fountain were desired, it obviously would not be located on a thoroughfare, but rather on one of the island-like plots separated from the traffic… This also explains why in each city and each public space the arrangement of the monuments is different, for in each case streets open onto the square differently; traffic follows a*

*different direction and leaves other points free.*
*(Sitte, 1945, 14).*

*Almost 100 years after Sitte's observations Edmund Bacon notes our contemporary ignorance of space as a preoccupation of designers that flies in the face of Sitte's realization of fountains placement: The basic ingredient of architectural design consists of two elements, mass and space. The essence of design is the interrelation between the two. In our culture the preponderant preoccupation is with mass, and to such an extent that many designers are 'space blind'.*
*(Bacon, 1974, 15)*

Today, when we look at many fountains and the spaces in which they are set, it seems the rationale for Sitte's findings has been forgotten.

Narrative form, as the interplay between masses and voids, is not the result of whim or whimsy. The mathematics associated with objects and spaces that have come to be known as strong orientation points was noted as early as the 1880s by Camillio Sitte (1945), again in the 1950s by Paul Zucker (1959), and even later by Cullen (1961), Ashihara (1970), and Lynch (1971). Spatial proportions, along with Barthes' philosophical interpretation of the Eiffel Tower, find the Parisian landmark set within three domains of influence: setting or context, the context's close-in composition, and the immediate surrounding's experiences. Based on "the human body and the range of human vision" Zucker notes that the close-in experience of an object is most favorable when the perception of an individual work of architecture is set in a vertical angle of 27 degrees within our vertical field of vision. He further notes that

*the beholder moves instinctively to a distance which provides for this angle. However, in order to fuse various architectural units with their surroundings into a total impression, which is the genuine task of all city planning, the eye can employ an angle of 18 degrees only.*
*(Zucker, 1959, 7)*

If that angle-distance ratio cannot be achieved, then the space feels tight and confining.

*Before her, she sees the beginnings of the actual form of the town as it emerges from among the trees, but the path is still tortuous, and the form is merely suggested, except for the bell tower of the church, which will shortly emerge as a powerful architectural statement, suggesting the character of the experience which lies ahead.*
*(Bacon, 1974, 53)*

In Barthes's terms, the immediate experience of a landmark shifts the landscape narrative from one of anticipation and context setting to that of an emerging personal experience or with luck the meeting of one's expectations. As with the Eiffel Tower, this transition in narrative occurs not only as you get closer but also when you enter into it, when you become part of the landmark or point when it seemingly disappears.

*And as the scene unfolds the church tower, the climax, is at last revealed. Due to [enclosing buildings] the angle of the road it performs the last act of closure before we enter the wide main street where everything is revealed. This is the final of the successive acts of closure which formed a series of dramatic visual events in a coordinated sequence which provides, on a delightful domestic scale, a model piece of townscape.*

*(Cullen, 1961, 110)*

Figure 3.12 Hatch Memorial Shell, Charles River, Boston, MA. Photo Credit: Susan Pease/ Alamy Stock Photo.

Figure 3.13 Columbia River, George, Washington. Courtesy of Photographer, Anthony Fabrikant.

Another recognition by designers of a narrative form related to points is that of the progression one experiences along a journey. The power of punctuation to provide a rhythm to a text or a poem is felt in walks and journeys through the landscape. A trip gains its identity from identifiable points along the way. Squares, plazas, alcoves, and intersections are all intermediate points along a journey that are more than rest and renewal. At a larger scale, this form of narrative is experienced in a road trip such as along the Virginia to North Carolina Blue Ridge Parkway, the 2,200 mile Georgia to Maine Appalachian Trail, or the 2,650 mile Mexico to Canadian border Pacific Crest Trail.

As intermediate anchors, those special points along the route provide a place-based rhythmic sequence of happenings linking (or imparting a continuity) to places as they help build people's sense of identity with the larger setting, e.g., garden, park, town, city, wilderness trek. In *Design of Cities* (1974), Edmund Bacon provides a wonderful exploration of the interplay of the mass-void relationship of obelisks and their gathering spaces as they relate to a pilgrims' journey. He details the meandering route taken by pilgrims visiting seven churches in 16th-century Rome. Using an engraving of the time, Bacon describes "the phenomenon of undirected, undifferentiated movement" and further states that "[b]ecause it is meandering, it would induce in the pilgrims no clearly organized sequence of purposeful architectural impressions" (Bacon, 1974, 136). Bacon goes on to describe how, with the 17th-century introduction of a series of obelisks through Sixtus V's plans for Rome, there is produced "an awareness of the structural relation of functions in what appear to be a chaotic distribution of independent functions" (Bacon, 1974, 139). Landmarks are the punctuation in a narrative's progression through space and time. As physical objects, they are the accents in an unfolding message. They are the physical features accenting people's experiences as they pass by, enter into, and participate in the associated spaces. The spaces associated with landmarks provide for pauses or rests in an evolving story. Pivotal spaces along a route provide for activities different from the linear character of the activities associated with moving forward. Implied landmarks are the voids we know as plazas, piazzas, or the intersections of walkways. The physical, observable nature of implied landmarks is their tendency to represent recognizable moments in a temporal progression. Such landmarks are found in Sienna, Italy's Piazza del Campo and its semiannual festival, Palio delle Contrade and the horse race; the annual Fourth of July gathering at the Hatch Memorial Bandshell on Boston's Charles River; and the annual Folk Festival on the Mall in Washington, DC. At a greater frequency, people enjoy weekly concert gatherings during the summer along the Columbia River Gorge in George, Washington. Yes, there is a town called George, Washington. And nearby, believe it or not, is Martha, Washington. The bowl-like landform overlooking the Columbia River Gorge in Washington State is home to a dozen or more big-name concerts each summer. These locations, like that of Michael Pollan's personal hideaway space under a forsythia bush in his childhood backyard (1991, 7–8), easily take on the double duty of points of initiation and destination, as well as geographical locations in time. Pollan distinguishes between a child's and an adult's perceptions when he notes,

> *I say that no grown-up knew about it [this space] because, in an adult's picture of this landscape, the hedge row runs flush against the fence. To a four-year old, though, the space made by the vaulting branches of a forsythia is as grand as the inside of a cathedral, and there is room enough for a world between a lilac and a wall. Whenever I needed to be out of range of adult radar, I'd crawl beneath the forsythia's arches, squeeze between two lilac bushes, and find myself safe and alone in my own green room.*
>
> *(Pollan 1991, 7)*

Personal and public affairs begin and end in sites of hiding and sites of history: first loves and lost loves, triumphs and catastrophes, are places never forgotten. Points as spaces or voids are places that highlight or accent our lives, as opposed to those spaces of more mundane meanings in which we gather on a regular basis. Both magnitudes of space are meaningful, some are the turning points in our lives while others are the stability of regular everyday associations on which we rely for a sense of security and orientation: sidewalk café, gym, parking lot, post office parking lot.

## SUBTLE MESSAGES

Points in the landscape are not always as blatant as a public square announced by a spire or tower. Another family of points includes those that are

AMERICAN
SOCIETY OF
LANDSCAPE
ARCHITECTS

Figure 3.14 Lines express a message. Courtesy American Society of Landscape Architects.

simultaneously implicit and explicit. In the realm of communication, this is like reading what a text has to say while simultaneously reading "between the lines." Or, you may hear what is explicitly said by someone while becoming increasingly aware of that person's "hidden agenda." In basic design a point can be implied in the twists of a collection of lines. A similar example is found in the American Society of Landscape Architects logo of a few years ago. Precisely located expansions of certain of the logo's explicit lines convey the implied lines of a landscape.

Landscape examples find people gathering under the trees that blend with the surrounding landscape to worship at the 236-acre open-air Cathedral of Pines in Rindge, New Hampshire. Crowds regularly gather for an evening of musical entertainment under the trees of Columbia, Maryland's Pavilion. From the air, in plan view, these points covered by their respective tree canopies are not evident. At eye level though they become more explicit. Even on the ground, the continuation of the vegetative canopy of the Pavilion's trees out across the streets and to other wooded areas provides a continuity to the greater place, differentiated more by shadow patterns and the activity areas delineated on the ground plane. There are landscape narratives to be read as plan views from an altitude of 30,000′. But for us, and for those for whom we design, landscape narratives most readily occur on or close to the ground at eye level.

These measurable (observable) attributes of points in space provide for the locational aspects of here and there, distance, right-left, up-down, and a sense of orientation. Visitors to and residents of Seattle find help in orienting themselves with the city's Space Needle and Mount Rainier. In their singularity, points are statements: the object or space itself, with its explicit meanings, is associated with attributes like size, shape, material and finish, location (placement), and directional alignment (Greene, 1962; Merleu-Ponty in Rosenthal and Bourgeois, 1990, 401–409).

A bench or seat reads "sit here," "sit alone," "sit with others," "be comfortable," or "don't stay long." Its placement and orientation says, "look in this or that direction." In conjunction with a scenic overlook you are set up for an outward experience of a long or panoramic view. Or, set within a small urban park or garden alcove, you are provided a more introverted experience of foreshortened, enclosed views. A single bench, set near the base of a basalt outcrop with water trickling down it, can say, "Sit here and take time to look inward." As statements they are not complete narratives, a fact not recognized in the papers that comprise volume 2(3) of *Landscape Review* (Lincoln University Press, 1996). While the various *Landscape Review* authors recognize individual statements from a landscape-as-text point of view, they miss seeing the greater continuity that comes with the statements flowing in a landscape-as-narrative progression.

Figure 3.15 Seattle, WA Space Needle and Mount Rainier in background. Shutterstock ID: 776152702. Contributor: Rudy Mareel.

*...the use of landmarks involves the singling out of one element from a host of possibilities, the key physical characteristic of this class is singularity, some aspect that is unique or memorable in the context.*

*(Lynch, 1960, 78)*

Advertising uses the principles of placement of text, images, and corporate logos to grab your attention and direct it in a particular direction across a page or computer screen. Lynch's landmarks provide for much the same behavior relative to finding your way (Lynch, 1960). The more readily identifiable (Lynch says "readable") the word, object, or landscape feature, the more meaning-full your relationship with it. The less readable the objects the more disoriented or lost you feel. "Anyone experiencing even momentary disorientation and lack of recognition of immediate surroundings has experienced the uncertainty of being 'lost'" (Golledge, 1999, 5). When new to a place you test the place's readability, through a trial-and-error dialectic, and eventually acquire an understanding of how to read the new, to you, landscape vocabulary, to read the new shape or form of art, or to appreciate the previously unfamiliar object in the landscape. Orientation is acquired. A fundamental goal of life is to maintain a sense of orientation (Rosenthal and Bourgeois. 1990; Bloomer and Moore. 1977). However, one's sense of orientation is a complex of three realms of situated knowing: functional, socio-cultural, and that of the deeper meanings coloring your life. Orientation derived from a functional narrative pertains to one's location or knowing where you are relative to where you are going or where you have been. This spatial orientation along a path or route links where you are going with where you started from. Within the socio-cultural realm of narrative, orientation reaffirms or clarifies who you are in your life (Foster, 2009). You have roots. You know where you come from and have some idea of where you want to go relative to your future: a blending of the social history of which you are a part and your emerging personal biography. The message content of a design can spark within you aspects of culture/history, kinship/family, place experiences, and beliefs and spirituality.

As the number of points within or across an area increases, their distinctiveness is diminished. Their message changes from accents within a greater context to becoming the context. In the massing of points, an emerging pattern gives rise to rows, edges, or enclosure. One tree or building in a field may be a focal point or ornamental accent. Multiplied one after the other, and depending upon their layout, a grouping of trees grow into an implicit or imagined line with space between their canopies. Then, with the addition of more trees packed closer together, they become a seemingly explicit, solid line. Lines of trees with canopies overlapping are hedgerows, buffers, or screens. A line of residences moved closer and closer together till their exterior walls become the common walls of townhouses or row houses. A scattering of small retail shops along a street easily

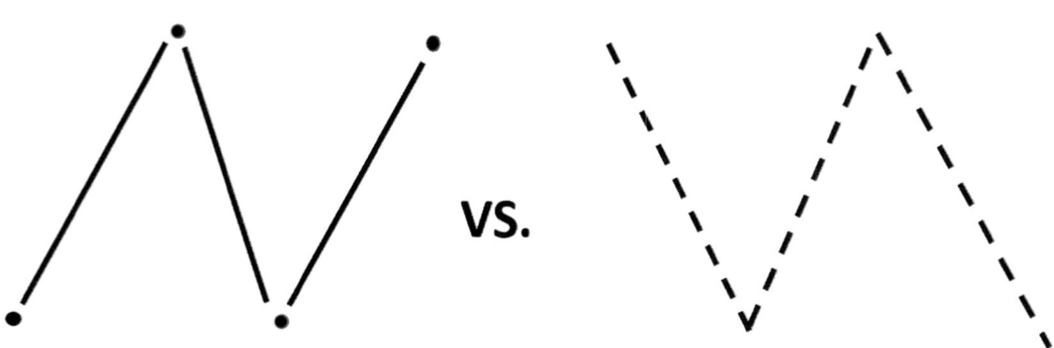


Figure 3.16 Implicit and explicit points. Drawn by author.

Figure 3.17 Brick buildings and tree-lined street in Back Bay, Boston. Shutterstock ID: 89093767. Contributor: Jorge Salcedo.

multiply into a strip development. Multiplied outward in various directions they become land use areas. One pine tree can be ornamental, while millions are identified with the culturally distinctive region of New Jersey known as the Pine Barrens: an ecologically unique region spanning more than seven counties (McPhee, 1968; Berger and Sinton, 1985).

The multiplication of individual words into sentences and then into paragraphs works to strengthen the context and content of a well-written narrative. In a sense, planes, like paragraphs, give body, substance, boundaries, and direction to a message. In Fitz Hugh Lane's painting, *Norman's Woe*, individual rocks make up the curve of the waterline that separates the sweeping ground plane of the beach from that of the ocean.

*The curved recessional of the western shore measures the depth of space in the right half of the picture, the left recedes by planes from the screen of vegetation thence to the rocks at the waterline...smaller rocks chart lateral distances across the inlet.*

*(Conron, 1985, 392)*

## SUMMARY

Space, time, progression, and context are paramount among the building blocks that structure the languages with which designs are made and through which their narratives are communicated. Learning landscape, architectural, or interior design means learning about what, so very often, isn't seen or heard, e.g., where people are not gathering. We often become so captivated with things, the word, the mark on a paper or canvas, or the object in the landscape, we forget our relationship to it and its relationship to us. Whether figure or ground, mass or void, object or space, a point like a solitary word on a page does not exist as an individual entity. Points exist in contrast to their surroundings, and in relationship to us once we recognize their presence. Points are context dependent, yet interrelated in their relevance to each other. Points in whatever form provide a complex of implied bits of information to those who experience them. Within the realm of people's sense of identity the appearance (or smell, sound, touch) of an object provides for a You-It or I-Thou relationship. If the relationship is readily identified as a part of your personality, your psyche, or profession the sense of identity is personal. The story, piece of artwork, or spatial environment expresses, and to a degree is, a part of you. If it holds meaning for you, if you identify with it, then aspects of your heritage and personal biography come into play. This is the foundation of Potteiger and Purinton's *Landscape Narrative*

Figure 3.18 Fitz Henry Lane (1804–1865), The Western Shore with Norman's Woe, 1862, oil on canvas. Collection of the Cape Ann Museum, Gloucester, MA. Gift of Isabel Babson Lane, 1946 (1147.9).

(1998), a blending of history and heritage in a site, a temple, a community's landscape. These are critical aspects of meaning and message delivery explored more completely in the upcoming chapters on narrative.

*When there is no interpenetration between the forms of monuments and the [Negative] N-space, which is the reverse space, and when the two balance each other beautifully, monumentality becomes unique and its quality enhanced.*
*(Ashihara, 1970, 37)*

As designers, we are wrong to consider a point as a static singular entity. In landscape architectural design we identify points in terms of their aspects of location, orientation, size, shape, and more. Within the realm of human psychology there is again a complex of characteristics that we identify and interpret in order to read points both in and of themselves and as moments in a larger progression. Good design is an orchestration of sequential sensual experiences. As John Simonds concludes, in his 1961 Epilogue to *Landscape Architecture*, a good designer is one who "...plans not places, spaces, or things – *one [who] plans experiences* (my italics). The places, spaces, and things take their form from the planned experience" (1961, 225). So, in whatever form, chose your points well. They could be the start or end of something wonderful.

# 4

# Lines

Ever wonder on the myriad qualities that can make up lines? Ever wonder on the multiple possibilities those linear qualities can bring to a landscape design? If you have, but not yet read Theilgaard Watts' observations (1975, 198–201) of a forest path that becomes a dirt road, that in turn through a series of modern-day modifications becomes an urban sidewalk, then you are in for a treat. With no philosophy. No mathematics. Through just good old personal observations, she sees, in the actualities of linear features, what the likes of J. B. Jackson sees as the connectivity of places, Susan Langer sees as sequence and rhythm, and Ursula Le Guin sees as implicit and explicit boundaries. For Jackson, "The Path, as I see it, has the prime function of introducing new life to the city, of bringing the city into touch with the outside world" (Jackson, 1970, 102). Langer sees that new life in how words are strung together; "But, words have a linear, discrete, successive order; they are strung one after another like beads on a rosary…" (Langer, 1980, 80). Among that myriad of qualities I came to understand that make up linear features Le Guin shined the brightest light with her opening to *The Dispossessed* when she wrote,

> *There was a wall. It did not look important. It was built of uncut rocks roughly mortared. An adult could look right over it, and even a child could climb it. Where it crossed the roadway, instead of having a gate it degenerated into mere geometry, a line, an idea of boundary. But the idea was real…Like all walls it was ambiguous, two-faced. What was inside it and what was outside it depended upon which side of it you were on.*
> *(Le Guin, 1974, 1)*

Figure 4.1 Q'emiln Park, Post Falls, ID. Photo by author.

Figure 4.2 Trees in a line, Lexington, KY. Photo by author.

DOI: 10.4324/9781003286981-5

A critical association between points and lines is found in all cultures. The Middle Eastern aspects of streets and squares are provided for us at a variety of scales of movement by Zucker (1959), Rudofsky (1969), Bacon (1974), Benzel (1998), Whyte (2001), and more (see Anderson, 1986). Among the examples, Benzel shows us how the structural relationship between point and line are "the two basic organizing elements of the Roman house – the main room (atrium) and the corridor – correspond to the two basic organizing elements of the city – the main square (forum) and the street" (Benzel, 1998, 83).

*For change has always happened in the margins, across human history, and it's happening there now. Seismic shifts in common life, as in geophysical reality, begin in spaces and cracks.*
*(Tippett, 2016, 1)*

*I have learned that people in the center are not going to be the big change makers. You've got to put yourself at the margins and be willing to risk in order to make change.*
*(Tippett, 2016, 33)*

Linear features in architecture, in landscapes, and in interiors provide two forms of dialog. One form is of information delivery, or that of sharing as brought about by connections. We are allowed to connect with others through paths, walks, roads, rail lines, and airline routes. Each works to sequentially deliver, in one way or another, information and experiences. The other aspect of dialog is generated along the interface of two or more abutting areas, in the contrasts or harmonies along the boundary lines of abutting areas. Kuhn (1970) talks about new ideas coming from the periphery of what we know. He addresses how old thinking is labeled as the status quo at the center and how revolutionary ideas come from those who travel to the outer reaches and experience new kinds of knowledge. We read a similar thought regarding new ideas, technology, music, and the like coming from the outer reaches in what sailing Captain Littlepage says in Jewett's *Country of the Pointed Firs* (1956) of his travels to little known regions.

*Most man-made structures or places have meaning only to man, and only as he experiences them. They are revealed to man by lines or patterns of circulation that lead him to, through, over, under, or around them... We thus realize that the circulation pattern is a major function of any project, because it establishes the rate, sequence, and nature of its sensed experience or visual unfolding.*
*(Simonds, 1961, 145)*

Functionally, paths, walks, and the like allow for the sharing of goods, commerce, and information. China's Great Wall both separated enemies and delivered soldiers and munitions. So too was President Dwight D. Eisenhower's plan for the United States Interstate Highway System for the delivery of equipment, munitions, and soldiers. Unlike Eisenhower, who saw roads as part of a defense system, Benzel notes that socially, to William H. Whyte (2001), streets and roads are the "rivers of life, pathways to the center of the city, and places where we come together" (Benzel, 1998, 99). In instances of homes with porches and park walkways lined with benches, streets and paths act to bring the "show" to the audiences, as they bring the walkers to the variety of shows along the route. Spiritually and psychologically, linear features are entwined with developmental experiences and with the passages toward spiritual and psychological growth: death marches, marriage walks, approaches to rites of passage. At a more mundane level, J. B. Jackson (1970; 1994) deals with the social development of our society and various adolescent rites in numerous articles about strip malls, hot rods and cruising, fences and paths, and roads in the landscape.

The idea of growth, particularly the step-by-step growth of a line, is indicative of the narrative process. In its physical nature on a page, as oral speech, and as that which is heard and read, the telling of a story is sequential and additive: word by word, sentence by sentence. On foot we construct an ever-increasing understanding of what we see, hear, smell, and touch: an orchestration of our senses' abilities to collect data cum information. As the breadth and depth of our knowledge of a place grows, it either confirms what we expected the place's vocabulary to read like or we piece together the vocabulary, based on what we know, in an effort to identify the place's syntax. To some extent learning to read landscape narratives occurs every time we experience a new place for the first time. Our acts of coding, encoding, and decoding bring into play the design principles of rhythm, repetition, and gradation all in support of learning over time and the growth of meanings embodied in the place.

Rapoport notes, we don't know if the meaning content of environments comes about in the same way as it does in language. No matter. Written words, and words as sound symbols are social constructs. They provide us the means to communicate (Rapoport, 1982, 48). In terms of organizing objects and spaces, designers have a lot of latitude in orchestrating what can be seen, heard, smelled, touched, and the sequences by which they are presented. We have the potential to structure the delivery of messages. Barry Lopez notes the Eskimo's fear in regard to this power, the power to (re)write a place's narratives:

> *...what the traditional Eskimos fear most about us is the extent of our power to alter the land, the scale of that power, and the fact that we can easily effect some of these changes electronically, from a distant city... They call us, with a mixture of incredulity and apprehension, 'the people who change nature.'*
> *(Lopez, 1987, 34)*

It seems they were well aware of the emerging Anthropocene Era, and with just cause.

> *The Anthropocene Epoch is an unofficial unit of geologic time, used to describe the most recent period in Earth's history when human activity started to have a significant impact on the planet's climate and ecosystems.*
> *(https://www.nationalgeographic.org/encyclopedia/anthropocene)*

The implicit and explicit aspects of linearity are not exclusive to the content of a landscape design. Whether we feel we live in a four-, five-, or six-dimensional world, we are linear beings. We move about for particular reasons. Our actions are goal oriented, and I would say that personal safety and satisfaction are at the root of our actions. For this reason, and because we are most immediately concerned with our personal well-being, I argue that function and use are easier to design into a narrative and read than those aspects of landscape narratives that communicate feelings, emotions, and the deeper values making up our collective unconscious. Functional messages are the physically featured vocabulary we author and read first.

Figure 4.3 University of Guelph, ON, Canada Photo by author.

Figure 4.4 Hole-in-the-Ground, South of Cheney, WA. Photo by author.

Figure 4.5 Providence Urgent Care, Spokane Valley, WA. Photo by author.

The landscape lines that communicate function and use are explicit. In them are reflected a great degree of the causality and probability that drives our actions. My thinking, that of Bob Scarfo, is: "If I travel in this direction then I will get to…"; "If I step here instead of there I will be safer"; and "Mixing these ingredients will produce a good salad dressing" (see Chatman, 1980, 44). Our thinking, as are most of our personal concerns, is colored by our expectations (Barthes, 1979; Goodman, 1954, 14; Rapoport, 1969; 1982). People give more attention to the "channels along which the observer… moves… [Pathways are] the predominant elements…and along these paths the other environmental elements are arranged and related" (Lynch, 1960, 47). Even when the goal is to meander, the stroller doesn't want to find herself brought to an exit with the only choices being to leave or turn around and back track. In this situation the stroller's expectations to wander are disappointed. Motion along a path requires periodic inspections of what's up ahead, "Am I still going where I intended?" When I arrive, the edges of the area I'm in become more important. "Have I arrived where I expected to?" Still, "Am I in the right place?" Two aspects of our movement through space are found in narratives: linearity and sequential unfolding of anticipated experiences. We move along a landscape's story, and at times slow to reflect more deeply on our immediate location, situation, and more distant surroundings.

Linear features found in written text, basic design, and landscapes are the repetition of points. Words multiply across a page to form sentences; points move through space to inscribe lines. Whether words in a sentence, dots on a canvas, or objects in the landscape, the motion, sequence, gradation, and connection of the respective points is guided by organizing principles. In language the form and character-giving principles are grammar and syntax. In the arts, they are basic design principles (Dondis, 1973; Wong, 1972; 1977) or the expressive rules of aesthetics. In society they are a group's social norms such as guided by religion, cultural and family values, laws, regulations, canons, building codes, zoning regulations, and the like (Yamin and Metheny,

Figure 4.6 Farm Laneway, Maxville, Ontario, Canada. Photo by author.

Figure 4.7 Multnomah Falls, OR. Photo by author.

Figure 4.8 Radium Hot Springs, Hwy 93, British Columbia, Canada. Photo by author.

1996). While variety comes from how the principles, codes, or rules are interpreted, there remain certain fundamental similarities in the resulting linear products.

Lines demarcate boundaries, separate areas, and give shape. Linear features act as boundaries and edges (Lynch, 1960, 62–66, 99–100, 111). Paths direct us through and bring to us a sequential progression of information. The power of our information-gathering progression rests in the qualities of comprehension, cohesion, and continuity of the informative statements along the way: Lynch's legibility and readability of built environments (1960). Keeping the messages simple, easily read, in no way negates the depth of what they can express, e.g., Vietnam Veterans' Memorial. That developmental progression of words to sentences to statements that become a narrative is not haphazard (Canemaker, 1999; Kooser, 2005; Rae, 1996). The ordering of information is grounded in the socio-cultural grammar and syntax of a place and a people. The result is a sequence of linear experiences that are progressive (sequential) and developmental (additive). We move through a landscape as pattern seekers searching out informative patterns to which we've been socialized to recognize and use. We look to find individual statements building on each other in the same manner that individual words build on each other to become sentences that add up to a story or narrative.

As we move through a landscape, building, or interior our senses reach out to points along our path. Piaget calls this "taking aim" (Piaget and Inhelder, 1967, 155). Personally I like Gibson's idea that our senses are "aggressive, seeking mechanisms" (1979, 97–98). Lund and Merleau-Ponty see our information-gathering processes as "exploration" and "becoming" (Lund, 2012, 236). In each instance, each of our "taking aim" moments occurs along a continuum.

*The construction of a straight line may be defined as the linking of distant points by*

*the interpolation of a series of points along a direct path… [they go on] to outline the development of the concept of the straight line through the act of "taking aim" or "sighting"….*
*(Piaget and Inhelder, 1967, 155)*

These taking-aim moments could also be considered moments of transition from what I take in with this step, and then the next step and on to becoming more of who I am, where I am, when I am. In pursuit of our daily lives we move from place to place for various reasons. The fact that many of those reasons and accompanying expectations are shared by most people works to a designer's advantage. While moving or standing still our eyes scan the surroundings in an effort to recognize things that we can identify. As we build a pattern of recognized points along that continuum the points become joined. Even in collectively developed vernacular environments they are not individual, discrete objects and voids. Each is part of and contributing to a larger whole. What structures the points, as meaningful objects and voids along our path, are the purposes for which the paths and the areas through which they pass were built and our brought-with-us reasons or purposes for using them. As we traverse a landscape we interpret the functions of walkways versus entryways, streets versus boulevards, and intersections versus forking paths. We do not, however, make these interpretations in a vacuum; they are not disconnected moments in time or places in space.

*The geometric line is an invisible thing. It is the track made by the moving point; its product. It is created by movement – specifically through the destruction of the intense self-contained repose of the point. Here the leap out of the static into the dynamic occurs.*
*(Kandinsky, 1979, 57)*

Points, as lines, are not without context unless, as designers, we purposefully toy with the users, hopefully in a playful way. Linear features are part of the greater continuum, we call life as purposeful motion, and our getting from here to there and then doing what we need to once we get there. As parts of a continuum, what phenomenologists call the "act of becoming," what we sequentially read builds on what was expressed by the previous points to formulate complete statements: entry, exit, ascend, descend, turn here, walk around me, stop, slow down, speed up. All contribute to a narrative with a particular theme.

In the realms of written and oral language we take aim at the individual words we see and hear. Based on our learned communication skills we expect the words to flow into identifiable and comprehensible patterns (Habermas, 1981; Forester, 1985). Our expectation is that the information will build from word to word and make sense on the way to meeting our expectations. In landscapes, in buildings, and in the spaces within and between buildings those statements as edges, boundaries, paths, walkways, and roads each tell us something in how we might interact with the delivery and the receipt of information, goods, and services and, hopefully, meet our expectations.

Critical to our understanding of movement systems is that their location, orientation, and surroundings all work to organize the order and magnitude of what they transmit. Information delivery is why the laying out of circulation patterns is so critical to a design's success. The organization or authoring of information as oral or written expression is a linear progression across space and over time (Aminzadeh et al., 2016).

*Life is a continuous flow of experience; each act or moment of time is preceded by a previous experience and*

*becomes the threshold for the experience to come. If we acknowledge that an objective of life is the achievement of a continuous flow of harmonious experiences, then the relationship of spaces to one another, as experienced over time, becomes a major design problem…in which no single part can be considered except in relation to what immediately precedes or follows it.*
*(Bacon, 1974 19)*

A design that expresses a particular character starts with a sequence of experiences that introduce and build that narrative's character. The rate at which the character-building experiences take place may be quick, or slow: traveling a college campus or walking by a 12′×12′ front yard of a rowhouse in Alexandria, Virginia. The rate often depends on available space, the speed of the landscape participant, and aspects of frequency and rhythm which as a designer you can, to a great extent, control. Keep in mind qualities such as rate, duration, and frequency require markers by which people measure speed, time, and change. As Lynch notes, a path's directional quality is felt through a gradient, or "regular change in some quality which is cumulative in one direction" (1960, 54), that could be land uses along an urban road or sidewalk. Lynch also notes that as people move toward the center of a city they encounter increases in the number of signs and symbols, particular land uses, and age of the buildings and landscapes (54).

*But by formulating topic sentences…a writer…greatly improves clarity and unity – clarity because the reader need not infer…which of various points touched on is the key to the rest of the paragraph, unity because stating major points clearly and explicitly helps…the writer to weed out irrelevancies and digressions.*
*(Guth, 1961, 143)*

*A street is perceived, in fact, as a thing that goes toward something. The path should support this perceptually by strong termini, and by a gradient or directional differentiation, so that it is given a sense of progression…a common gradient is that of ground slope…a progressive thickening of signs, stores, or people may mark the approach to a shopping node; there can be a gradient of color or texture of planting as well; a shortening of block length or a funneling of space may signal the nearness of the city center.*
*(Lynch, 1960, 97)*

## PHYSICAL ATTRIBUTES

Lines do not occur in nature. In fact, like points, lines do not exist. They are without width or height so they cannot actually have length. Mathematicians and designers would find it difficult to agree on what defines or constitutes a line. In mathematics lines are one dimensional. Yet they're invisible. As design elements they must be given width or height in order to be visible.

For scientists, engineers, and mathematicians to produce a graph both the points and the lines must take on a certain dimensional quality. Rather than getting into the philosophical depth of Kandinsky's thinking (1979) let's agree with Lynch that as edges or boundaries lines are readily observed where two different materials or textures meet (1960, 100). Paths, roads, and other linear surfaces accommodating movement are in distinct contrast to the areas around them. Linear features are considerably longer than they are wide.

*Line – As a point moves, its path becomes a line. A line has breadth but no depth. It has position and direction. It is bound by points. It forms the border of a plane.*
*(Wong, 1972, 7)*

*The geometric line is an invisible thing. The track made by the moving point; that is, its product… Here the leap out of the static into the dynamic occurs.*
*(Kandinsky, 1979, 57)*

Edges, boundaries, and to an extent circulation routes are typically made of different materials than their adjacent areas. In drawings lines are our way of visually representing what in reality is the separation between objects and spaces or textural changes in a surface. Betty Edwards' Exercise in Edges and Contours (1999, 81–96) makes this fact clear. You are asked to stare at one of your hands while not looking at your other hand and the pencil it is holding, or the sheet of paper on which it is about to draw. Your eyes are to start anywhere and then follow a line around your hand, its fingers, their creases. As you visually study your hand you simultaneously reproduce, with your pencil, the movement of your eyes with a single pencil line on the paper. What you draw as lines on paper are actually representative changes in lighting over the continuous yet changing surface of your hand and its contrast to its background. The pencil lines are abstractions of the changes in texture and lighting.

Figure 4.9 Pure Contour Drawing by student Kim Hoang. Edwards (1999, 89).

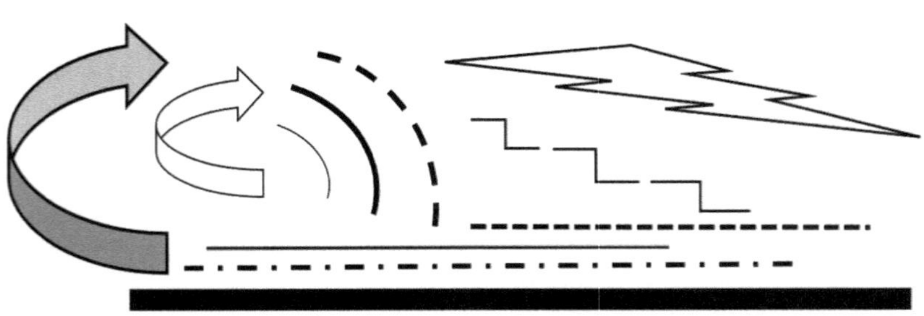

Figure 4.10 Line variations. Drawn by author.

With each (conceptual) line we draw on paper or build in the landscape we automatically append to its length the added dimensions of width or height. The graphic designer Wicius Wong recognizes the kinds of features designers see as lines. He notes that

> *three separate aspects should be considered in recognizing a line: the overall shape... general appearance...curved, bent, irregular; the body – As a line has breadth, its body is contained within two edges. The shapes of these two edges and the relationship between them determine the shape of the body...tapering, knotty, wavy, or irregular; [and] the extremities – These may be negligible when the line is very thin. But if the line is quite broad, the shapes of the extremities may become prominent...square, round, pointed, or any simple shape.*
> *(1972, 9)*

## CONTEXT-DRIVEN SHAPES

If a line is quite broad then we tend to see it as a path, sidewalk, road, boulevard, and the like. With regard to urban landscapes, Lynch notes the importance of the destination and origin points [or ends] of paths to people's sense of orientation in the city (1960, 54–55).

> *People [in his study] tended to think of path destinations and origin points: they liked to know where paths came from and where they led. Paths with clear and well-known origins and destinations had stronger identities, helped tie the city together, and gave the observer a sense of his bearings whenever he crossed them.*
> *(54)*

Figure 4.11 Jet as point; contrails as lines inscribed by points. Photo by author.

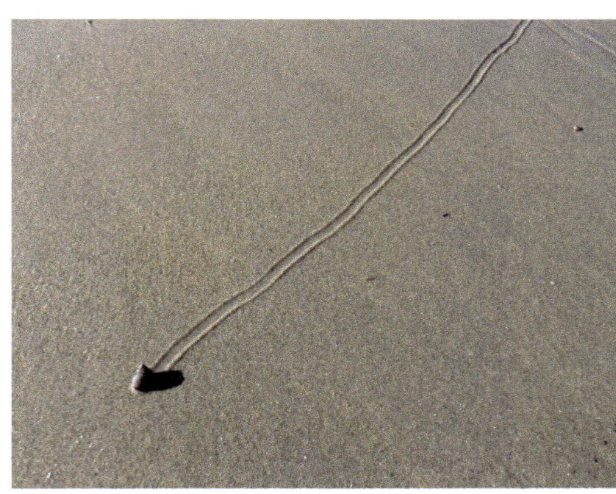

Figure 4.12 Snail racing after the receding tide. Photo by author.

Kandinsky (1979) carries Wong's thoughts on determining the shape of a line's body further when he notes the interdependency of a line with its surroundings. He recognizes basic line forms as horizontal, vertical, diagonal, angular, and curved as reactions to adjacent forces (55–61). We are helped in our understanding of the nature of lines when we consider his conceptual line as a "track made by the moving point" (57). An excellent example of such "tracks" made by moving points is seen in jet aircraft contrails or the track of a small snail chasing an ebbing tide. As tracks of points in motion, lines are the products of one or more forces acting on them. Kandinsky sees straight lines, whether they are horizontal, vertical, or diagonal, as the result of singular forces acting on a point which in turn is propelled in one direction "with the tendency to run in a straight course to infinity" (57). Angled and curved lines are the products of two or more forces acting on a moving point. With his idea of forces acting on lines, we are brought to the static world of a point as a pause, a stop, or a place to step outside the world of motion, gather together, and take a break from that dynamic world of life between points. Kandinsky's view allows us to see the relationship between the character of a linear feature and its adjacent forms: sharp or sudden bursts of energy versus continuous uniform or graduated forces. We see this in geologically young landscapes. Single pulses of energy, or repeated bursts of action on a moving point produce one or more angles in a line. Much more real than Simonds' river analogy (1961, 158), Theilgaard Watts, in her chapter on Canyon Story, provides on-the-spot observations of pressures acting on not only a river but also the evolving shape, over time, of the canyon it runs through (1975, 38–52). See Figure 7.9 in Chapter 7, Narrative.

Figure 4.13 Thomas Cole (1801–1848). View from Mount Holyoke, Northampton, Massachusetts, after a Thunderstorm – The Oxbow. 1836. Photo Credit: Image copyright © The Metropolitan Museum of Art. Image source: Art Resource, NY.

Motion for Kandinsky is of two types. One is visual. Your eyes scan an area or follow a linear feature across your field of vision. The other form of motion is of the body or object. Built and fixed linear features traverse and connect separate areas. Visually, paths and edges move across your field of vision. While such features are fixed in place, the act of your moving your eyes to visually traverse them contributes to your anticipation of actually traversing them. This is Kandinsky's other form of motion, our actual movement along a path, road, or walkway.

When linear features gain form and character from their social and ecological surroundings they appear to fit with those surroundings. When walks, roads, and the like are forced into their surroundings, cutting across contours, they do not "fit" their topographical contexts. Landscape features forced into their settings do not fit their contexts. As a result, they typically require continuous upkeep to maintain their shapes' integrity. In this case a site's maintenance is higher than it need be. In the four-dimensional world of everyday life, Kandinsky's forces are likely noted as things that are in your way (rock outcrop, poorly located street sign), gusty winds (zig-zag sailing course), extremely loud noises (you turn away), and the sudden appearance of very bright lights (you wince). Watch the trajectory of a hockey player just before and then after taking a body check. Or, imagine yourself walking along a beach, or through the woods, the path you walk is likely related to the location of rocks, trees, fallen branches, gullies, and earthen mounds. The amount of change and sweep in a linear path is seen to relate to the size of the object causing the change in the arc of your path. In design jargon, the paths visually "fit" their contexts (Figures 4.11 and 4.12). Excellent examples of working and not

Figure 4.14 Spokane, WA, Rockwood neighborhood's topography gives shape to the Olmsted Brother's road alignments. Photo by author.

working with natural forces and features are seen in John Reps' *Town Planning in Frontier America* (1969). The early towns and villages of a more "organic" collective personality as seen in the map of 1722 Boston, MA are laid out with roads respecting hills, ponds, and rock outcrops (177). Those towns made up of a more intellectual collective personality as seen in the map of 1748 New Haven, CT display lot and road layouts made up of straight lines and 90-degree angles; topography and landscape features be damned (160).

Like water finding its path, points in motion not only react to exterior forces but also continue in their altered direction due to momentum. Simonds' use of a river analogy helps us understand pedestrian movement and momentum (2006, 228). As a river flows by and around a protruding embankment the water courses toward

Figure 4.15 A contrived pathway along the Spokane River that reacts to no immediate features or forces. Photo by author.

and cuts into the opposite shoreline. Simonds relates this "negative current" to pedestrian "traffic oriented to the opposite bank...by force of momentum" (209).

As artist-designers both Kandinsky and Wong's thinking relate to two- and three-dimensional design and art. Even so, their arguments for external forces giving form to lines work in the realms of architecture, landscape architecture, and interior design. Similarly, Bacon (1974) and Benzel's (1998) thinking is that linear features respect their surroundings; they gain their shape by reacting to their contexts. As landscape architects we would say that such features are ecologically sensitive as seen in much of Olmsted's residential and

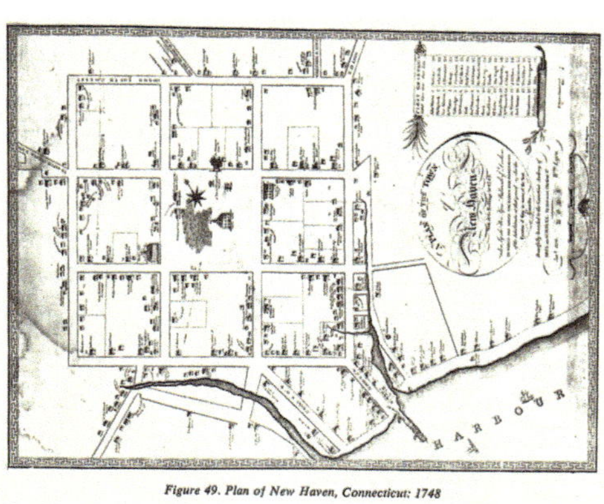


Figure 4.16 1748 New Haven, CT. Library of Congress Geography and Map Division; G3784. N4S1 1806.L9 and Courtesy John Reps' Estate.

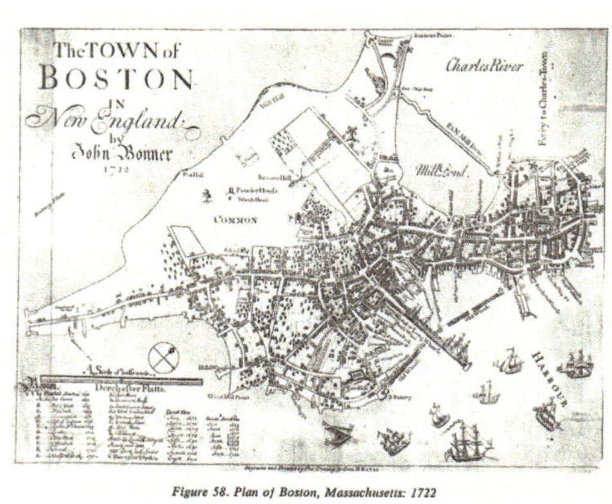


Figure 4.17 1722 Boston, MA. Library of Congress Geography and Map Division; G3764.B6 1725.B6 and Courtesy John Reps' Estate.

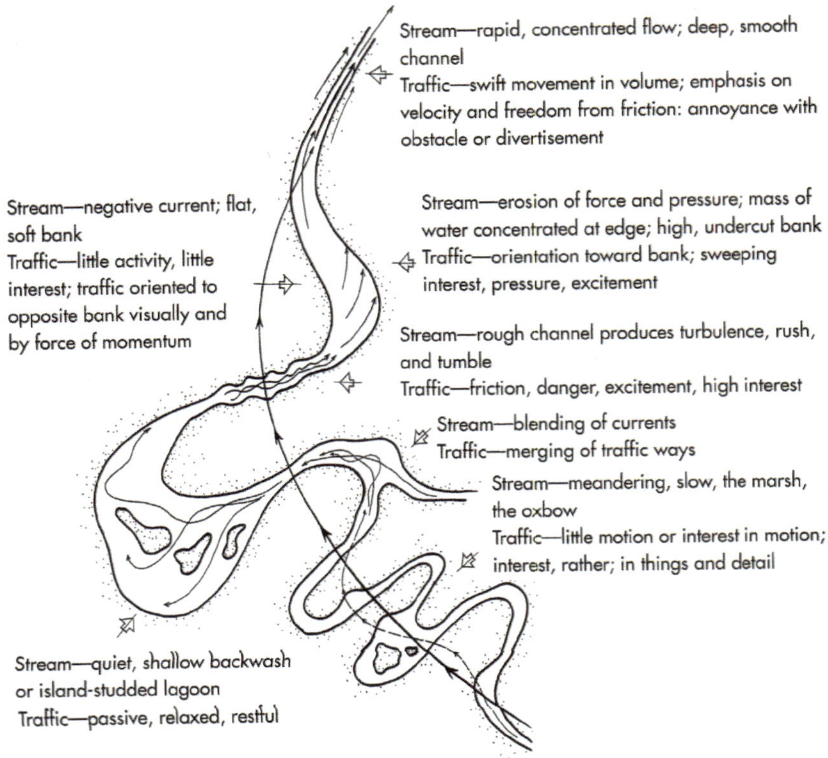


Figure 4.18 Simonds' stream and pedestrian traffic analogy (2006, 4th ed., p. 228). Courtesy McGraw Hill.

park walkways and roadways. However, when we explore linear features used by people, we also find they gain their shapes from forces from within. People and their needs as related to their surroundings also give form to the paths they travel (Bacon, 1974; Benzel, 1998; Cullen, 1961). A fun description of forces shaping Boston's street patterns, based on the belief they arose from cow paths, is found at https://www.bostonmagazine.com/news/2018/03/06/boston-streets-cow-paths/ (last viewed November 27, 2021).

Edmund Bacon's elaboration of a continuity of experiences along a route of travel, rhythms in time, and their relationship to natural phenomena brings the forces Kandinsky sees acting on two-dimensional lines into the four-dimensional world of people's daily lives. Bacon states that at a macro-scale "movement systems must be related to natural or manmade topography; they must take into account the nature of terrain and the natural features and structures that are a part of it"

Figure 4.19 Implied line of benches and explicit walkway edge line. Rose Garden, Manito Park, Spokane, WA. Photo by author.

Figure 4.20 Tree-lined Manito Blvd, Spokane, WA. Photo by author.

(1974, 35). Within a more personal framework he states that "the actual design of each system must relate to the tempo of the movement it is to accommodate as well as to the general nature of the surroundings...pedestrian movement systems require interest, variety, and impressions of rapid change" (35). Between the two magnitudes of motion, what we can call public and personal, he relates the character of circulation systems to carts, cars, trucks, and masses of people. Through a wealth of examples, we come to see Bacon's reasoning for believing that good urban design is based on how well pedestrian circulation is designed.

*I believe that when the [Market Street] project [in Philadelphia, PA] is completed, it will have been decisively demonstrated that the significant architecture actually is contained in the people street and its movement systems, and that the architectural expression of the several buildings along it, each of which may be developed by a different owner with a different architect, hopefully of a high architectural order, nevertheless will be subordinate in importance to the people street.*

*(Bacon, 1974, 295)*

Without explicitly stating it Bacon refers to the importance of the street's spatial character responding to people's needs and wants. In exploring people's needs and wants as related to landscape, architecture, and interior spaces, we quickly begin to realize the ways we must consider those who will use the spaces we design (Norberg-Schulz, 1980; Rapoport, 1969; Relph, 1987; Tuan, 1974). This realm of consideration is the crux of Chapter 7 which deals with how a narrative unfolds.

The idea of an even deeper personal reaction to one's environment is explored by Solnit (2000) in *Wanderlust: A History of Walking*. She sees

*The rhythm of walking generates a kind of rhythm of thinking, and the passage through a landscape echoes or stimulates the passage through a series of thoughts. This creates an odd consonance between internal and external passage, one that suggests that the mind is also a landscape of sorts and that walking is one way to traverse it.*

*(2000, 6)*

As designers, landscape architects can play with walkers and strollers' pace and rhythm. "Thus the changing rhythm of walking indicates an alteration in one's relations to the outside world, social as well as physical. Thus walking is a metaphor that indicates finding direction in life..." (Lund, 2012, 229). Walking actively contributes to the collection of information over space and time and therefore inescapably contributes to an individual's "becoming" who they are. "If one's path through life reflects the becoming of a person, it is clear that walking can be analyzed as a metaphor for one's position in life as well as society" (Lund, 2012, 229). Often my most creative moments, or periods of creative thinking and problem solving occur during my early morning dog walks. I refer to those periods of freer thinking and where my mind takes me as "Free-Range Thinking." I was so happy years back when I learned Charles's Darwin used his daily walks to discover answers and direction to difficult problems he faced while writing *The Origin of the Species* (Young, 2015).

> *While working on his famous theory of evolution, Charles Darwin took a daily walk, to exercise his mind and body, to think, and to observe. Walking the same route each day from his house, through shady woods and back along a privet and hazel hedge-lined field, Darwin referred to this as his 'thinking path'. Working on his theories and ground-breaking book 'On The Origin Of Species', Darwin thought as he walked. If he had a particularly challenging problem to consider, he would pile 3 or 4 stones on top of each other and knock one off with a stick on each lap, returning home only when all stones were cleared.*
>
> *(Wildlife Wonder, 2020)*

## FIGURE-GROUND SUBSTANCE (MASS – VOID)

As noted, lines, like points, can be explicit, a continuous visually apparent edge, boundary, or path, or implicit, as seen in a line of trees, street lights, or benches along a garden path. In design jargon lines are virtual or conceptual, positive or negative, masses or voids, and figure or ground (Cullen, 1961; Laseau, 1980, 102; Wong, 1972, 11). Lines constitute physical edges, explicit boundaries, or they are implied, and yet felt, as noted in this chapter's earlier quote from Le Guin's, *The Dispossessed* (1974) or the invisible-to-most turf boundaries fought over by neighborhood gangs (Ley, 1974, 230) and noted in the March 2018 KCRW news: "Gang Borders Create Invisible Walls in Los Angeles" (Avishay, 2018). Implied lines, those we believe we see in the spaces between points, are envisioned due to our mind's desire to recognize patterns (through each of our senses). Our pattern-seeking nature allows us to connect the dots and see lines where there actually aren't any. We've all experienced a sense of connection through our other senses. Particular smells, sounds, and tastes connect many of us with other places, people, and events in space and time: the taste of grandma's turkey dressing, the sound of a bicycle bell, and the smell of lilacs or sage connect us with other times and places. The connecting of temporal points, or points in time, is the basis of the work done by Ron Lee Fleming's The Townscape Institute,

Figure 4.21 Turning eyes from looking straight ahead to the right and seeing what Cullen calls "Netting": shifting view from trees as wall to trees as frame, Spokane, WA. Photos by author.

http://www.townscape.org/projects.htm (viewed December 2021).

*Points arranged in a row may evoke the feeling of a line. But in this case the line is conceptual and not visual, for what we see is still a series of points.*

*(Wong, 1972, 9)*

There is a gray area between the implied conceptual lines that appear to connect a sequence of objects, and explicit, actual lines that are unbroken across an area. Spatially, the strength of an implied line is related to the relative positioning of the viewer to the objects being viewed or read. The intensity of implied lines relates to the reduction in space between the linear sequence of the objects, as the angle of view becomes more acute. When the spaces reduce to the degree that the objects seen in perspective seemingly touch or overlap, you have the strongest of implied lines. Looking down the length of trees alongside the road that approaches the French chateau at Vaux-le-Vicomte, the acute angle of view allows the tree trunks to present a solid wall directing your vision straight ahead. At any moment you can turn your head to the right or to the left and see the adjacent fields framed by the same trees that a moment ago made up a linear visual barrier directing your view to the road ahead. When acting as an expert witness in a lawsuit over the existence of an allegedly illegal hedgerow, I was presented with a real mind teaser by the defense attorney: "At what distance apart will these plants not constitute a hedge?"

Figure 4.22 Approach to Vaux-le-Vicomte in Maincy, Ile-de-France. Google Maps Copyright 2022 Google Maps/GoogleEarth.

The intensity of the connection between points in time (this moment with another in your past) rests on the importance of the attributes of the physical objects, or momentary experiences, to the person experiencing them. A particular lacey material may evoke memories of a tablecloth used at annual family gatherings. The sight of a knotted rope over a swimming hole connects you with your first escape from Mom and Dad. The smell of sizzling bacon, the sound of a foghorn, and the crunching of new fallen snow under foot trigger memorable connections for each of us. What times and places in your life did they connect? A number of place-time moments have stood the test of time for me: the turpentine, nostril-burning smell around the refineries flanking the New Jersey Turnpike; the nectar-sweet aroma around the Necco Candy factory (many years ago) in Cambridge, MA; Mom's spaghetti sauce; and the lemon and mint herbs' aroma by my parking space at home.

## DEFINING AREAS

When lines as edges

> *are not used or considered as paths… [or as boundaries they] may be barriers, more or less penetrable, which close off one region from another; or they may be seams, lines along which two regions are related and joined together.*
>
> *(Lynch, 1960, 47)*

Many "seams" provide for the interface between two areas, that zone of adjacency that results in social interaction: property lines, demarcation of a park's activity areas, and the signals we set out to mark and enclose our personal spaces. Gordon Cullen (1961) sees Lynch's "seams" in a variety of forms, all of which provide for a sensed separation of two areas and yet penetrability: a low sitting wall between a plaza and major walkway, the narrow streets of many British and European cities and villages actually working to join both sides of the street. Cullen describes other seams in terms of pedestrian ways (54), juxtaposition (60), immediacy (61), and intimacy (69).

Figure 4.23 A seam: Woornerf, London, UK. Shutterstock ID: 1066971254.

Seams come about and are sustained through people's actions and the degree of respect those actions show for what we consider *here and there, us and them, theirs and ours*. Our spreading out of books, pencils, computer case, and ballcap at a library table works to keep most people away. Wolves have another way of marking their territory. I can't help but wonder what category of line the fence in Robert Frost's poem, Mending Walls, is. The final line of the poem says "Good fences make good neighbors." Does it do that by separating or joining? In Lynch's language, is it an edge or seam?

While most edges do not stand out in people's minds the way paths do, they are important organizing features and in no small way influence people's lives. Throughout history the political boundaries between Germany and France literally moved all over the map. The Cartesian thinking that surrounded and constricted Native American lands was more lines on maps than on the ground. Yet, what those map symbols represented, however, led to people dying and to the devastation of numerous cultural landscapes. For some, lines aren't seen:

> *close to the meeting of the territories of the three Monroe [Street gangs]. The site which had been intended [by city planners] as a playground was… well suited as a battleground, and in a short while [after being built the*

*playground] was under gang control.*
*(Ley, 1974, 230)*

Despite the very real yet invisible turf boundaries making up West Chicago, eight young men who came to form the first all-Black high school rowing team share their lives, and those of brothers and sisters lost, in the documentary, "A Most Beautiful Thing" (Cooper, 2020). As a youngster, I was very impressed by the storyline of a television show, the name of which I can't remember or find now. The storyline focused on two warring factions arguing over who rightfully owned or controlled the land covered by a thick pencil line on a map that had originally been drawn to demarcate the boundary between their holdings.

The last type of implied line is one of connection rather than separation. Lines of sight join the viewer and the viewed: views of Mt. Rainier and the Cascade Mountains are as much a part of urban Seattle as they are part of the residents' psyche. The idea of connection is found in Japanese gardens in the use of nearby vegetation to frame, and in doing so borrow, and thus bring into the garden, distant landscapes. Originally a Chinese practice, the term *shakkei* meant borrowed landscapes or borrowed scenery (Itoh, 1981, 15). Adopted as a Japanese landscape practice, it was called *ikedori*. Gorden Cullen's term is netting (Figure 4.21): "this serves to link the near with the remote" (1961, 39).

*We had been driving through the northeastern corner of Poland, a country where frontiers march back and forth to the abrupt commands of history. The same fields of wheat and rye moving in slow waves with the rhythm of the breeze had been Lithuanian, German, Russian, Polish.*
*(Schama, 1995, 23)*

Lines of sight, as presented earlier, are an extremely subtle yet powerful aspect of a place's character. What is seen, what is brought from a distance into an experience, is as important as how the sightline is defined. Consider characterizing the sightlines associated with New England, Nebraska, Montana, and coastal rainforest landscapes. How would they differ? When relating characteristic sightlines with particular regions it isn't difficult to imagine why the long-time resident of Kansas can't understand her sense of claustrophobia when visiting New England. Being used to regionally characteristic views that extend in all directions to the horizon, she feels enclosed in New England where the average sightline through tree-laden, rolling hills probably averages one-quarter to three-quarters of a mile, with occasional opening of a mile or two. Then there is the almost con-

Figure 4.24 Kashan, city plan with bazaar route seen as horizontal line moving through open areas. Courtesy Nader Ardalan and Laleh Bakhtiar (1973, 92) and University of Chicago.

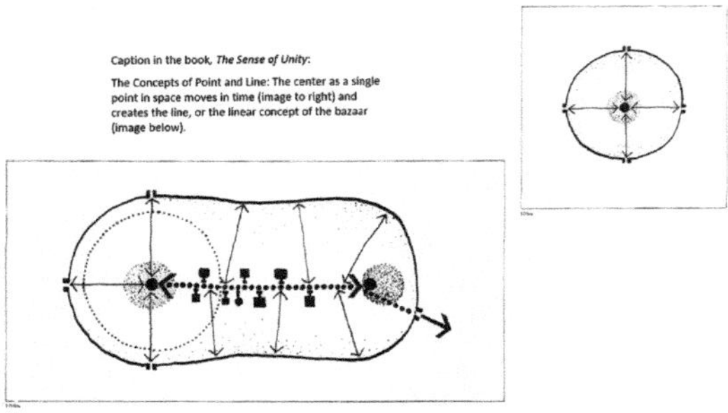


Figure 4.25 Point-line concept found in the bazaar (Ardalan and Bakhtiar, 1973, 89). Courtesy Nader Ardalan and Laleh Bakhtiar and University Chicago Press.

tinuous summer tree canopy overhead. Recently, a Pacific Crest Trail hiker used to the wide-open west coast panoramas referred to his Appalachian Trail experience as hundreds and hundreds of miles in a green tunnel.

> *In time, as the linear system of the bazaar grows, and the residential areas (mahall-ha) grow from the main spine, the old skin is shed and the new layer is created.*
> *(Ardalan and Bakhtiar, 1973, 93)*

The concept of a point moving through space and inscribing a line is wonderfully evident in Islamic cities. The pedestrian spaces

> *maintained the concept of a center as a single point in space that moves in time and creates the line, or the linear element of the bazaar (1973, 89). This reorientation towards a moving point introduced a more vital planning concept which, even today, accepts the growth and change as natural phenomena of existence.*
> *(Ardalan and Bakhtiar, 1973, 89)*

As a result, the bazaar also exhibits the linear characteristics of growth and direction.

> *The bazaar traditionally begins at the palace precincts, which symbolize the spiritual head of the body, and grow cellularly in an apparent natural pattern in the direction of its symbolic heart...going on, then, to the opening of one of the city gates.*
> *(Ardalan and Bakhtiar, 1973, 93)*

This cellular growth of an Islamic city is an honest, genuine, example of the socio-cultural production of space. When I first read Ardalan and Bakhtiar's account the designer in me had two thoughts. Bear with me, what I am about to say will seem quite strange at first. My first thought was that as a designer of others' living environments I was bound to the same ethical responsibilities as when humans domesticated the cow. "And when man domesticated them, he took on the responsibility of nature" (Bronowski, 1973, 26). Think about it. Maybe we are taking on the responsibilities of people's parents or grandparents. Whatever our relationship we do accept the responsibility to provide for and contribute to the health, safety, and welfare of people of all ages and abilities. Let that simmer in your mind for a while, we'll return to that responsibility in Chapter 10, Ethics in Landscape Architectural Design Practice.

That realization of our responsibility as designers led me to my second thought. When I design an environment that might have otherwise occurred at the hands of hundreds of people over many years in response to their needs and wants, I am ethically bound to at least be aware of those personal and social dynamics. In taking on a design project, I accept, as it were, the responsibilities of individuals, social groups, and a culture evolving through its unfolding history. I am duty bound to respect their cultural heritage.

Ardalan and Bakhtiar, like Bacon (1973), provide numerous examples of pathways, processional corridors, roads, and open spaces that grew in a linear fashion due to changing social and environmental pressures, or what we might consider Kandinsky's forces. As a result, each author's analysis of linear and pedestrian movement systems expresses the dynamic ebb and flow of cultural motion indicative of growth and change. Changes over time are not just related to the linear character and forces acting on the lines: they are not restricted to basic design or landscape features. The character of what is contained in sentences making up pieces of literature often changes with the re-editing and re-publication of subsequent editions. Early editions of E. A. Poe's short story, The Landscape Garden, make no mention of the Picturesque or the Beautiful. Later editions note both artistic styles. Walt Whitman altered *Leaves of Grass* many times over almost 37 years and eight editions. We might say that our responsibilities to those left to live with our designs should carry with them a certain, within limits, flexibility.

To this point we primarily explored lines as horizontal linear features: edges, boundaries, and paths flat with the ground. The earlier example of a line of trees flanking the road approaching Vaux-le-Vicomte and Lynch's mention of signs and buildings does open us up to linear vertical features in the landscape as single points or sequences of points seemingly set along a line. Introduced as repetitive features they delineate and frame space and help measure one's speed. As stand-alone features, vertical linear features make bold statements as points in space and time as we've mentioned. Church spires announce public gathering spaces at their bases. Train yard switching towers, air traffic control towers, and custom's house buildings set at the edge of urban harbors announce the location of people with the power to guide movement, sell and trade goods, and manipulate, and tax. They announce authority.

Figure 4.26 CN Tower, Toronto, Ontario, Canada. Shutterstock ID: 776152702. Contributor: jessica.kirsh.

While particular skyscrapers symbolize economic power and corporate egos (Domosh, 1996), they also often act as landmarks enabling people to orient themselves as they approach and enter into a city or town. The Seattle, Washington and Toronto, Ontario space needles and the Eiffel Tower each provide the same services of orientation, sense of place, and announced gathering. At a smaller scale, similar features can be used in parks and plazas, and at an even smaller scale pillars and cairns can be used in gardens.

Markers of gradation, and Lynch's variant called scaling (1960, 55), occur along paths or a sequence of nodes or vertical features or landmarks along the path. Nodes are spaces in which people gather, reorient, and transition from one direction to another. They can be intermediate areas or final destinations. Such spaces are road intersections, piazzas, plazas, town squares, or a sequence of small parks. They are central business districts and ethnic districts (Little Italy, Barrio, Chinatown), and transportation hubs. While

Figure 4.27 Arc de Triomphe, Paris, France. Contributor: Eric Isselee. Shutterstock ID: 1867822423 Inc.

Figure 4.28 Dupont Circle, Washington, DC. Photographer Julia Armstrong, Aero-Imaging, Inc.

nodes are areas of use (intersection and gathering), they are not typically considered linear features. Many are, however, often associated with the vertical feature Lynch labels as a landmark.

Imagine approaching a community you've never been to before. Imagine the signals you're looking for that will first get you to a gas station and then a hospital. The shapes that express one or the other, even at a distance of a mile or more, bring with them very different collections of contingent and anticipated forms. There isn't a cause-and-effect order linking specific forms and uses but there is a collection of contingent objects and spatial forms that you piece together as a pattern that expresses the likelihood you will arrive at the hospital and the gas station. This piecing together of parts into an identifiable pattern that says gas station or hospital is based in what Kooser refers to as "a reader's collective associations with words" (2005, 112).

Taking from what we've dealt with thus far, it is important to touch on the potential we, as designers, hold to the authoring of a design's narrative. For many, I'm sure the idea of introducing causality and contingency into the creative process rubs the creative side of your personality the wrong way. New design students so frequently believe that creativity is the freedom to sprinkle things about a site, or structure, or interior until

Figures 4.29.1 and 4.29.2 The Watts Towers in Los Angeles, CA. Once condemned now on the National Historic Register. Built mostly of found objects by Simon Rodia between 1921 and 1954. Left: Shutterstock ID: 98082746, Contributor: James Michael Dorsey; Right: Shutterstock ID: 1400348636. Contributor: Walter Cicchetti.

they look right with no agreed-upon basis for what defines what looks right. If you feel, like some modern authors, that causality is too strict a relationship between story (narrative) and discourse (expression) you may find more to agree with in Jean Pouillon's concept of "contingency," as the dependency of word-images for their "existence, occurrence, character, etc. on something not yet certain" (Chatman, 1980, 47). As we move about a landscape we often refer to Pouillon's "not yet certain" anticipations or what we expect to encounter up ahead. People have the ability to connect the most divergent of events and items in an effort to make sense (and identify a pattern). I remember a creative-thinking problem that asked us to list the similarities between a watermelon and a spoon.

Our tendency to pattern building (seeing faces or animals in clouds, shadows, or ink blots) works in a landscape architect's favor but even then there are limits as to what can be brought together and in what order. With even a general knowledge of why people approach, enter, traverse, or stay in a site the designer acquires insight into the user's anticipated landscape vocabulary and experiences. Once identified we have a context from which to derive contextually appropriate messages and begin to author a narrative. We have a basis on which to employ Pouillon's contingencies and author a narrative's contents with considerable assurance that the story to be read will be read and will be understood, at least at the level of functional messages and users' health safety, and welfare.

The acts of authoring and reading landscape narratives are those of mediation and interpretation: "…the ordinary reader is not trying to recreate the author's original meaning" (Wolff, 1981, 107). I agree with Wolff's statement that "The reader therefore fixes the [anticipated] meaning in the act of reading," but the author and designer-as-author must work to guide those anticipations and limit the reader/user from straying too far.

In basic design and landscape design the final narrative belongs to the observer-participant or user-as-reader. Done well though, landscape design can display an explicit, easily read, narrative of function and use. More challenging, but not impossible to express, are those message items that communicate feelings, emotions, history and heritage, and national pride. Aesthetic narratives leave the greatest room for flexibility between what the author-designer intends and what the user-reader constructs. Keep in mind, landscape aesthetics are not those of a passive art observer viewing a painting or listening to a concert, landscape aesthetics are derived from a choreographic involvement of at least four, if not five, of the landscape participant's senses as that participant/reader moves through an environment.

## BEGINNING TO THINK IN TERMS OF NARRATIVE

*The comparative study of the mythologies of the world compels us to view the cultural history of mankind as a unit; for we find that such themes as the fire-theft, deluge, land of the dead, virgin birth, and resurrected hero have a worldwide distribution - appearing everywhere in new combinations while remaining, like the elements of a kaleidoscope, only a few and always the same.*

*(Campbell, 1969, 3)*

Well-articulated, functional messages expressed by landscape features either work or they don't; with degrees of ease a gateway allows access or it doesn't, a walkway serves people's movement or it doesn't. The ease of people's movement through a place is provided by a certain degree of control over the delivery and the unfolding of the message sequence and the opportunities to slow and reflect or become reoriented in the narration. Granted, what the user sees and what the sequence allows to happen do offer a certain latitude in the user's unfolding narrative. Still, the freedom afforded the user-as-reader is not so loose that she can, in a sense, think what she wants. This logic is found in the structure of oral history, or storytelling. The

*order of the presentation need not be the same as that of the natural logic of the story. Its function is to emphasize order - emphasize certain story events, to interpret some and to leave others to inference, to show or to tell, to comment or to remain silent, to focus on this or that aspect of event or character...each arrangement produces a different plot, and a great many plots can be made from the same story.*
*(Chatman, 1980, 43)*

*Literally, syntax means the business of constructing sentences: it defines the way of putting words together so that they have meaning, carry information. In this context the idea of syntax of urbanism conveys the belief that the elements of the built environment can be put together in such a way that they have collective meaning. The buildings and spaces and all impedimenta of the city comprise the vocabulary. Here [in Syntax of Cities] it is being proposed that the vocabulary be used according to a syntax which draws a positive response, not only in the verbalistic sense but also from the deeper, non-rational levels of the mind. Urbanism has the potential to speak through images and symbols which unite urbanized humanity....*
*(Smith, 1977, 17)*

*The origin myths of the world's mythological systems differ too; but in all except the most rarefied the conviction is... that the living universe is the handiwork or emanation - psychical or physical - of some father-mother or mother-father God.*
*(Campbell, 1969, 85)*

*[The I Ching is comprised of] family members in their various roles: the strong father, the devoted mother, the arousing eldest son, the dangerous middle son, the resting youngest son, the gentle eldest daughter, the intelligent middle daughter, and the joyful youngest daughter...[which*

*in turn represent] the seasons, parts of the body, points of the compass, plants, animals...*
*(Wing, 1979, 15)*

The order in which landscape features are presented is guided, to a great extent, by syntax as the rules, principles, and processes that govern the structuring of sentences, specifically their word order – the sequence in which words are put together to make sentences, and make sense. Like word choice and sequencing, the collection of landscape features expressing a place's proposed functions, activities, and destinations will make or break a place's readability and the users' experiences. Consider a parking lot. What do you anticipate and as a result look for in order to safely arrive at a parking slot? What do you anticipate finding when you decide to look for a café? A high school versus elementary school? Urban center versus town or village center?

*An ordinary language like English is a system which allows us to create an infinite variety of one-dimensional combinations of words, called sentences.*
*(Alexander, 1979, 185)*

*A pattern language is a system which allows its users to create an infinite variety of those three dimensional combinations of patterns which we call buildings, gardens, towns.*
*(Alexander, 1979, 186)*

Semantics is a branch of linguistics concerned with meaning – relationships between signifiers (the words or landscape objects) and what they stand for. With some understanding of people's likely goals as they participate in a particular place a designer is presented with a basis for the formulation of the syntax of a place. That grouping of what I consider the family of spatial forms, their defining colors and textures, and ancillary defining objects is drawn from what the user-as-reader will most likely seek to identify. Consider your expectations as you approach a marina versus a used car lot. The degree to which a design realizes the syntax sought by the user contributes to the comfort the user will likely experience. The greater the discrepancy between what is expected and what is found the more the reader must work to piece together the syntax of that place, the vocabulary of the landscape narrative, before he or she can move about comfortably. Finding a discrepancy between the anticipated and the actual is like reading a mystery which may be the design's objective.

The better we understand "who done it," the more readily we can identify a landscape narrative's signals and patterns. Increasing the gap between the syntax sought and the syntax found presents the designer with the responsibility of anticipating the extent to which the user-as-reader will have the time or the inclination to figure out and become familiar with the design's syntax and therefore its vocabulary. The more unique, and unusual, a design's form and content, the longer a person needs to come to understand the rules of the place. This brings to the designer the added responsibility of providing, through organization and arrangement (of objects and spaces), the means by which the user can come to recognize, identify, and interpret the unfamiliar spaces, objects, and sequences that make up the design.

Context and grammar are inescapable in landscape narratives. Sites are situated in established geological, geographical, and climatological locations, each of which brings certain conditions or contingencies (Forman and Godron, 1986; Watts, 1975). Each site comprises larger contexts dependent upon smaller and smaller subordinate contexts and vice versa. Each has a particular history, and a current status as seen through the tinted glasses of people's cultural value systems (Conzen, 1990; Cosgrove, 1984; Cronon, 1983; Rapoport, 1969; 1982). In an ecological, or grammatical sense, each of these interdependent contexts provides for rules that, if followed, allow a design to "fit" or "work." If not followed then a design fights with the social and environmental dynamics of its setting and requires constant upkeep.

Christopher Alexander speaks to the interconnectivity, and interdependence, of patterns or what we read, when brought together, as contexts.

> *What is most important about this sequence [of larger to smaller patterns] is that it is based on the connections between the patterns...In short, no pattern is an isolated entity. Each pattern can exist in the world only to the extent that it is supported by other patterns: the larger patterns in which it is embedded, the patterns of the same size that surround it, and the smaller patterns which are embedded in it.*
>
> *(1977, xii–xiii)*

> *There are an enormous number of people out there with invaluable information.... all you have to do is pick up the phone...And it may also turn out that in searching for this one bit of information, something else will turn up that you absolutely could not have known would be out there waiting for you.*
>
> *(Lamott, 1994, 147)*

Figures 4.30.1 and 4.30.2 Sea wall vs. ocean, San Padre Island, TX. Photos by author.

In a good design, all the components can be read and interpreted. This calls upon a designer/author to research a topic before writing on it. To do so for a design is to become informed of the character of a design's elementary parts and the overall characteristic dynamics of which it will become a part. To violate the grammar of a place is like building seawalls to save beach-front homes, only to have to regularly replace both the walls and the homes, e.g., Ocean City, Maryland, San Padre Island, Texas, and Whidbey Island, Washington provide great examples. To work within a place's ecological setting a good designer not only works with the local natural dynamics but also becomes aware of social meanings associated with the respective points, lines, planes, sequence, and rhythms of the place's elementary features.

## POWER OF LINES

The power of linear features in a landscape resides in their ability to bring about connections both as paths and edges or what Lynch calls seams (1960, 47). While most people consider edges or boundaries as separators, take a moment to consider the kinds of concentrated political and economic power associated with boundaries. The boundaries to the north and south of the United States receive tremendous amounts of attention, money, rhetoric, and political time and energy. Neighbors argue over property lines, cities and citizens litigate rights-of-way, and politicians and religious leaders debate national boundaries. By the very fact, as Lynch states, that edges and boundaries separate different activity areas and land uses our attention is also brought to bear on the contrasting differences of surface materials and adjacent landscapes or building character actually brought together by the edges, or seams.

*Edges are the linear elements not used or considered as paths by the observer... They are lateral references rather than coordinate axes. Such edges may be barriers, more or less penetrable, which close one region off from another; or they may be seams, lines along which two regions are related and joined together. These edge elements, although probably not as dominant as paths, are for many people important organizing features, particularly in the role of holding together generalized areas, as in the outline of a city by water or wall.*
*(Lynch, 1960, 47)*

Edges may act as seams, linear features "along which two regions [or activity areas] are related and joined together" (1960, 47). Seams are penetrable "to some depth with the regions [or areas of differing activities] on either side...[they constitute] a line of exchange along which two areas are sewn together" (100). Edges and seams are more connectors of space than time. As such they remain rather static. Maybe we should consider edges as features of potential energy.

However, a more important and vital form of connection comes in the form of paths, walks, roads, and other forms of circulation. They may act as seams, boundaries, or edges, but they are characterized by motion. As such they bring into play the power to connect time, and space. Also, as stated earlier, people's movement across space and through time enables the additive quality of sequential progression. These qualities are not abstractions. As designers we must think of physical features and their qualities in terms of the sequential experiences they foster and the experiences and functions participants are there to fulfill.

The power of these linear experiences is twofold: that of the user and that of the designer. From the user's viewpoint motion enables life; life is learning by being brought to new experiences. The sequential experiences by which we arrive at (or are brought to) statements have a lot to do with the degree to which we enjoy what we read, watch, and listen to as we traverse a landscape. A smooth ground plane going straight ahead can provide a clear concise order to messages. A straight, uncluttered motion contributes to users' efficient and comfortable movement. Too much though would be boring. Less obvious, more subtle variations in ground plane and direction of pathway call for slower pace, even occasional stops to take in the surroundings and become reoriented before continuing on. Then there is the withholding of information or messages, breaking or altering of an expected linear sequence to introduce a degree of mystery or levity; the need to circle back or backtrack moments of "aha," and decision points asking you to decide "Do I go this way or that or maybe both?"

In that a landscape narrative, in order to be read, calls for motion, a designer's authoring of compositions takes the form of a planned orchestration or choreography of others' movements (Aminzadeh et al. 2016). The primary and secondary sequences designed into a site have a lot to do with how comfortable people are in the execution of their daily activities. There are the experiences we label as "having a good day." Our experiences went smoothly. There were no unexpected detours or delays. Understanding a different kind of comfort has become the focus of psychologists, physiologists, medical doctors, and child development specialists regarding individuals with intellectual and developmental disabilities: autism, attention-deficit/hyperactivity disorder, dementia, Alzheimer's and Parkinson's disease, and the elderly and infirm. We now know that walking to school through green environments helps kids with ADHD remain calmer and more focused longer than when they walk through concrete environments (Kuo and Taylor, 2004). There are a growing number of related residential projects in Britain, Denmark, Canada, and the United States for diverse populations needing as much attention by designers as by medical and health-care professionals. Examples include (all the following sites were viewed on March 1, 2022):

Dementia Village, Denmark: https://www.dailyscandinavian.com/dementia-village-denmark/
Dutch "dementia village" offers beer, bingo, and top-notch health care: https://www.businessinsider.com/inside-hogewey-dementia-village-2017-7
Canada's Langley Village: https://www.thevillagelangley.com
Building Ohana, Spokane, WA: https://www.buildingohana.org/village-in-planning
Japan's Villages & Towns: https://www.alzheimers.net/11-02-16-japanese-city-confronts-dementia-through-community-care/
UK, Outdoor Spaces for Adults with Autism: https://www.researchgate.net/publication/256547367_Green_Spaces_Outdoor_Environments_for_Adults_with_Autism
Housing Design for adults with Autism: https://www.researchgate.net/publication/228448997_Living_in_the_Community_Housing_Design_for_Adults_with_Autism
Portland's Memory Garden: a multisensory garden for those suffering from memory loss. https://issuu.com/joelgrogan/docs/portland_memory_garden_history_fina

## A CALL TO ACTION

Consider what we take for granted each time we read a poem or experience a landscape. Imagine yourself along with a hundred other people and no mutually agreed upon sounds, smells, textures, and visual signals with which to communicate: no memory, nothing to recall, no ability to recall, no basis of recording or capturing a thought or observation, storing it for others, and later sharing it with them. Without individual memories built into a larger collective social memory how would you formulate an idea? And, even if you were able to conceive a thought, without mutually agreed upon sounds, symbols, and signals, or combined sensations, how would you communicate that thought? Oral history would be nonexistent, and the transmission of culture would not occur. An alphabet would be so many pebbles randomly scattered across the sand. In such an environment, how could we safely walk, drive, or ride a bicycle anyplace, let alone find our way back?

One note before moving on. The above paragraph pretty much defines dementia and Alzheimer's disease.

> *Dementia does not make us crazy. It simply forces us to interact with the world and people around us using a more limited set of skills than we have been accustomed to employing, and that in turn causes predictable emotional reactions.*
> *(Cornish, 2019, 136)*

If in fact, as Judy Cornish states, there is a degree of predictability to individuals living with intellectual and developmental differences, I believe it is our responsibility as designers to study and interpret those differences as they can be supported by designed environments.

I took this moment to share these thoughts with you because I agree with Cornish's distinction between treatment, which assumes a cure, and care or strength-based care for the whole person. We cannot give the medical and pharmaceutical industries full responsibility for those whose cognitive skills are changing. Cornish's recognition that people developing Alzheimer's and dementia appear to retain and in some cases hone their intuitive skills is, I believe, the doorway through which architects, landscape architects, and interior designers have a strong potential to design supportive living environments. What we consider a loss of cognitive abilities is more a shift in perception, or the manner in which one reads their surroundings, than a loss. While millions of people wait for medical and pharmaceutical solutions to regain what is lost, my hope is that professional design programs will promote research into what and how cognitively impaired individuals read and interpret the messages and stories expressed by their surroundings and in doing so bring about an environmental form of care for the whole person.

Design guidelines, and therefore the basis upon which to author appropriate architectural, landscape, and interior design narratives are limited. It is up to us, as designers, to develop interdisciplinary partnerships to research and provide environmental design criteria that may be applied to inclusive personal and intentional community environments. It will be a challenge. But not only will those being designed for benefit but so will the greater public by allowing them to participate in the greater community.

As designers, we give up the luxury of being ignorant of so many things we previously took for granted. As designers, we need to become familiar with the fundamental social attributes that structure the languages we use (words, forms, spaces) to compose a design thought, to hold that thought, and to communicate that thought as built form so that it can be comprehended by those for whom we design. We also need to recognize how communicating personal and cultural appreciations through designed spaces involves the dynamic interplay of time, space, and the participants. This thinking goes to the heart of a response I gave my Dad years ago when he questioned my becoming a landscape architect. "Why four years of university just to plant bushes?" I responded, "Dad, I don't know what landscape architecture is but I am sure it is much more than just planting bushes."

# 5

# Planes

Planes are the outcome of points moving through space to become lines and lines, in turn, moving through space to inscribe planes. Earlier the mathematical absence of points and lines was introduced even though as landmarks and edges designers would argue they do exist. Now, with the introduction of planes, we have the three primary landscape features – ground plane or floor, vertical plane or wall, and overhead plane or ceiling or sky – which when brought together provide us with measurable spaces or volumes.

In the world of geometry planes, like points and lines, have no thickness, they are two dimensional. Planar surfaces are measured in terms of area, not volume; this is a simple distinction, but as evidenced in their language, it is one not easily grasped by many design students or, for that matter, practitioners.

Planes are characterized by length and width (or breadth). They are described (or measured) with reference to their location and their direction, for those that are elongated. Compact planes, such as a circle, square, or star shape, are more directionless and typically noted with regard to their location. The repetition of planes across a space to provide a front-to-back, side-to-side, or top-to-bottom structure, or the use of two or more planes in combination introduces the sense or actuality of volume. Our consideration of the part planes play in The Analogy is helped by a simple reminder of the vocabulary distinguishing the two-dimensional character of area forms from the three-dimensional character of volumetric forms.

Figure 5.1 Rainbow Road, Reykjavik, Iceland. Photo by author.

Figure 5.2 Three planar features: horizontal water, vertical rocks, and overhead sky – Banks Lake, WA. Photo by author.

| *Area* | *Volume* |
|---|---|
| Square | Cube |
| Rectangle | Box or Cylinder |
| Circle | Sphere |
| Triangle | Cone; Pyramid |
| Trapezoid | Trapezoidal Prism |

A more complete overview of the two- and three-dimensional figures mentioned here is found in Ching's *Architecture: Form, Space, and Order* (1978).

DOI: 10.4324/9781003286981-6

As to stretch your mind a bit, let's build upon the reading-writing and basic design metaphor. Added one after the other, sentences are brought together through a grammatical process into paragraphs that physically take up a two-dimensional area on a page, a computer monitor, or phone or tablet screen. While some paragraphs can be single sentences, appearing longer than they are wide, others can be equal on all sides, or take up a whole page. In basic design planes rarely occur alone, in much the same way paragraphs come together with other paragraphs to make a chapter or a volume. Although, having said that about single planes, all paintings are actually single planes, but made up of multiple overlain planes to provide a sense of perspective. The planes they house are illusory. The sense of depth provided by a series of planes in a painting does not diminish their capacity for narrative, particularly when they include human figures. People depicted in paintings "can act as a surrogate inviting the spectator into the picture." In reference to Asher Durand's tribute to Cole, which included Cole and the poet Bryant in conversation in the painting, the spectator "Becomes a third party to their discourse" (Novak, 1980, 185). Once we give our attention to the surfaces of planes that work as a path, a walkway, or a roadway, we begin to consider the sequence and positioning of ground, vertical, and overhead planes which bring with them the greater substance or content of the narrative we are authoring across a site.

Figure 5.3 Kindred Spirits (1849). By Asher Brown Durand. Oil on canvas. Courtesy Crystal Bridges Museum of American Art.

The easiest way to understand what I mean by planes as related to sketches and paintings is easily experienced where you live, work, or go to school. Get a thin-nibbed marker. Pull a stool up by a window. Select a distant object you see through the glass, then draw it on the glass with your marker. The glass is acting as a picture plane in the perspective you are viewing. While there are discernable foreground, middle ground, and background planes, their objects and spaces are all captured on the one plane that is a canvas, sketch pad, computer screen, or the pane of glass on which you outlined your selected object.

*Nikola Tesla, the technological genius [who invented...the florescent light and the A-C generator, 'could project before his eyes a picture, complete in every detail, of every part of the machine'...Further, he claimed to be able to test his devices in his mind's eye 'by having them run for weeks – after which time he would examine them thoroughly for signs of wear'. (McKim, 1972, 8)*

In most instances, when two or more different planes are used in a landscape design, a distinction between two- and three-dimensional thinking is called into play:

*...a difference in attitude. A three-dimensional designer should be capable of visualizing mentally the whole form and*

*rotating it mentally in all directions…thoroughly explore the play of depth and flow of space, the impact of mass and the nature of different materials.*

*(Wong, 1977, 7)*

Today we have graphic software capable of providing fly-throughs. A shift in perception, from two-dimensional sketches to the sculptural objects they represent, is difficult, but not impossible to make. Some people can go so far as to imagine four dimensions in their mind's eye and run machines, check them for their efficiency, and locate wear (McKim, 1972, 8).

Those who have difficulty imagining three dimensions need to sketch objects and then build them. The more easily we can picture three dimensions, the closer our conceptual abilities are to including four-dimensional thinking: the time involved in moving through the environment we are imagining or designing. Computer graphics and the ability to generate a "fly-through" help both three-dimensional thinkers and non-designers. Moving through a proposed designed environment allows the viewer to identify gaps in the progression and flow and begin to include even more considerations associated with the sequencing of spaces, the movement through a site, the growth of plants, and the phasing of projects over time.

The practice of three-dimensional thinking is readily helped by doing quick sketches on trace paper. It's a cheap, readily available learning technique. Taking the next step of transforming a two-dimensional representation into a three-dimensional model or sculpture quickly illuminates the shortcomings of a sketch in the early stages of a design and helps us in our ability to see a finished object. This is where you come to see how storyboarding helps find gaps in one's design thinking. Another good exercise to help hone our visualization skills is to recall a common interior or exterior space and draw it from memory. Once done, then visit the space with your drawing and check your accuracy. The ability to envision three dimensions in a two-dimensional drawing is the reason I believe three-dimensional modeling and sculpture studios are excellent foundation classes for would-be designers. Seeing what you are looking at or envisioning is key to being able to bring design ideas to reality. Betty Edwards provides a wonderful foundation, or first step to learning to see. In *The New Drawing on the Right Side of the Brain* (1999), Edwards doesn't so much help you learn to draw as she helps you learn to see. The better you can see what you are looking at, or envision what you want to bring into existence, the more accurately you can draw it for others to see, whether on paper or computer. I also believe that with the ability to clearly see what you're imagining comes the ability to write more clearly. Many of you will experience this improved clarity in the interconnectivity of your seeing-writing-drawing when you reflect your written narrative off your storyboard, and again when you interpret your storyboard into your final design.

*Drawing is a curious process, so intertwined with seeing that the two can hardly be separated. Ability to draw depends on ability to see the way an artist sees, and this kind of seeing can marvelously enrich your life" (Edwards, 1999, 2). "Most people never learn to see well enough to draw (2)…[but] as many of my students have told me, their lives seem richer because they are seeing better and seeing more" (7).*

Time, the fourth dimension, works in favor of most landscape designs because the dominant landscape medium is plants. They grow in size, area of coverage, and seasonal character. The temporal aspect of architecture, landscape architecture, and interior design is most applicable to the visual testing of people's movement through a project, neighborhood, downtown, or region. Even more difficult for students to grasp than seeing three dimensions in their two-dimensional drawings is the role time plays in their designs. Even though there are computer programs to help reduce the difficulty of seeing beyond two dimensions and over and around changing elevations and through plants' growing periods, students will likely continue to have difficulty considering the influences on their designs of changes associated with sunrise to sunset (spectrum of colors, light levels, shadow movement, angles and patterns); sunset to sunrise (artificial lighting, phases of moonlight, changes in temperature and moisture levels); seasons (birth, growth, maturing, and death of vegetation; micro-climates; changing forms of precipitation). These are cyclical processes, periods of transition, that are so much a part of our daily lives that they are frequently taken for granted, even by designers. This is why it is so important that we consider the message content of our work from at least two, somewhat antithetical, viewpoints: coming and going, setting up and breaking down, entering and leaving, awakening and going to sleep.

*Astronomical time is really spatial... best represented by symmetrical space... of cardinal directions modeled after the stars, particularly the North Star and the sun. It is a diagram of the daily path of the sun, and more importantly, of the seasonal cycle.*

*(Tuan, 1978, 8)*

While everything about us exists in at least four dimensions, we regularly consider, and therefore, reinforce life as two dimensional. We live in a four-dimensional space and inhabit a life realm that is a dynamic state of equilibrium, or, as the phenomenologists would say, a continuous state of becoming. Yet, we seem intellectually ignorant of the spherical and cyclical aspects of life. We erroneously refer to the "paths" of the sun and the moon as though, for all intents and purposes, the sun is in motion. Well, it is in motion, but in relative terms it is stationary. It does not inscribe a path, nor does it "rise" or "set." The earth revolves. More accurately, the earth's horizon rises and falls relative to the sun. Daily the fact that many environmental designers conceive of life as a two-dimensional still life, and I realize

Figure 5.4 Mandala. Shutterstock Item ID: 671473225. Contributor: Katika.

Figure 5.5 Yin-Yang. Shutterstock Item ID: 1457918075. Contributor: mrjohnblack.

that is redundant, is reinforced by our work and by much of what surrounds us. In our study of mythologies and religions we are as limited in our perception of the volumetric aspects of life as we are in our design awareness. We rely on two-dimensional representations to grasp and grapple with life. "A geometric figure may serve the same purpose of harmonizing the opposites, and of such figures the most important is the circle or the mandala" (Tuan, 1974, 17) and the yin-yang or Tai Chi symbol. These two-dimensional figures represent and embody the multiple dimensions of life and relationships. Note Tuan's reference to two-dimensional figures. One striking difference between earlier cultures recording of life and our current use of geometric representations for four-dimensional cycles is that our predecessors lived with them, where guided by them, and orally shared their beliefs as they learned about them. The dynamic spatial differences between primitive and modern cultures are excellently provided in Lera Boroditsky's Ted Talk, "How Language Shapes the Way We Think" (see back in Chapter 2). We, however, sit at tables and work out our visualizations on the two-dimensional surfaces of paper and computer screens as if they are separate from us.

Even during the heyday of environmental psychology, and today's academic and media blitz promoting ecological awareness, two-dimensional thinking continues to be reinforced. Many environmental psychologists strengthened the foundations of design in the 1960s and 1970s. Gans (1962), Sommer (1969), Hall (1969; 1981; 1983), and Downs and Stea (1973) typically expressed their work in plan view or as imagined in plan view. Ecological literature also uses plans, and cross sections, and the less-dimensional medium of graphs and charts. The presumption is that viewers automatically interpret plans, sections, and charts into dynamic, four-dimensional environments. Their diagrams and sketches repeatedly represent human behavior from above and not from eye level where the circle becomes a sphere, the square a cube, the triangle a cone, and the ellipse an ellipsoid. This is a lot to expect from a designer, let alone a layperson. In addition, land planners see "land use character" as two-dimensional colored areas on a map (e.g., yellow represents residential land use) and not what the land use character looks like at eye level. So a community settled in 1845 and another a few miles away settled in 1962 both display the same color representing residential land use even though one is 117 years older and at eye level looks nothing like the newer residential neighborhoods.

*Personal space refers to an area with invisible boundaries surrounding a person's body into which intruders may not come....Personal space is not necessarily spherical in shape, nor does it extend equally in all directions.*

*(Sommer, 1969, 26)*

For Hall (1969; 1983) and Sommer (1969; 1974) "proxemics" and "personal space," respectively, are, if anything, a semisphere when we look at a person standing erect. Yet, the authors do not recognize height; therefore, the spatial nature of life remains two dimensional, or flat. This was all the more concerning when I saw the change that took place with my landscape architecture students between first and third year. A review of my video tapes showed first-year students miming the spaces they designed: walking across the front of the room as they described a path, lifting their heads as they told of the sunlight through the trees, lifting a foot as they stepped across a bit of water (Scarfo, 1989). My concern surfaced when I saw those same students three and four years later no longer miming the experiential content of their designs, but presenting the future life worlds of people through the two-dimensional language of geometry: axis and focal point being their favorites. The dangers of thinking one and two dimensionally are made humorously evident in Edwin Abbott's *Flatland* (1991) written in 1880. Later, in 1965, certain corrections with regard to both the social and mathematical thinking in

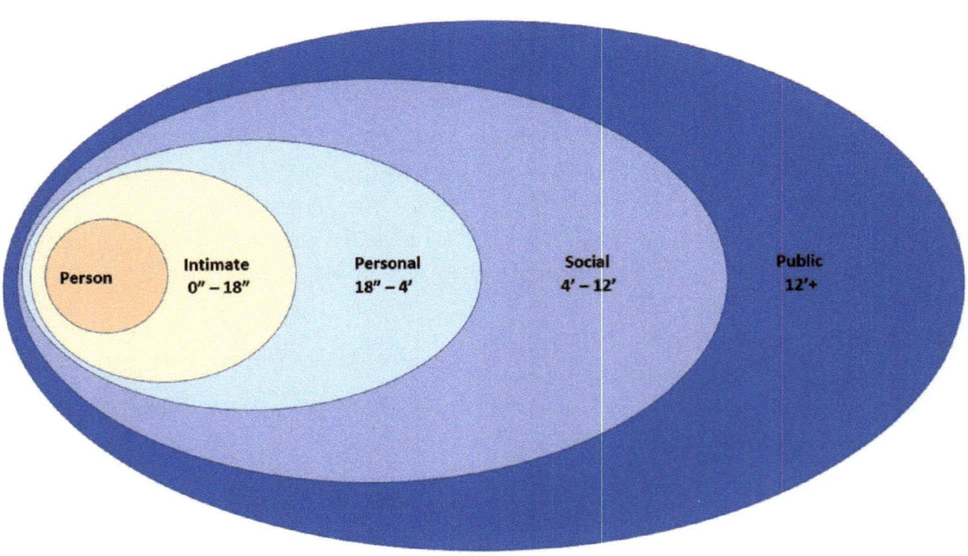


Figure 5.6 Proxemics. Drawn by author.

*Flatland* were made in Dionys Burger's *Sphereland*. As professionals whose work gives form to the life worlds of others, landscape architects need to be comfortable with the fluidity of volumetric spaces that are people's lives in space and time.

> *Proxemics is the term I have coined for the interrelated observations and theories of man's use of space as a specialized elaboration of culture.*
> *(Hall, 1969, 1)*

## GROUND PLANE: LIFE AS MOVEMENT

> *The other method of analyzing verbal landscapes is grounded on the concept and the techniques of literary setting. How can a place in which events "take place" be represented in the time-flow of language? That, in its broadest sense, is the problem of setting, and it is a problem shared by anyone who would render a landscape in language – explorer, naturalist, tourist, and literary artist alike.*
> *(Conron, 1974, xxi)*

Fiction offers us a way to counterbalance our tendency toward a still-life mentality. Our four-dimensional world is recognized by writers and those who study literature (Conron, 1974; Lutwack, 1984; Smitten and Daghistany, 1981). Spanning more than 300 years, John Conron's analysis of American literature, poetry, and paintings identifies four ways in which authors reproduce "a coherent and detailed landscape" (Conron, 1974, xxii). John sought to show how details in literature and poetry can "be arranged so that they will not only accumulate, but also connect to each other as parts of a living picture" (1974, xii). His findings built upon three- and four-dimensional spatial relationships: fixed position, physical movement, metaphoric frame of reference, and dominant impression.

1. "*From a fixed point of view*: I stand at the eastern edge of a field, facing a tree in the middle of the field…here the landscape is organized from where I locate myself in terms of a coherent visual logic: foreground to background, left to right" (xxii).
2. "…organize the landscape around physical movement: I walk toward the tree; wet empty furrows ripple under my feet; the tree balloons as I approach it. Movement thus energizes the heart of the sentence, the verb, and consciously or no it activates description, it suggests that the landscape is animate rather than just inertly there" (xxii).

"Two other techniques play off the visual with cognitive logic…" (xxii)

3. "Landscape is arranged in terms of a *metaphoric frame image*: the furrows in the field are like corduroy; the field is a flat, slightly mussed piece of grayish-brown corduroy on which the tree sticks like a burr…" (xxiii).
4. "…*a dominant impression* arranges the significant details of a landscape in terms of its effects on the perceiver ('impression' means literally a mark on the mind). If I am a painter or a hunter or a tourist, I will naturally ascribe different values to the field. If I am cold, I will describe the field in images which emphasize the chill I feel: gray light, rigidity, starkness, even loneliness" (xxiii).

Conron admits "there are other ways of organizing verbal landscapes, but they depend much less on visual than on cognitive logic" (xxiii). You will find Conron's spatial analysis of literature is best applied later here in Chapters 7 and 8, when you reflect the narrative you are asked to write against the storyboard of that narrative you produce. Another beneficial activity for you to consider, as we get closer to thinking in narrative forms, is to review various cinematographic techniques, e.g., fade in, fade out, jump cut, and flashback to name a few (Beaver, 2009).

*The simplest model of man's existential space is therefore a horizontal plane pierced by a vertical axis. On the plane man chooses and creates centers, paths and domains which make up the concrete space of his everyday world.*
*(Norberg-Schulz, 1980, 40)*

In architecture, landscape architecture, and interior design the ground plane is the most important of the three planes. On it rests the walls, or vertical supports, that in turn carry the ceiling or frame the sky. Even so, Francis Ching's (1978) way of introducing planes exemplifies a problem encountered by most designers wishing to learn about them. The floor is all too often taken for granted. Ching's graphics do deal with ground planes and their importance in meticulous detail. But, the drawings and text that introduce the concept of plane, and precede his graphic analysis of the ground plane, deal with vertical planes. In the various sketches' ground planes are left vague, at best. Little of the contemporary design literature deals with the ground plane in more than ambiguous terms (Downing, 1841; 1969; Guilpin, 1792, 1802; Repton, 1794, 1803; Ruskin, 1859; Vitruvius, 1960). Similarly, little of the design literature deals with interior rooms' ceilings, often confusing them with roofs. More on that later.

*Light gives life to a space. Manipulation of light and shade gives the urban designer power to transform spaces of stone, bitumen, and concrete into tapestries of sunlight and shadow…Intricate shadow patterns on pavement*

*and building walls create architectural richness from even the crudest materials.*
*(Arnold, 1991, 44)*

I'm not sure if there is a connection with the limited attention given ground planes in design literature, but too few landscape designers seem to take advantage of the potential message content that can be conveyed by a ground plane to the people walking or gathering on it. The works of Burle Marx and Dan Kiley come to mind as exceptions, along with the admission that in most landscape designs the planting beds do get a significant amount of attention. Planting beds are the very essence of the landscape architects' design medium. Their shape typically communicates directional messages through framing.

Figure 5.7 Burle Marx – Portuguese stone pavement mosaic in the Carioca square. Alamy Stock Photo ID: 2HEFT4N. Credit: Donatas Dabravolskas.

They demarcate one area from another, and in doing so provide a sense of direction and enclosure. Planting beds along the edges of walks and gathering spaces modify a sequence of messages by directing or gathering the user-readers' attention. If we think of walk, path, and plaza as nouns, then how their surfaces are textured (scored or inlaid) and how their defining edges are articulated can be thought of as adjectives modifying their message forms and activities. As a homogeneous mass of brick, wood, concrete, or stone, any modification to the character of the ground plane is left to the ground plane's patterning, or to the shadow patterns being cast on its surface. Shadow mosaics can give a ground plane a particular identifiable character that provides a space with a uniqueness contributing to its sense of place. The application of shadow patterns can also save money.

*[U]nexpectedly, the cost difference between brick and asphalt paving can buy enough*

Figures 5.8.1 and 5.8.2 Leaf shadows: left (in daylight) and right (at night under lights). Photos by author.

*immediate tree cover to justify using the most disparaged material of urban civilization – asphalt paving – which, with trees, assumes the richness of a Spanish tile floor that changes with the seasons.*

*(Arnold, 1991, 83)*

## PHYSICAL ATTRIBUTES

Many design students tend to take for granted the character of the ground. Although nature provides plenty of examples of ground planes from the seemingly level to precariously steep, many young designers have yet to see the possibilities in such variations and use them to enhance the character of their work. Ground planes are another design feature we are too close to, we have grown to take for granted. Individuals new to design frequently miss the opportunities provided by the topographic nature of the ground in that they often think of the ground as flat and without character. This is what I call a plan-view mentality; after all they are looking at a flat piece of paper or a flat screen. In many design projects they even go so far as to argue for the removal of bothersome topographic and vegetative features in an effort to begin a project on a plain plane. However, when asked to do a design on a flat site, those new to design quickly realize there is *too much* freedom, too much opportunity. There is no natural starting point around which or with which to grow a design. Like being asked to write something, anything, "Give me a sample of your writing." I'm sure you have memories of staring at a blank piece of paper or computer screen, your mind wandering as it searches for a starting point – that first word. Even when given a topic to write about, students rarely ask "what's the emphasis of what I've been asked to write on?"

Forty years' teaching landscape architecture showed me that students approach writing and design in much the same way. So often they expect their first draft to be their last. No surprise, they also expect their first design ideas to comprise their final design. They miss the opportunity to utilize, in a sense to be guided by the topic itself, or what the topographical context provides every site design project. The narrative design process introduced shortly overcomes these problems. With guidance, you will be asked to summarize a site's sense of place in terms of the likely users and the client's wishes; develop a one-, two-, or three-page narrative of someone experiencing the eventual design; interpret the narrative into a storyboard; blend narrative and storyboard out loud to find weak spots, and move to interpret your storyboard into a final design.

Landscape sites offer opportunities to respond in a variety of creative ways. That is why designers learn to do inventories, to find what resources are available to work with and in what portions of a site the work may be done. Do the ground plane's rolling contours suggest soft meanders, or does the rock-strewn bare surface indicate a harder colder approach to a solution that reflects the probable users' expectations? How does all that blend with probable users' expectations? Ground planes' patterns can suggest behaviors such as passage through, stop, turn here. They may be a solid, emphatic statement, "Walk this way," or orchestrated with changes in views and directions and rates of movement into a dance across a site. For an excellent example of how a designer's work can control a sequence of movements and message gathering by a landscape participant, see the experiential description below of the Japanese Katsura Imperial Villa teahouse garden produced by one of Katherine Benzel's students (1998, 171–178).

Benzel's student gives us an example of the two most critical qualities of paths, their ability to move us through an unfolding sequence of bits of information and to structure the sequential nature of that delivery. Shared through the drawings and words of her former student Junko Miyata, she writes

*If we visited the Katsura Imperial Villa built in the middle of the seventeenth century in Kyoto, our*

*experience would unfold in the following manner. The first path to the teahouse, made with stones positioned to facilitate an easy, leisurely stroll is enclosed by a wall of trees on both sides to close down the view. A lantern at the end and stones off the path indicate a place for us to turn. As soon as we turn, the space opens up and we unexpectedly encounter a significant view of the teahouse across the pond. Since the teahouse is not immediately available to us, its appearance heightens our anticipation. We take another path to continue our journey around the pond. The new path is made of stepping-stones spaced far enough apart to require us to look down and concentrate on where we are stepping. It is harder to walk on this path because the spacing between the stones and their height and unevenness make it impossible to assume a normal walking rhythm. We cannot look around but must focus on what we are doing to avoid tripping or falling. Relieved to follow a stone that directs us off the path, we pause to view an island in the pond. After we resume the path, a thick white, rough, monolithic stone bridge stops us in our tracks. Considering our physical and mental exertion, the view of the teahouse right before us is dramatic. The bridge, however, appears to be out of agreement and scale with the teahouse because it is so raw and inharmonious. We walk over the pond on top of it, paying attention because it is very narrow and there are no railings, and again we are forced to think about our efforts. Once we step down on the other side, we find a water basin on the path near the house for symbolic cleansing and purification. The entrance to the teahouse is straight ahead and through a small high window that we must climb up to and crawl through on our hands and knees. Crawling into the house forces us to experience humility. The sequence of transition, contrast, emotion, self-awareness, and time prepares us for the tea ceremony.*

*(Benzel, 1998, 172–176)*

Figures 5.9.1–5.9.4 Drawings and description of the Katsura Imperial Villa Tea Garden are those of Junko Miyata from Katherine Benzel's *The Room in Context* (1998, 173–175). Courtesy of McGraw Hill.

*A visible walking surface (sidewalk, path) is a fundamental provision for the promotion of pedestrian movement. On-site paths can be seen as connectors to off-site paths. According to Gibson's (1979) ecologic theory of perception, "surface" is the provider of possibility for movement. Visible connections and walking surfaces are key elements in Lynch's concept of legibility, in which the pedestrian uses visual cues to gain an understanding of the environment and organizes it into coherent patterns.*
*(Zimring et al., 2005, 189)*

The ground plane is the basis, the foundation of a landscape design. Across the ground plane the interrelationship of foreground, middle ground, and background areas does a lot to set the stage for the type of experience(s) a design provides. Much like the grammatical format of a paragraph the reader need not even be aware of the paragraph's basic narrative structure for it to be enjoyable or add to the reading experience. On the contrary though, the designer/author must be acutely aware of the narrative's structure. With the topic sentence in first place a paragraph's message is introduced or set up. Moving toward a landscape we refer to this as a first impression of which there may be several as the participant moves from one area to another. The shape of a ground plane greatly influences how the narrative is delivered.

A level plane without obstructions allows you to easily see where you are going. A destination is even more pronounced when the ground plane raises up toward you, unless the design is on the other side of the hill. With the ground plane raised the perspective is foreshortened, a path is more easily read than when it is laid out on level ground. The topography of an ascending ground plane brings our attention to the rate of message delivery. Generally, at a normal pace there is a fairly uniform rate at which the message units come into identifiable (readable) range. However, site size and character can influence the rate and rhythm of message delivery.

Two other factors influence the rate at which details occur across a given space. One is what Forman and Godron refer to as a landscape's grain (1986, 216), and the other is the path's context which adds to or diminishes the coherency and clarity of an evolving message. By grain, they refer to a landscape area covered by similar features that account for a coarse, medium, or fine grain. A mountain range, what with the "average diameter or area" of a mountain, is a coarse grain. A landscape of dry shrubs and small trees is a fine grain. A coarse-grained landscape would likely afford longer views, sight lines, than a fine-grained landscape.

In general terms, a landscape's flat, concave, or convex form makes a difference in one's sight lines and the direction in which a landscape narrative unfolds. A landscape that climbs, or curves upward into your field of view, presents more of the context (middle and background) of the evolving message, more quickly. A landscape that falls away requires you to discover the narrative as you approach the crest of the ground's curvature as it reveals what is below where you are. Interestingly, with respect to message delivery, this kind of topography presents its messages from the top down.

In contrast, a concave landform allows our eye to sweep over more ground area more easily, more readily know what's ahead, and what's around.

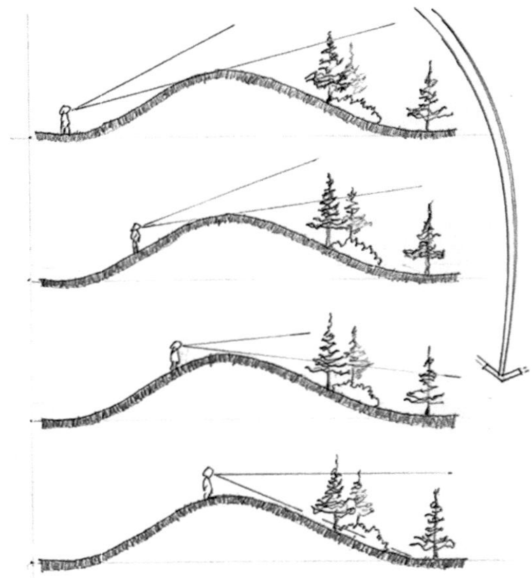

Figure 5.10 Message delivery sequence as you move toward a hillside. Drawn by author.

You are allowed to become more comfortable and spend more time identifying the immediate messages. Information comes not only "from the ground up" but also through what makes up its greater context. This more explicit, more open manner of disclosing messages typically reduces the need for the user-as-reader to "work" for the messages and eventually the narrative.

In the act of ascending a hillside or stairway, the physiological act of climbing is part of both the mood-generating and the information-gathering processes. The landscape's context is being filtered (or modified) by the person's inner context, or state of mind. The user's arrival at a destination had better be equal to the effort made to get there or a certain disappointment ensues. Years ago I was impressed by the way the act of ascension was used in the old Masonic Temple that was then the office of Sasaki, Dawson, DeMay in Watertown, Massachusetts. You ascended nearly three flights of red-carpeted stairs set between two walls. The first two flights, each bringing you to a landing, were straight. Two-thirds of the way up the stairway it took a right-hand turn. At the top was the receptionist's station set in front of a large wall. In this case, the visitor's destination was located over what would be the crest of a hill, which, in this case, was the top step. Climbing the stairs, the person was unencumbered with visual messages, other than the crimson red carpet at her feet. As she turned to climb the last flight, the backdrop, the receptionist, and then the counter/desk (barrier) at the destination came into view, from the top down. As the visitor climbed, and was allowed to consider why she was there, I'm sure the crimson red, carpeted stairs only added to her arriving at her destination out of breath and with an elevated heart rate. She had ascended the heights, earned the right to speak to the receptionist, but at a disadvantage. Gasping for breath while trying to make a good impression, Sasaki, Dawson, DeMay and their staff, whether intended or not, had a certain sense of power over the person before she even introduced herself.

Think of the types of communication associated with straightforward dialogue and that of anticipation, then relate each to a representative ground plane. Think of the types of writers who would associate with each emotion and ground plane: mystery, poetic, meditative.

In 2006 John Simonds spoke to "abstract [spatial design] qualities....of a series of varying volumes, each designed to induce a predetermined response" (2006, 159).

Think, too, of the rate and sequence of information disclosed as we walk over a ground plane that curves downward, away from us. Tension is produced in the landscape especially when the obvious existence of views that can inform us of a destination and how to get there are blocked and yet to be discovered at points (times) along the pathway. The removal of views of your destination occurs when the ground plane is convex. In a convex landform, the ground plane causes people to ask for more information "up front." The "falling away" of the ground plane relative to the location of the reader removes a portion of the topic sentence, the essence of the place over the horizon, hidden by the lip of the downward curving ground plane. As we crest a slope or stairway, background messages are typically presented first, then middle ground messages are presented as we move forward. As we move toward the peak of a curving landform, middle and background information is provided for us from the top down.

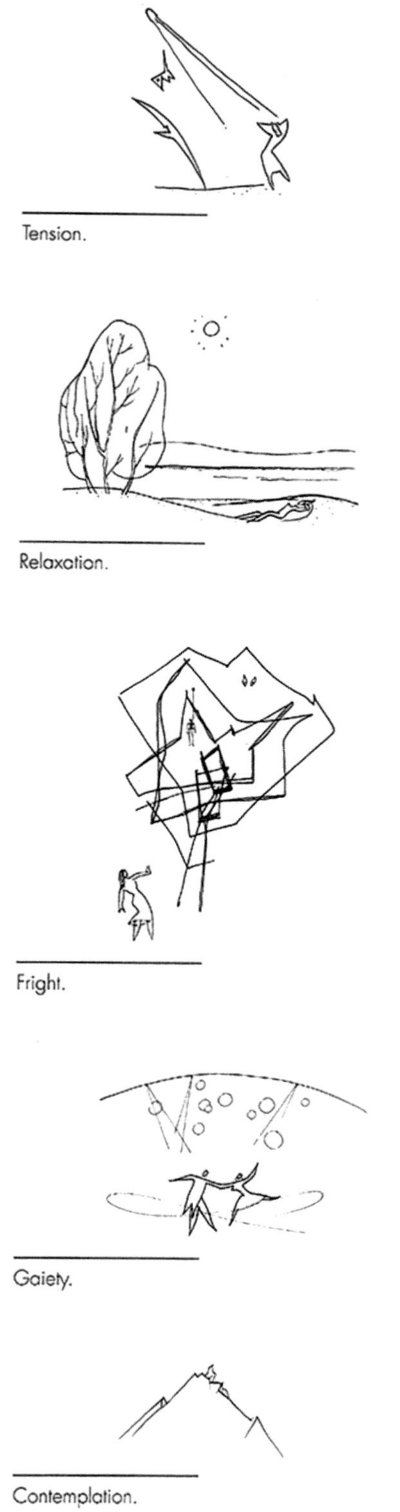


Figure 5.11 Lines and objects indicative of emotions. Simonds (2006, 159). Courtesy McGraw Hill.

In a sense, information rises into view, remember, the tops of distant objects and spaces appear first. The upper portions of the remaining visual information appear first and then progress downward to the ground. If you think about this from a landscape point of view, the tops of tree canopies will be contributing to a user's first impressions, or they will supplement a first impression gained by walking up the slope. The tightness of the plane's curvature determines the distance from the viewer to the information that comes into view. In an undulating ground plane, such as sand dunes or a moraine landscape, information is intermittently presented and hidden. Warped ground planes also introduce the aspect of varying sightlines. A designer could provide bits and pieces of a narrative, the culmination of which would either satisfy what the landscape reader was anticipating or surprise them. Keep in mind, a designer-author could also use sounds and aromas to influence the reader-user's expectations before they reach a narrative's main visual messages.

As a regional, community, or neighborhood feature, a place's predominant sightlines play an integral part in contributing to its local character and greater context. I grew up outside Boston, Massachusetts. Growing up I took the characteristic New England sightlines for granted. While vegetation was a contributing factor, the rolling land itself made for typical New England sightlines of one-quarter to three-quarters of a mile. An excellent similar example of such a landscape, without the tree cover, is the Palouse region of Washington State. Although topographically similar, the interesting difference between New England's and Eastern Washington State's landscapes is the absence of trees across the Palouse. With few trees across the Palouse, the early morning and late afternoon sun seems to appear sooner, and linger longer, and depart more quickly than in New England. Easterners start and conclude their days with lower levels of light for longer periods of time as the sun has to climb higher before it shines over and through the tree canopy. In a lot of the Palouse and the high desert of eastern Washington State, the sun is in your face from the moment it breaks the horizon to the moment it sets. In eastern Washington sightlines are longer than in New England and overlooks are broader and deeper in their panoramas. In fact, high points and overviews are a major feature of the eastern Washington landscape character.

Figure 5.12 New England village. iStock ID:824654196. Photo: Kirkikis.

Years ago, in a humorous gesture to a readily identifiable sense of place, someone recognized an aspect of Kansas' extended, flat landscape and seemingly forever sightlines. That aspect became a bumper sticker that was divided evenly across its length with a horizontal band of blue resting atop a horizontal band of white. It read "Ski Kansas."

Another example of people's sense of and comfort with a regional landscape character

Figure 5.13 "Palouse" Southeast Washington State. Courtesy of Spokane, WA. Photographer Dan Price.

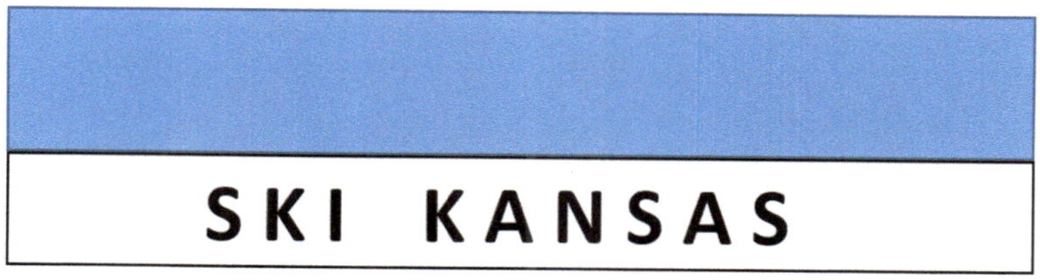


Figure 5.14 SKI Kansas. Drawn by author.

occurred several years ago when two friends from the Plains States came home with me to Massachusetts. We arrived at night, so it wasn't until the next morning that they had their first introduction to the limited sightlines of the New England landscape. For two days they were obviously uncomfortable. I could see it in their faces, they could feel it in their bodies. But we couldn't figure out exactly why they felt so uncomfortable. When we visited the wide-open beaches in Ipswich and walked over the last dune to enjoy a view of the Atlantic Ocean we all suddenly knew what had been happening. They'd been fighting feelings of claustrophobia. The New England landscape's reduced sightlines contributed to their feeling closed in. Once they crested the last dune and looked out across the ocean they could see, what was for them, a typical horizon at a distance of about three miles and foregrounded by the beach's flat open landscape. Their facial muscles then, finally, and obviously, relaxed.

The ground plane sets the stage for a narrative. Whether naturalistic or modified, a landscape's ground plane is an ecological composite of natural, cultural, and social dynamics: a place's cultural geography (Yamin, and Metheny, 1996). Even the most built upon site is still a composite of interactive, and to a great extent, interdependent relationships between what's built, what's built upon, who's doing the building, and who lives there using it. If what is built works well with the natural dynamics of a site and setting then it requires less maintenance than if the designed environment fights the local ecology. If the environment complements the cultural and social dynamics then less social maintenance is required in the form of laws, regulations, codes, and the people to enforce them. This is a fundamental aspect of sustainability.

Figures 5.15 and 5.16 Seawall, San Padre Island, Texas, 1978. The seawall reminds me of Ernest Callenbach's statement, "Nature bats last." Photos by author.

As presented above, within the realm of message delivery, the nature of the ground plane contributes to the landscape text's frame of reference or setting. Whether flat and open, concave and enclosed, or convex and foreshortened (but interestingly enough with high points and long views), the ground plane sets the character for how people move across the land. On flat terrain the typically open views can be easily enhanced or diminished with the selective placement of vertical planes. Vertical planes either work with the circulation pattern provided along the ground plane or work against it, giving the narrative a message structure that contrasts the topographical character of the place. Later, as we think about the movement of the landscape's reader we will gain a broader understanding of how ground planes contribute to a landscape's narrative form and content.

## VERTICAL PLANES

*In our experiences of the built environment, action settings constitute our social worlds… When we enter a built environment, we immediately survey and apprehend its character, figuring out, usually with a glance what we can and should do, discerning all the norms that govern people's conduct there.*
*(Goldhagen, 2017, 199)*

With regard to narrative, vertical planes and upright objects provide eye-level messaging features. At eye level, vertical planes tend to fill more of our cone of vision. We read them in an effort to acquire more explicit information along our narrative path and as such they take on a more subtle, yet influential, power. Vertical planes tend to hold and guide our immediate attention. The reason is simple, as vertical planes they are in our face rather than at our feet. Their addition to a ground plane immediately divides the larger space so to provide the overall narrative with a sequence of smaller message-delivery areas. Vertical planes separate spaces. As foreground features, they more emphatically frame and direct how we may move about a space. They structure both flow of movement and message delivery.

Watch the person delivering the mail along your street. Lawns, driveways, and low planting beds are all walked through with the aim of traversing the shortest distance between front doors and mailboxes. This speaks to vertical features efficiency and their power. But, install a fence, or watch what happens as a wall of wild grasses along a property line grows to a height of 4′ during the summer. The growth of taller vertical masses not only redirects people's attention, if not movement, but also introduces a more striking demarcation of foreground, middle ground, background, and framing, or stage right and stage left. Verticals direct, enclose, and enframe. As solid features verticals define landscape rooms. A fun landscape quality to play with is how the sequence and character of landscape messages change over time. As vertical landscape features grow, they narrow the setting provided by a ground plane, intercepting our sight, taking away, as it were, the more distant ground plane, the narrative's greater context. With regard to narrative, the segmenting of space with verticals enhances the sequential character of the messages being read. Since verticals can control the visual sequencing associated with a place, they can also, to a great extent, control the sequence in which the messages are read. They contribute to the message rhythm and comprise a lot of what we so easily call the character of a place.

*In his discussion of cross-sectional cut lines, Wester notes, "The area at or near the cut line is designated as foreground, the main message or emphasis area as middle ground, and the horizon or context area as background".*

*(Wester, 1990, 65)*

Interesting is the positioning of the viewer and the viewed to both graphic communication and message delivery in the landscape. While Lari Wester's (1990) focus is the development of good graphic skills, his approach makes a wonderful contribution to our understanding of a landscape's message delivery capabilities. A viewer's position relative to what's viewed is critical to delivering the desired message(s). Throughout *Design Communication for Landscape Architects*, Wester relates the viewer's proximity to the viewed, the message area, scale of drawing and detail contained in it. Since Lari works off the viewer's cone of vision all these qualities easily transfer to the development of message delivery in a landscape. Lari's work provides an excellent, and fun, foundation on which to build a landscape's message delivery capabilities. A good example is his introduction of axonometric drawings. He wants to be sure the student realizes "that simply knowing how to plot an axonometric is *not* the objective here. Learning to control and adapt the process for communication is the objective" (Wester, 1990, 79).

*From the stage to the city is more of a leap in scale than symbolism. Built environment has the capacity to transmit messages which range from the conceptual and verbal to the abstract and inchoate. These intimations do not consist of objective information but rather reflect an internal mental assembly of symbols and images. The message can be 'read' and understood because the mind is versed in the language.*

*(Smith, 1977, 16)*

Positioned across a ground plane, vertical features provide more information for the reader than the ground alone. They increase the author/designer's intended message delivery by adding detail to the ground plane's message and greatly reducing the reader's wandering eye. Message continuity along a continuum calls for the designer to give greater attention to syntax – a strong interrelationship between the media delivering the messages and the messages delivered. The channeling of motion and sensory perceptions helps align the information being perceived and interpreted. Reading of the text becomes more refined and directed, we might say more pronounced, than that found

Figures 5.17-5.19 Providence Sacred Heart Hospital Healing Garden, Spokane, WA. Photo by author.

on an open ground plane. Vertical planes may restrict wandering or provide for a multiplicity of planned opportunities to wander about a site. For example, the seemingly free wandering about Spokane, WA's Sacred Heart Hospital's Healing Garden, was a planned part of the overall narrative. Broken up, with openings at differing intervals and heights, permeable vertical planes allow (even call our attention to) specific views of things to come. They can foreshadow destinations and at times the routes to be taken. Vertical planes can draw our attention backward, strengthening our sense of beginning and return. Openings in vertical planes to the sides of our line of travel enhance information regarding where we are, where we have been, and where we are going. Their spacing also influences the sensed rate at which we are moving. What they allow and disallow as views provide for subplots and diversions to the main messages along our journey, flashbacks and flash-forwards, reminders, and hints.

The strength of vertical planes lies in their ability to demarcate, divert, and direct both movement and views. While he doesn't use the word plane, Lynch's "district" is more a demarcated place than a thing. Although it is "conceived of as having two-dimensional extent...the observer mentally enters 'inside of'...[this place of] some common, identifying character" (1960, 47). Two facts need to be noted with the introduction of Lynch's definition. First, Lynch's "observer" is not passive but an active participant in the district, reading and interpreting information. Second, the participants' ability to enter a district calls for the existence of edge boundaries that are pronounced so one knows when thy have entered and exited. Or, with sightlines back to the edge, participants can know how far they have come into a district. However, Lynch goes on to note the edges need not be solid and can act as seams allowing movement between districts of activity areas.

*In general vegetation is the manifestation of living reality. But vegetation has also forms which are less friendly or even frightening. The forest, thus, is primarily a "wilderness" full of the strange and menacing forces. Bachelard writes: 'we do not have to be long in the woods to experience the rather anxious impression of "going deeper and deeper" into a limitless world. Soon, if we do not know where we are going, we no longer know where we are'. Only when the wood is of limited extension and becomes a grove, it remains intelligible and positively meaningful.*

*(Norberg-Schulz, 1980, 25–27)*

The experience of entering into a place does evoke the simultaneous sense of (1) having left a place and (2) having passed by, over, or through a feature of demarcation: a line of low-growing groundcover on both sides of a walk; an opening in a stone wall, a picket fence, a plaza, or an alley between two buildings; or under a canopy of trees, a bridge or trellis. In moving toward a vertical surface, fence, building facade, or hedge row or barrier planting, we quickly scan and read our options. Are there openings to move through the mass confronting us, or must we turn to move along its facade? As verticals become increasingly distant from our viewing position, their influences transition from that of framers of scenes and directors of movement to that of destination points or backdrops. At a distance they act more to provide hints, message seeds, as it were, of what's to come or what to move away from. Close in they help us understand how to carry out those hints.

Gordon Cullen notes the framing aspect of verticals as "netting...the devise of framing brings the distant scene forward…into our environment" (1961, 39). Netting is experienced more to your right or left providing momentary glimpses as you move along a narrative progression. Remember the Japanese practice, *shakkei*, of visually bringing distant landscapes into a garden or borrowing the distant scenery (Bring and Wayembergh, 1981, 180). In "serial vision" Cullen recognizes verticals that undulate to reveal a series of contrasts that bring a place to life. Serial vision enables that progression of a spatial sequence of messages that make up a place's narrative or story.

In the case of verticals that frame distant scenes, the reader is more a viewer than a participant in the scene viewed. They can, but do not typically, walk through the framing elements until further along the narrative. This is echoed in Conron's description of the first of four ways an author "produces a coherent and detailed landscape" (1974, xxii). Allowed to walk through the framing features – entering Alice's mirror as it were – allows the design participant to cross the line that separates individual views, or a narrative of stop-action scenes, to that of a progression of scenes or of serial experiences, or Conron's second way an author introduces the reader to a coherent landscape (xxii). The change from moving up to or passing by a framed view to moving into that which was previously framed shifts the reader from being a passive viewer/reader, as in a third-person narrative, to that of being an actor living the scene as a first-person narrative. The serial nature of an unfolding narrative is dealt with further in the upcoming subsection on volume and sequence of volumes and then in Chapter 8, Authoring a Landscape Narrative.

## BASIC ATTRIBUTES

Both Hall (1969) and Ashihara (1970) relate the interrelationship of viewer and viewed to vision and eye-level views. Hall more strongly relates the mathematics of his proxemics to the cultural contexts in which they are applied.

*Proxemics is the study of human use of space and the effects that population density has on behavior, communication, and social interaction. Proxemics is one among several subcategories in the study of nonverbal communication, including haptics (touch), kinesics (body movement), vocalics (paralanguage), and chronemics (structure of time). (see https://en.wikipedia.org/wiki/Proxemics, last viewed March 1, 2022)*

As the number of planes defining a space increases so does the sense of enclosure to the extent that eventually the space becomes enclosed. Even a patio surrounded on four sides by 3″ tall groundcover becomes either a stepping off platform or a terminus to the walkway that cuts through the groundcover. Such a patio area is, in a sense, a gathering space open to its surroundings. By contrast, an enclosed space, surrounded by solid vertical features, is inwardly focused and to varying degrees cut off from its surroundings. Vertical planes most readily eliminate the greater visual context of a space and to a lesser degree the audio and olfactory contexts. Context in this instance was built up prior to the landscape reader's arrival at the enclosed space and to an extent may be maintained by surrounding scents and sounds. A wonderful example is Thomas Church's residential design in, I believe, Woodside, California on a hillside south of San Francisco. Having arrived at the circular turn-around at the front of the house you are drawn to a narrow, cinder-surfaced walkway that takes you toward the side of the house. You walk between two tall hedges flanking the pathway. There is nothing to see except the path in front of you and the tall vegetative hedges immediately to your right and left. Yet, there is a

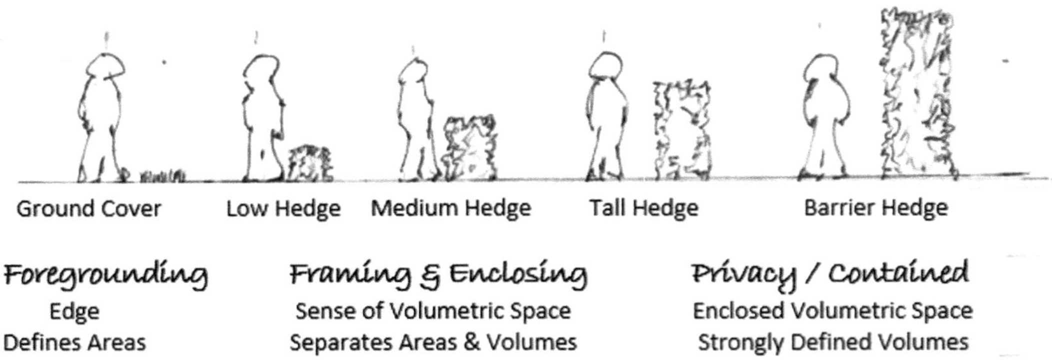


Figure 5.20 Enclosing vertical features. Drawn by author.

quiet sound, a hint, of water. But where? You are given the answer through a narrow break in the hedge to your left. One step forward and there it is, a small fountain, a kind of birdbath. One more step and then it is gone. The enclosed walk continues around the house, finally dropping away as you arrive at the backyard, and wham, a panoramic view to the north, up the peninsula to San Francisco and the Bay Area.

## AMBIANCE

Surface textures and colors mediate one's sensed depth of a space. The finishes on manufactured materials and the leaf structure, density of overhead canopy, and color not only enhance or reduce a space's ambient light but also express aspects of the place's character. A place may appear airy and both visually delicate and light with regard to weight. The weight of a space, in a darker location, is typically the result of both the enclosed space and what is enclosing it. Dark spaces appear to be heavier than light ones, a dark space's air can seem heavy. But, in the mid-summer heat a darker space is often a welcome relief: lower temperatures combine with less radiant energy on your skin. Some of both aspects can be brought together by manipulating a space's enclosing planes' surface textures and colors. Many urban courtyards and small parks receive little if any direct sun light. Yet, as University of Toronto landscape architecture students found in their video-taped study of urban spaces, a number of downtown Toronto spaces not receiving direct sun still showed people casting shadows. Sunlight was being reflected off adjacent buildings' glass surfaces.

While vertical planes are readily apparent as landscape features, most rely on ground planes for support. By thinking of the ground as a foundation on which to locate plants or built objects is to take the opportunity to enhance a place's story. By blending the message delivery capabilities of both the horizontal and vertical surfaces a place's story becomes more substantive. Verticals also connect the ground plane with the overhead plane. In this way the story line is diversified and as a result more interesting. The cathedral-like presence of elms and London plane trees can raise you to the Heavens, while small verticals can bring the grandeur and beauty of clouds down to earth. Or, like the tea garden in Kyoto (Benzel, 1998, 175–176), the horizontal ground plane's varied surfaces require you look at your feet in some places and allow you to look at the garden in others. Working in tandem with vertical plantings and foreshortened and elongated views, people's behaviors are choreographed in a way that directs them to selectively "read" a landscape's narrative.

## SETTING TO DETAIL

As a final note in the introductory ideas related to planes, we are continuously influenced by the location of planes relative to us and our reading of them. Designers can argue for the application of design principles to a landscape narrative from three points of view: the global, the local, and the personal. In the process of reaching out and acquiring information, people's eyes scan the space ahead. Taking from Rubin's theory of serial recall (Rubin, 1995, 176–179), a distant, panoramic body of information provides the larger setting of a more close-in narrative. That larger setting is global. As people move toward the global text it is constantly becoming more local. As we walk from one outdoor room to the next, along an urban street, or through a garden, we moment by moment review increasingly detailed textual information. The distant, global, information becomes increasingly specific, or local, and then more personal. In the act of a text becoming more detailed our message acquisition shifts from the ground plane to the vertical planes in front and then around us. As the shift occurs the ground plane grows in its subtlety unless it is broken and uneven. As stated earlier, the ground, or base plane, provides physical support for vertical elements that mold the form of landscape spaces. The extent to which it supports the eye-level text is critical to a design being considered good or great. However, that support is usually quite subtle.

*Hall's proxemics can be used as a guide here in regard to the degree of detailed information that can be or is collected. (1969, 113–129)*

As long as we see the ground plane as uniformly unobstructed it becomes a more subtle influence as it becomes local to the reader. The reason is seen in many of the quick-production methods of one-point perspectives (Reid, 1987, 159–163; Wester, 1990, 96–99). Due to our cone of vision, which has a vertical span of 60 degrees, what is viewed without being distorted is seen beyond 20′ away from us. In most landscape narratives, anything closer than 20′ has already been read. What is close to the viewer, up off the ground, and inside one's cone of vision, is what has the viewer's immediate attention. Interiors are different. In most residential interiors people consciously look at the ground plane, particularly if its tonal character changes frequently. It is a matter of personal safety with regard to the potential of changes in heights of flooring materials or changes in elevation. If you think for a moment and react to my statement with, "I don't do that." stop and ask yourself why, where, and when. How are interior floors typically handled? Which ones do not continuously call for you to check on their being level and unobstructed; which do often call for your attention? Commercial and institutional interiors, those of churches, retail shops, and government office buildings, would fluctuate between the two; some having the space for the ground to transition between global and local, and some maintaining a local-personal approach to the delivery of narrative messages.

If you ever visit the Governor Calvert House in Annapolis, MD and you don't watch where you're stepping you may suddenly find yourself standing above and looking 10′ or so down through plexiglas flooring into an old heating system known as a hypocast "used for heating a greenhouse used to grow semi-tropical plants and flowers" (Archeology in Annapolis, http://www.aia.umd.edu/calvert.html). Fascinating as it is it is still startling to suddenly find yourself seemingly standing on air.

Vertical planes can open up to you, or reject and push you away or cause you to angle off to the right or left of your line of movement. How blunt,

Figure 5.21 Plexiglass Floor, Governor Calvert House, Annapolis, MD. Photo by author.

Figure 5.22 Vietnam Veterans Memorial. Shutterstock ID: 121748179. Contributor: MDart10.

or in your face, the messages are rests largely on the plane's shape, visual and acoustical character, and how it relates to the other planes around it. The messages are also influenced by the material of which the plane is made, color, texture, apparent density, and scale. Looking straight at the edge of two planes joined along a vertical edge or corner and spreading to your right and left places you in or near what Ching refers to as "a field of space" as it relates to the two outstretched planes (1978, 137). The interior corner of the Vietnam Veterans Memorial is softened by the fact it is 125 degrees 12 minutes. Planes joined along an edge present a harsher field of space than when presented by a single curved plane. The influence of enclosing planes is relative to (1) the depth and width of the inscribed space, (2) the height of the plane(s), and (3) the resulting distance-height ratio of the viewer to what is being viewed. More background can be found in Ashihara (1970) or Edward Hall's "proxemics." (1969). When you are at the lowest (and tallest) part of the Vietnam Veterans Memorial, 10 feet below the top of the wall, it is quite surprising how quiet it is. I remember just hearing birds. You're left with the deceased veterans' names and you reflected in their memories. The absence of city sounds makes the experience all the more solemn, solitary, and powerful.

## OVERHEAD PLANES

Yi-fu Tuan provides a wonderful summary of the bases on which much of our interactions with our environments rest.

*Human time also has an important psychological aspect. Life is lived in the future – close as the next meal and distant as the next stage of a career… one's back is to the past and ahead lies the future (Tuan, 1978, 9)…Looking up at the night sky is contemplation, for people have few practical reasons for turning their vision skyward. Since looking forward is habitual, people must make a special effort to convert [the sky] into an aesthetic experience.*

*(Tuan, 1978, 12)*

In many reference books that deal with overhead planes, landscape architecture students must interpret what is given. Ching's graphics beautifully depict several hundred overhead planes as a result of the attention he gives to the ground

Figure 5.23 Sky as an overhead plane. Photo by author.

and vertical design features in his perspective and elevation drawings (1978, 110–193). Even so, much of the potential information regarding overhead planes remains implied. Even in *Sense of Unity* (Ardalan and Bakhtiar, 1973), where the articulation of overhead features is explicitly dealt with, there remains a wealth of information in graphic and photographic images left for the reader to interpret. In that we rarely consider overhead planes, we are not attuned to thinking about what they have to offer. Overhead planes and ground planes are primarily found at our peripheral vision. In design that means they subtly influence what we see ahead of us: patterns above and below sandwich what is up ahead adding to the sense of direction, supplementing information, and complementing the character of the messages being read.

As the designers of environments people inhabit, we seem to have slipped into an apathetic appreciation of what's overhead. Yet, in that so much of what I've been asked to design included trees, my understanding of the nature and the characteristics of tree canopies has always been an important aspect of what I considered in the design of a space. Although I'd like to believe that I don't think that has been the case. Only in the latter part of my career my background in history and narrative form provided me with an appreciation for context or for what is, at any given moment, at the periphery of what I am looking at. Like so many designers, for many years I took overhead planes for granted, most likely due to little attention being given the design capabilities and principles of overhead planes in my classes and books. Thus far, many of those writing about overhead planes missed an important distinction. While sources do mention ceiling and sky, overhead planes are typically explored as ancillary features of walls and trees. Numerous architectural and landscape architectural sources refer to roof or roofs as though they are addressing overhead planes. What's taken for granted is that a roof is viewed from afar, not typically experienced from underneath as a ceiling or overhead tree canopy is. Ashihara (1970) believes that unlike interior architectural space "delimited by three planes, a floor, a wall, and a ceiling," exterior space, regarded as "architecture without a roof," is

> *delimited by only two planes – a floor and a wall; in other words, it is space created through the use of only two dimensions...this makes the floor and wall planes all the more important determinants in this type of design.*
>
> *(14)*

I disagree with his observation. What is overhead has an inescapable influence on exterior space and vice versa. The articulation of space at eye level and its sense of enclosure relate to the seeming height of the sky as a ceiling or overhead plane and how it is viewed. In the same way corners are implied by structures around but not enclosing a plaza or town square. So too the heights of vertical planes imply the proximity of a space's ceiling. The 12′×12′ air shaft of a tenement house is a hole above which the sky-ceiling is inconsequential compared to that of a 12′×12′ interior courtyard of a one-story residence. The effect of the ground features silhouetting the sky is influenced by the makeup of the sky, its color, and characteristic clouds. "In general we may say that the sky is as

Figure 5.24 Piazza del Campo, Sienna. Shutterstock ID: 1946041681. Contributor: Diego Mariottini.

large as the space from which it is seen" (Norberg-Schulz, 1980, 40).

*The grass was very green by the roads and by the streams; the shadows of the mountain slopes were very dark on either hand overhead, and the sky with snowy flanks and tips was high up.*
*(Lawrence, 2007, 123)*

*We climbed up the gradual incline for a long time. The slopes above became lower, they began to recede. The sky was very near, we were walking under the sky.*
*(Lawrence, 2007, 121)*

In all the attention Ching (1978) gives his analysis of space, spatial definers, and principles of order that cover the 387 pages of magnificent drawings, only six pages are devoted to Overhead Planes (130–135). Of the six, the first page provides an excellent introduction to the nature of overhead planes. The remaining pages deal as much with roofs as they do with ceilings, and even then the ceiling is addressed as structure rather than design feature. Herb Greene's work (1962), while aimed at helping us appreciate lived space, focuses on roofs as though they are something we live with by living under them. As in most of the literature, roofs are the focus of sculptural appreciations seen from a distance, not as appreciated from underneath. When under a structure's roof we typically appreciate the interior space's ceiling. Equally cursory attention is given by Norberg-Schulz (1980), although he does address the importance of the sky more than Ardalan and Bakhtiar (1973) and Sitte (1945). In each case their respective books' indexes are of little help. The reader must go hunting. Even those who have explored the body metaphor seem unaware of there being any importance associated with what's up and how high. Simonds two pages on "The Overhead Plane" adds detail to the work of others. While not addressed directly, a number of Simonds's sketches, throughout *Landscape Architecture*, provide insight into the use of overhead planes (Simonds, 1961, 89, 111, 121, and 150–153).

*It is only when we visit places very different from our home that we suddenly experience the sky as "low" or "high"... seen from an extended open plane, the sky becomes a complete hemisphere, and when the weather is "good," its appearance is all-embracing and truly grandiose. In places with a pronounced surface relief or rich vegetation, however, only a sector of the sky is seen at a time. Space contracts, and the landscape becomes intimate or even constricted.*
*(Norberg-Schulz, 1980, 39)*

Keep in mind, overhead planes are typically perceived at the periphery of our cone of vision. Although seen in less detail than what we, at any moment, are focused on, what is perceived is no less important to the character of a place's main messages.

## MACRO TO MICRO

In landscape architecture, the sky is inescapable and ever changing. The sky is one medium found on a landscape architect's pallet, broadened when you consider daytime/nighttime, changes in lighting with the four seasons, and the weather. The myths and religions of most every culture connect their existence on earth with the sky through mountains both real and imagined as in the form of pyramids (Norberg-Schulz, 1980, 51) and the "world mountain" or ziggurat (Ardalan and Bakhtiar, 1973, 89), trees (Campbell, 1974,190–194), and obelisks (Bacon, 1974). The sky is the earth-home's ceiling, nurturing with sun and rain and chastising with lightning, hail, tornadoes, and heat. This, seemingly little considered design feature, has directed the life and growth of numerous cultures. The sky's contents, the sun, moon, stars, comets, have directed people's sense of time, fired their imaginations, and provided them with their gods. Yet, looking up is a child's past time, all too often lost to adults.

This will sound strange, but clouds were seemingly not really "discovered" until the Dutch painters began exploring them.

> *There is an evolution in our painting of skies and clouds over the last 500 years…In the late 1400s…clouds appear more as symbols of clouds than observed… A theatrical backdrop…into the 1600s… not yet painted as a result of direct observation…By the mid 1700s…Finally, we have clouds that feel vaporous.*
> *(Dunlop, 2020)*

Among those who studied clouds are Ruskin (1859, 45–59), William Blake (in many of his writings and paintings), Olmsted, and the Hudson River School of Painters.

> *The practice of using trees to make urban spaces…[is like using] words, they can be organized to create confusion or poetry.*
> *(Arnold, 1991, 47)*

The intriguing, yet brief, four-paragraph introduction to "roofing" found in the *Poetics of Gardens* (Moore et al., 1988) is elaborated upon in Arnold's *Trees in Urban Design* (1991). Where Moore et al. have a limited view of a landscape's overhead plane, "the ceiling of the garden does not so readily fall within the designer's grasp since its basic element is the sky" (33), Arnold recognizes the results of the interplay between the tree canopy and sky in the branching structure of a tree and its shadowed-light patterns, and the leaf density and magnitude of its resulting sense of enclosure. In combination the works of Arnold (1991), Moore et al. (1988), and Ashihara (1970) provide an introduction to the attributes of overhead planes in the landscape. The pragmatic slant of their views is drawn from deep within the being of our existence, as recognized by Norberg-Schulz's existentialist interpretation of the sky in various landscapes (1980).

> *In the desert areas of North Africa and the Near East, the cloudless blue sky gives emphasis to the infinite extension of the land…[on] the plains of Northern Europe, instead, the sky is usually "low" and "flat"…and the feeling of being under an embracing dome is usually absent.*
> *(Norberg-Schulz, 1980, 40)*

Three types of overhead planes define outdoor spaces: manufactured, vegetative, and the celestial as it is framed by either manufactured objects or vegetation. Manufactured overhead planes are

Figures 5.25.1–5.25.3 Overhead tree canopy densities. Photos by author.

those of wood, plastic, plaster, steel, and the like. Their potentials as designed forms and articulated finishes are found in a wide variety of sources not because they get explicit attention as overhead planes but because their part in shaping and unifying space is inescapable.

As features critical to capturing borrowed scenery (referred to as *shakkei*), Itoh presents the importance of the overhead eaves of the main building of the Jiko-in garden as more important than the posts of the verandah. He describes the need for the overhead eaves in recognizing that their absence "would allow too much of the sky to be seen, thus the view of the plain would degenerate into nothing more than a commonplace sight" (1981, 47). Itoh's multiple references to the posts, and the absence of photographs of the framing eaves, allow the reader to too easily miss the role overhead planes do play in spatial design and in landscape narratives.

All three types of overhead planes are characterized by the physical attributes of height, space, density of structural members, opacity or translucence, and the degree to which they contribute a sense of scale of actual or perceived space, shadow patterns, sound, ambient light and temperature. The character of an overhead plane that results from the interplay of these attributes adds to or detracts from the continuity and flow of a landscape narrative at eye level. This realization becomes all the more critical as we admit to how little attention designers give to what is happening overhead. In buildings, and more so outdoors, ceilings are not "relatively free of the functional restrictions that are the natural lot of the floors and walls…" (Moore et al., 1988, 33). Tree canopies, as outdoor ceilings, typically contribute to the ambient light and shadow patterns on the ground. Those patterns can either complement

Figure 5.26 The sunlit intersection up ahead signals cross traffic. Photo by author.

or contradict the landscape narrative as expressed through the verticals. The interjection of dramatic increases of light at intersecting paths and roads throughout the older parts of New England and the Mid-Atlantic States calls out to people using the heavily shaded paths or roadways. As, in a sense, the light at the end of the tunnel illuminated intersections and gathering areas, also signal destinations, and eventual arrival; changes in lighting can warn pedestrians and bicyclists of cross traffic ahead (personal safety): be alert a decision is coming up, better slow down (reorientation). Changes in the height, span, and density of structural members (tree canopy's leaf density) can also communicate the above messages. Indoor and outdoor narratives in which attention is not given overhead planes potentially contribute to disorientation, missed decision points in a journey, and dangerous situations. The unplanned criss-crossing of light and dark patterns is the objective of camouflage, the breaking up of edges to hide the outlines of objects that tell a story someone doesn't want told.

*The arrangement of trees in definite purposeful patterns enlivens spaces. Our habit of considering geometric composition as static comes from the limitations of two-dimensional representation in drawings. Moving through a space, what appears in plan as a static row or grid, becomes a rhythm sprung from tree trunks and provides a visual syncopation that improves human comprehension of the space, just as metrical form in poetry guides measurement and understanding. The effect is also like a kaleidoscope with each twist compounding a fixed number of elements.*
*(Arnold, 1991, 44)*

Overhead vegetative planes are either a continuous or broken blanket of branches, twigs, and leaves. To varying degrees the sky is filtered out depending upon the trees included in a design and where they are in their growth to mature size and spread. Overhead, sky planes are situations in which the sky is framed and seen as a complete entity. Sky planes depend upon the dimension of the openings in a tree canopy. Arnold (1991) provides the best information and direction in the organization of both vegetative and sky planes. His primary focus is the appropriateness of tree size to setting and intended effect. His conviction, backed up with photographs, is that the wonder of urban spaces relies on the use of large trees planted closely together. Unlike many authors his arguments are made from and refer to eye-level observations. Sitte (Collins and Collins, 1986) also works from eye-level observations, which is to his credit, but he's most attentive to buildings' facades and to the ground planes that they surround and enclose. Arnold justifies his eye-level approach by noting the need to avoid being caught up in thinking in plan view: "...because of their ineffectiveness in conveying three-dimensional character...design and planting plans nearly always give the impression that trees are visual barriers at eye level because of the way the tree crowns are shown superimposed solidly on the ground plane" (Arnold, 1991, 74).

From an eye-level perspective Arnold also states, and I'm sure to the glee of architects, that "good architecture should not be softened or hidden by plants, but should be reinforced and sometimes embellished. It is not the nature of our cities and urban buildings to be soft" (Arnold, 1991, 41). Within a narrative approach to design, Arnold's generalization bears questioning, especially with the increased attention being given the benefits of green environments on the health and well-being of hospital patients and staff, children with autism or ADHD, and longevity of the elderly (Amick et al. 1995; Kearns and Gesler, 1998; Sternberg, 2009). Research findings from Asia, Europe, and North America show that the softening of environments through the introduction of greenery benefits the health and well-being of all age groups (Kuo and Taylor, 2004). Such environments also positively influence those who are autistic and enduring the trials of dementia and Alzheimer's (Jauchius, 2017).

Indicative of an overhead vegetative plane is its continuous mass; depending upon the trees selected the canopy appears in varying degrees of solidity. What I refer to as a sky plane is found when tree canopy and building, or buildings, frame an overhead view of the sky. A vegetative plane is characterized by the following:

1. Continuity over circulation routes (pathways or roads) or over gathering areas (plazas and courtyards): over both linear and rectilinear areas "trees in their best use reinforce, articulate, add scale, and subdivide these spaces" (Arnold, 1991, 49).

As you drive the older portions of Prince Frederick County, Maryland you know when a small commercial area is up ahead. All the older roads have continuous overhead tree canopies. Retail areas tend to have buildings up along the sidewalk and no trees. As a result, you see a brightly illuminated area in the shaded roadway up ahead. Similar landscapes are found on Cape Cod.

2. Enclosing space from above: here canopy height above the ground is important. An arcade of tall pines with branching starting 40′ or 50′ overhead provides a cathedral-like space, while the lower more horizontal branching of Honey Locust provides a more homey, personalized space. Thoughtfully manipulated, a tree's canopy provides each landscape with "a scale that is sympathetic to the movement and perception of the pedestrians" (Arnold, 1991, 37). An additional benefit, depending on the canopy's density, is that the temperature can be 5–10 degrees cooler than adjacent sunlit spaces. In the heat of a Georgia summer, the difference is readily felt as you walk from the older to newer part of the University of Georgia campus in Athens, Georgia.
3. Sense of depth: overhead enclosures provide a sense of depth through more than just the density of a trees' branch and leaf structures. A combination of tree species provides numerous plantations throughout the South with a most dramatic spatial orchestration. At Rosewood, in St. Amant, LA, the massive Live Oaks are under-planted with smaller, yet full sized, trees. The rhythm of ascending and descending spaces is a tapestry worthy of contemplation. This "space within a space," as Arnold refers to it, also occurs across a horizontal area. "Bryant Park in Manhattan is a good example of trees used to define an outdoor room in a way that complements the architectural mass of the New York Public Library…the branching height in the center of the grove is 25 to 30 feet above the ground, though branches at the edges of the grove hang down to within eight feet of the ground" (Arnold, 1991, 49).
4. Lighting: light levels relate to leaf density and height of crown,

*light gives life to a space… [and] gives the urban designer power to transform spaces of stone, bitumen and concrete into tapestries of sunlight and shadow…Intricate shadow patterns on pavement and building walls create architectural richness from even the crudest materials.*
*(Arnold, 1991, 44)*

*…the cost difference between brick and asphalt paving can buy enough immediate tree*

Figure 5.27 What shadow patterns do for asphalt. Photo by author.

*cover to justify using the most disparaged material of urban civilization – asphalt paving – which, with trees, assumes the richness of a Spanish tile floor that changes with the seasons.*
*(Arnold, 1991, 83)*

5. Tree crown: height of a tree's crown and its leaf density influence the sense of depth in the mass of the overhead plane; that is whether it feels light or heavy, oppressive or comfortable.

*The Honey locust, which is valued most in summer for its light shade, is least effective when not in leaf because of a naturally sparse branch structure. It is a tree that, when small, almost disappears in winter. The Horse chestnut, a tree used in many of the formal spaces in Paris, has one of the most effective winter forms, because of its thick twigs organized in a provocative symmetry.*
*(Arnold, 1991, 83)*

Note: a tree that "disappears in the winter" may be exactly what a particular design calls for.

*Trees provide more benefits to more people per dollar cost than any other material used in the construction industries. Costs of general building construction rose by more than 100 percent in the decade between 1968 and 1978. In this same ten year period, the installation cost of a tree in a city increased between 25 and 30 percent... the value of the same tree [then] continues to increase after installation, when all other materials depreciate.*
*(Arnold, 1991, 119)*

In urban settings, the act of bringing "down to earth" the overhead plane with vegetative cano-

Figure 5.28 Allee of Linden trees line a tree lined footpath through park. Shutterstock ID: 632149790. Contributor: dugdax.

pies introduces a pedestrian or human scale to urban canyons. Vegetative canopies or allées over urban sidewalks provide a transition from the scale of interior living space to that of outdoor spaces that allow people to remain comfortable within the scale of the walkway. A swath of light down a roadway, as echoed in the overhead opening between rows of street trees, reinforces the linearity of an environment devoted to circulation, therefore reinforcing the idea of movement with direction.

As an overhead plane, the sky encloses urban open spaces edged with (or bounded by) buildings or trees. In both linear and rectilinear spaces, the dimensions of the open space, the height of the bounding objects, and the position of people relative to edge features are all important factors. Hall (1969), Ashihara (1970), Arnold (1991), and Sitte (Collins and Collins, 1986) give us parameters that guide our design of open space experiences. Each echoes aspects of the other's theories: Halls' proxemics, Ashihara's 1/10th theory, Sitte's principles from picturesque paintings, and Arnold's critical observations. Each author's analysis is based on "observations," and their respective analyses are derived from the physics of the human eye's cone of vision.

## MECHANICS AND DESIGN

So much of the influence of ground, vertical, and overhead planes relates to the visual information provided us through our eyes. To be more precise, our cone of vision. Our bioptic field of vision is considered a cone, elliptical but still cone

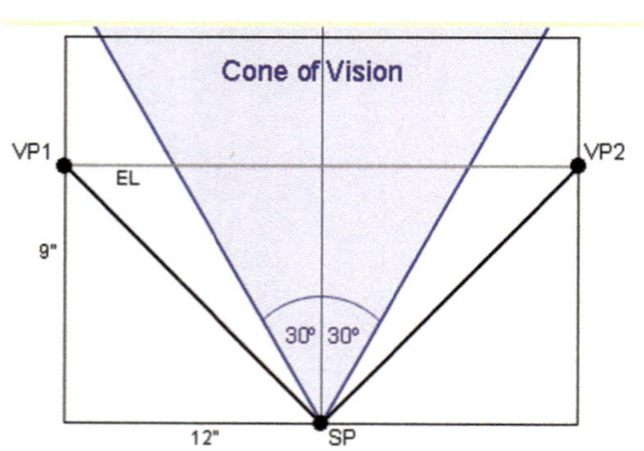


Figure 5.29 Cone of vision. Courtesy, the Art Department, Brigham Young University-Idaho

shaped with a vertical span of 60 degrees and a horizontal sweep of close to 180 degrees. With eye level always being the same as, or in alignment with, the horizon, the vertical sweep of our vision inscribes about two-thirds or 40 degrees of area above the horizon line and one-third or 20 degrees of area below it.

> *If part of the sky area is to be brought into the field of view, one can see a building as whole at an angle of 27 degrees [of the 40 degrees above eye level] when D/H=2 (D being the distance from the building to the observer and H being the height of the building).*
> *(Ashihara, 1970, 42)*

According to Ashihara, the critical point at which the quality of exterior space radically changes is at D/H=1. At D/H=1 the cross-sectional space between two buildings still provides us with a sense of balance between building height and open space. As the ratio becomes larger than one, the space becomes greater, to the point that at more than 4 "the mutual interaction [between the building and the space it defines] begins to dissipate" (43). Less than one leads to people feeling closed in, potentially contributing to a sense of claustrophobia.

These dynamics are interesting for two reasons. First, Ashihara's spectrum of feelings from claustrophobia through balance or comfort to a sense of standing apart is related to apparent sky-plane heights. If, as he says, when 27 degrees of the upper portion of our cone of vision provides for comfortable views of buildings, then a comfortable amount of sky is what is seen in the remaining 13 degrees. This places the upper edge of the vertical surface defining a sky plane at a 40-degree angle from the viewer. Second, if that is the case, then, when situations dictate open spaces between buildings of heights that result in a D/H greater

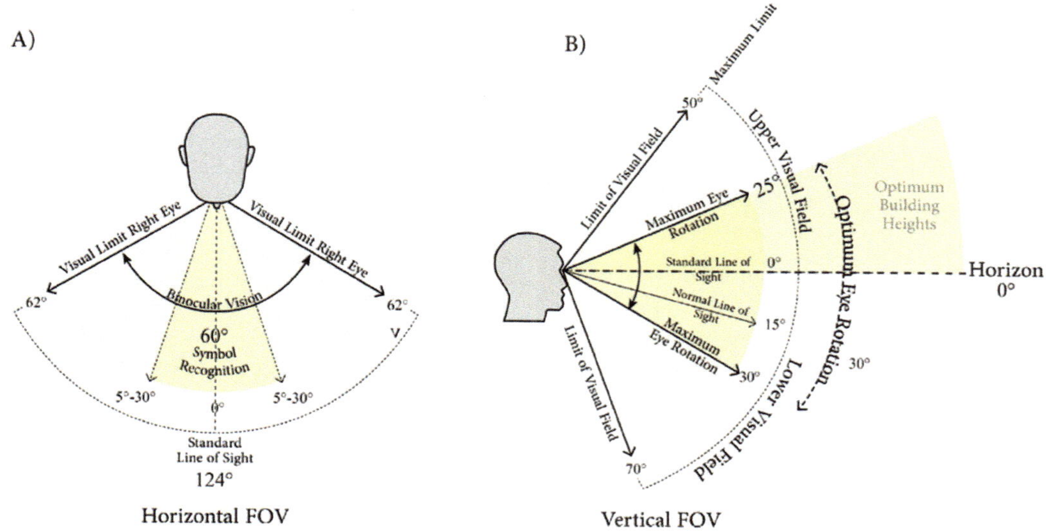


Figure 5.30 Human field of vision. Courtesy Tara et al. (2021).

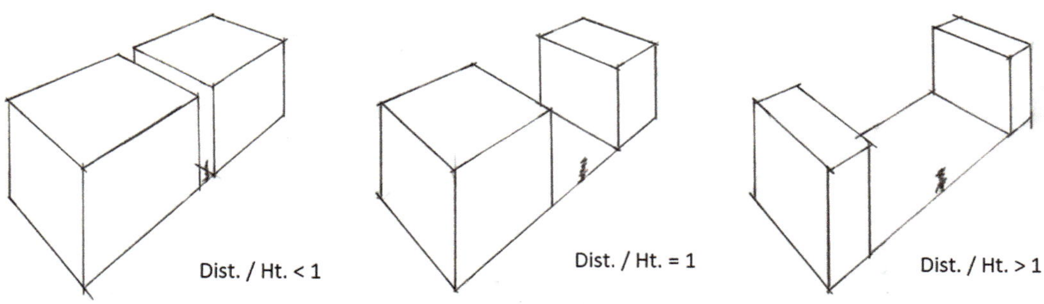


Figure 5.31 Sensed enclosure – alleyway experience of enclosing walls vs. comfortable plaza or patio vs. limited to no sense of enclosure. Drawn by author.

than 1 trees, or other overhead planes, could be used as intermediary comforters that "give urban spaces a sense of scale…to connect and extend the geometry, rhythms, and scale of buildings into the landscape" (Ashihara, 1980, 22) and heighten the spatial expanse situations where D/H is less than 1, the sky plane recedes. More and more you sense you are in a hole as the space between the vertical walls "tightens." The use of street trees in dense urban settings introduces a vegetative overhead plane that can, in screening a building's mass, reduce the sense of being closed in by walls. Urban street trees bring the perceived height to width ratio of the now tree-canopied human space into balance. This balance with the use of vegetation also works to soften the audio character of the space. In such spaces I would heed Arnold's frequent call for us to return to using larger rather than smaller street trees, even in tight spaces (Arnold, 1991, 59). On a field trip in an urban setting of D/H ratio less than 1, a student of mine commented, "I feel comfortable here with my friend the tree."

*When viewing a space from the ground, there is a [comfortable] feeling of enclosure when the surroundings are tall enough to intercept our entire cone of vision. A convenient rule*

*of thumb is that the observer should not be further away from the defining edge than three times the enclosure height.*
*(Arnold, 1991, 59)*

According to Ashihara a space-to-height ratio of 3:1 works well when you wish a building to be seen "as part of a group" (Ashihara, 1970, 42). If this is the case, then a mass planting that provides a comfortable sense of enclosure is one that is part of a larger "picture" and takes up about 18 degrees of your field of vision. Of interest here is the recognition that Ashihara's ratios are only related to buildings and their intervening spaces. He fails to note that a 3-to-1 ratio could mean a person is backed up to a vegetative wall or under a tree canopy that is in fact located in a larger space, a space that Ashihara would note at an even greater ratio.

## BRINGING THE PLANES TOGETHER

*...the apparent clarity or 'legibility' of the cityscape. By this we mean the ease with which the parts can be recognized and can be organized into a coherent pattern.*
*(Lynch, 1960, 2)*

Whichever of the three types of planes we start with in the authoring of a landscape text, the important factors are that the text makes sense, that there be a continuity struck between the locations and relative positions of information delivery, and that narrative form can be readily identified and properly interpreted by the reader (Lynch's "legibility"). This idea is helped by writing a first- or second-person narrative. Again, the fact your written narrative is through a person's or people's eyes greatly contributes to continuity in message delivery. The connection between narrating a person's experiences and the continuity of the landscape is almost a natural outcome of writing a first- or second-person narrative. If your narrative is disjointed, erratic, and confusing then you're approaching Alice in Wonderland, and just maybe that is what you want. But even then, there are limits. Go too far, negate a continuity across your messages, and the result is chaos.

*...a paragraph often corresponds to a major unit in the material treated. Such a paragraph may take up one limited aspect of a scene, outline a step in a process, recount a self-contained episode, or recreate a mood. It will only derive its unity from orderly presentation of significant detail. Often the sequence of details is inherent in the subject matter. Thus a paragraph may trace structural or geographic relationships (spatial order), follow the order of events in time (chronological order, or proceed from causes to effects (order of causation).*
*(Guth, 1961, 145)*

The development of a comprehensive narrative, a process dealt with in Chapters 7 and 8, calls design principles into play. While their application to the design of an environment structures a logic read through our senses, their application to a narrative, to messages and their sequence, structures the landscape's story and discourse and contributes to what we read and interact with as landscape users in particular environments.

The ground, or base plane, physically supports the vertical elements that give form and sequence to landscape spaces that fill most of our cone of vision. The strength of the vertical forms rests in the fact that they are encountered at eye level. Manufactured or natural, verticals can display a wide variety of characteristics: (1) as solid planes they may be opaque, translucent, or transparent; (2) as perforated planes they may have small to large "openings," some affording peeks and glimpses while others allow larger views (broader panoramas); (3) in form they may be uniformly flat or undulating, rhythmically repetitive or a fine, coarse, or broken texture across a landscape; (4) the planes' surface textures vary with their composite materials and the finishes of those materials, e.g., teak vs. exposed aggregate concrete or a vegetative wall. Textures have the possibility of being visual and tactile and as green walls, aromatic. Keep in mind such textures are complemented by temperature, and sound, if not also scents and colors. Texture related to the surface a plane may also involve the shadows and reflections cast by those planes; (5) surface colors also work to highlight, diffuse, bring closer, and recess; (6) surface depth, the "density or transparency of trees in leaf... is a function of branch height, spacing, and most importantly, tree type" (Arnold, 1991, 87). Two basic kinds of leaf arrangement: monolayered ("most of the leaves in a single layer are distributed around the periphery of the crown"), multilayered (leaves are more uniformly distributed in layers throughout the crown). The latter tends to cast more dense shade (Arnold, 1991, 87); (7) a vertical plane's height, the extent to which it fills our cone of vision, relates to factors of size and distance from viewer. The best understanding of this is gained by playing with Wester's (1990) or Reid's (1987) perspective techniques; and (8) the position of a plane, on grade, suspended in the air, or at or below eye level, as they influence movement, views, and air circulation.

*At every instance there is more than the eye can see, more than the ear can hear, a setting or a view wanting to be explored. Nothing is experienced by itself, but always in relation to its surroundings, the sequences of events leading up to it, the memory of past experiences.*
*(Lynch, 1960, 1)*

Figures 5.32.1–5.32.3 Vertical planes helped with vegetation – (left and center) walkway between parking garages; (right) parking lot, Cathedral of Our Lady of Lourdes, Spokane, WA. Photos by author.

Like a planar surface making up a room or landscape, a paragraph is a subunit of a greater narrative. It "tends to be a logical, rather than typological, unit. It reflects the way the writer has sorted out and organized material" (Guth, 1961, 142). As a subunit of a greater context a vertical plane or sequence of planes contributes to a more detailed narrative. Such design features serve an additive purpose, they enhance the developmental logic of both the progression of defined spaces moving across a landscape and to the sequence of people's experiences. In one setting, gravel placed in a walkway would appear to be more of a construction or sidewalk repair project than something that belonged there. Experientially, the placement of the gravel would likely break up people's movement about the space, slow their pace, and demark a location through the new sound underfoot.

With the coming together of ground planes, vertical planes, and overhead planes, we begin to piece together increasingly coherent spatial contexts leading to defined volumetric spaces. When there is a coherent orchestration of the physical characteristics of each of the three planes, when they work harmoniously, we are (1) presented with information of a uniform character that (2) gives designer-as-author determined direction through the prevailing narrative and (3) provides the user-as-reader with a sense of coherency and continuity. There is, for the user, a bundle of anticipated experiences that occur with a high degree of sequential logic.

The sequential units of information and direction in which they occur influence the user-as-reader on two levels, physical and emotional. In that the most readily observed messages express functional information, you interpret the information as it relates to the size, shape, and maneuverability of yourself and any associated bags, packages, children in tow, and vehicles. The character of the message units, their adjectives and adverbs, their finishes and craftsmanship, gives position to our greater contextual sense of geographical place and historical time, and give form to our more immediate sense of well-being, self-confidence, and a range of emotions (gaiety, consolement, affection, and more). A place's detailed finishing touches help a narrative work within the local ambiance or atmosphere that contributes to the design's sense of place and the participant's sense of being in that place.

As a component of a landscape's developmental logic, that is, the progression of its message parts leading to a particular conclusion, planes – like paragraphs – call into play a logic associated with such design principles as harmony, repetition, gradation, radiation, concentration, contrast, anomaly, and more (Wong, 1972). The idea of cohesion found in each of these design principles begins with the very physics and ecology of a landscape's planar surface. The elemental parts that make up a plane need to work together to sustain the integrity and character of the surface. As a physical feature, a landscape works within the parameters set by what is below and to the sides.

Doug Way's *Terrain Analysis* (1973) remains a classic expression of the interconnectivity that is ecology: a surface's parent or subsurface material gives rise to soil composition, the area's drainage patterns, the plants that will likely grow in that soil, and the animals most comfortable living in the particular plant communities. A soil's granular makeup, in turn, determines the potential maximum slope or angle of repose to its related landforms: "the limiting steepness beyond which soil grains will slip downhill" (Lynch, 1971, 161).

What falls on the ground also works within certain physical parameters. Most simply stated, rain on flat ground puddles and visually covers an area (basically a two-dimensional form). Rain falling on sloping ground runs off and visually forms streams, rivulets, and other linear features. The ecology of vegetative planes also relates to a soil's physics, solar exposure, and to the horticultural kinship of particular plant communities. There are ecological, as well as design principles to be addressed and employed.

*The continuity of thought and subject matter in a unified paragraph is often reflected in recurrent terms: the repetition or paraphrase of key words or phrases....*

*(Guth, 1961, 144)*

The application of design principles to a place's messages and meanings enhances or detracts from the ease with which the messages are read and understood. For example, imagine yourself approaching a designed landscape you never experienced before. Its vocabulary is unfamiliar. The designer of a unique environment has a choice. The uniquely different place can challenge the visitor, provide little introduction to the new landscape vocabulary, and have the visitor feeling confused, maybe even lost (see Lynch, 1960, 4). Without having provided the space and time in the design's approach to enable people to become acclimated, they likely feel disoriented, confused, and probably anxious as they work their way toward and through the new environment. Or, the design can gradually introduce new landscape vocabulary features in ways that allow the participant to experience and learn and read the new messages, thus feeling more comfortable in a new setting.

A close look at well-designed advertisements provides wonderful insight into how design principles are applied to the introduction and growth of messages. Blatant word messages are supported and reinforced by more subtle messages: precisely chosen words, subtle color palette, textures, shapes and orientations of the objects all in support of the main message which often isn't even stated: "Buy Me," "Use My Service," "Tell Everyone About Us."

Look at one or two landscape and interior spaces you take for granted. What do the spaces' messages say? Are the spaces' messages conveyed and supported as well as the messages are in advertisements? Is the floor pattern echoed in or contrasted by that of the ceiling? Do the shadow patterns work with the finish on your patio or sidewalks? Good examples are found in certain of A.J. Downing's country houses, a ceiling pattern in a library echoes that of the parterre garden outside the library (Downing, 1969, 399, 185). When standing in the library, what you see from eye level in the lower portion of your cone of vision is reinforced by what you see in the upper portion. What is the message conveyed by the parterre garden: authority, control, current or past aesthetics? And how does the narrative from the library (intellectual authority) unfold as you move to the parterre garden (control over nature)? Take time to analyze the messages found in a place in which you feel comfortable, and then in a place in which you feel uncomfortable. Your findings should illuminate a lot regarding the messages authored and the messages read. In advertisements the character of the product to be sold and the targeted clientele drive the design principles applied to the visuals and the sounds (background music) more so than the utility or need for the product itself. There is little that is attractive about a face covered with soap suds but convey it through growing gradation of flowing rhythmic motions of a provocative human nature and accompany it with equally provocative sounds reaching a crescendo and now you got something worth running out to buy. Well, at least a lot of people are left thinking that.

*The setting of a stage is but one aspect of the extended dialog between body and building... Upjohn's Connecticut State Capitol...corridors, halls, and stairways are faceted with body-sized articulations that welcome our presence....invite cozy leaning. Niches and bays occur on landings and in halls so that intimate groups can gather.*

*(Bloomer and Moore, 1977, 70)*

Great examples of design principles in action are also found in the cemeteries and the town halls of New England or the county-seat buildings of the Mid-west. Like paragraphs of varying length, intonation, and function, the introduction, rhythm, and gradation of planes introduce a place and help us transition from one place into another. In the case of a town hall, or county seat, we move from commercial to institutional, or from the privacy of our car to the public environs of the seat of government and, if we are lucky, or maybe unlucky, eventually into an elected official's private chambers. This sequence typically moves along a spectrum from the publicly grand to the intimately personal; monumental planes inescapably define large spaces, which transition to increasingly personal-sized planes and spaces. This gradation is likely accompanied by a sequential shift in size and finishes of the planes.

> *The old Governor's Mansion does have stairs and waste[d] space, which is precisely why it remains the kind of house in which sixty adolescent girls might gather and never interrupt the real life of the household.*
>
> *(Didion, 1979, 71)*

The finish of a plane's surface is critical. Contrast the sensual qualities of the materials and spaces defining older seats of government with that of almost prison-like character of Boston's "award winning" City Hall (late 1960s). Boston's City Hall spaces still echo the spatial transition seen in the older government buildings and their surrounding spaces, but the surface colors and finishes of the experience are cold and impersonal – rough brushed, gray concrete – not the warmth of wood and brick and the personal size of both interior and exterior spaces. The importance of the surface of such planes was strikingly evident for me at a recent AIDS Quilt display. The very same convention hall space went, within a 12-hour period, from an environment of echoing voices and squeaking pushcart wheels to a solemn, spiritually silent place where even with 500 hundred people walking about you could have heard a pin hit the floor. The activities associated with dimensioning the gray concrete floor and laying out the walkways reflected a dramatically different meaning-full environment than did that reflected in the brightly colored quilt panels that expressed hope, love, a devotion to life, and works to absorb sound.

I'm tempted, at this point, to steal from Gene Rodenberry and conclude this chapter with "Space, the final frontier...." Dealing with space as volume, or even more so as a series of volumes, brings us another step closer to better understanding serial vision as the four-dimensional world in which we live and are aided by the sequential messages that flow across space and time to become narratives.

# 6

# Volumes and Volumetric Spaces

Twelve years before the Americans with Disabilities Act was first drafted (1986), I found myself, during a New England winter, learning how to get about with a pair of crutches. In 1974, my sensitivity to indoor and outdoor spaces heightened daily as I learned to carry books, drawings, groceries, and the like. I was also, at that time, working to identify a thesis topic for my Master's in Landscape Architecture. My impairment and my search came together in my thesis, "Gray Matter: designing for the physically handicapped."

Back in the 1970s I was learning, as Benzel notes,

> *To design living environments that go beyond one moment in the story and reach life in the present while harmoniously connecting it to fulfillment in the future…[I was also learning] designers must develop a sense of spatial progression, perceiving design as an odyssey (a long series of adventures and experiences) as well as a specialty.*
>
> *(Benzel, 1998, 4)*

Figure 6.1 West entry to Duncan Garden, Manito Park, Spokane, WA. Photo by author.

Yet, at the time, many designers thought as Thiel did in the 1960s and 1970s, "Space experience in the broadest sense is a biological function, necessary to the continual adaptation of any organism to its environment for the purposes of survival" (Thiel, 1961, 35). Contrary to Thiel, and many others' thinking I thought the environment should be adapted to people's "purposes of survival" and not the other way around. I look back now and see how my early, and very conscious, efforts to "read" built environments continued to influence my emerging attention to the narrative content of landscapes.

What environmental designers see on paper or a computer screen and so often refer to as space is only two dimensional. Two-dimensional features represent area. Environmental spaces are three dimensional or what we call volumes. Space for those who design private residences, public open spaces, business structures and surroundings, walkways, and mixed-use developments is volumetric space. Volumetric space is that four-dimensional world in which people move about living out their daily lives, animals roam, birds fly, and fish swim.

DOI: 10.4324/9781003286981-7

*[The structure's wall] has however a window, an opening which makes us experience the inside as a complement to the outside.*
*(Norberg-Schulz, 1980, 9)*

Volumetric space is difficult to deal with. Our attention is too often drawn to the horizontal and vertical features that define a space more than to the three- and four-dimensional volumes that are the spaces themselves. Yet, the spaces in which we exist are the fundamental medium with which designers construct and through which they give character to our personal, social, and cultural environments. In those environments the visual and topographic stability of ground planes, the opacity and scale of vertical planes, and the degree or sense of enclosure of overhead planes as objects are what hold our attention and guide our movement. Volume, as four-dimensional space, is what isn't there and yet it is what is, at least, as important to the success of a design as the defining objects, if not more important. As Lao Tzu wrote in verse #11 in the *Tao Te Ching*,

*with a wall all around/a clay bowl is molded/but the use of the bowl/ will depend on the space/in the walls of the void. So advantage is had/ from whatever is there/but usefulness [in what we make] rises from whatever is not [there].*
*(1955, 63)*

What we make useful in landscapes, architectural structures, and their interiors is volumetric space or, being even more definitive, experiential volumetric space. Movement toward, around, into, and through volumes of space involves not only the aspect of time (the fourth dimension) but also our existence as we live life from moment to moment and step by step.

Our existence is accommodated as space-time (an area of geographical studies) in all the forms we give it: the apparent directional character of linear volumes and the apparent gathering power of compact volumes, e.g., as the plaza surrounding a fountain or a backyard patio gathering friends for a BBQ. It is, at this time, important to recall Henri Lefebvre's recognition of space as a social product both formed and informed by people's values, meanings, and practices (1979; 1994; Moreno, 2012). Landscape architecture is the blending of Lefebvre's recognition of space as a social product and Simonds' recognition that spaces exist to provide us with life experiences. Taken together, landscape architecture plays a vital role in the social production of experiential space. Recognizing landscape spaces as social products makes the subjective door into environmental design much easier to open. Lefebvre's work and that of his critics allow landscape architects to readily balance the objective and subjective, physical and social, sciences. Lefebvre and many more authors noted throughout The Analogy contribute to our deeper understanding and appreciation for "the art and science" that is landscape architecture.

The languages employed in the development of environmental designs (landscape, architecture, and interior design) both form and inform how designers come to think of, and appreciate, the portions of the world to which they give form, content, and character. Throughout the preceding

Figure 6.2 Thinking spatially with pop-up books. Amsterdam Holland City Skyline. Shutterstock: ID: 1544575148. Contributor: ShustrikS.

discussion of points, lines, and planes, even when considered as discrete objects, a sense of volume, particularly volumetric space, has been unavoidable. We know the spaces we design are a composite of four dimensions that accommodate people, animals, insects, vehicles, and dynamic ecological systems. Most newcomers to design have yet to be able to explicitly deal with volumetric space as readily as they do the objects defining and giving character to the spatial volumes. The question being, "Is character, for those who participate in our designs, associated with what is seen in plan view or seen at eyelevel?" You already know how I lean – "eyelevel." A fun thought at this point is that maybe we should take lessons from someone like Shawn Sheehy and instead of doing drawings we should create pop-up site plans or at least ask how the work of Sheehy and the late Professor Masahiro Chatani can be brought into landscape, architecture, and interior design processes.

In that so much of what landscape architects do is in plan view, volume with regard to design is a misnomer. Students, and many designers, maintain the erroneous idea of landscape volumes while so often seeing and describing designs in plan view. Doing so carries with it three assumptions. First is that designers and their clients are able to imagine the plan view objects' heights above the ground level, proximity to things above and below, and sense of enclosure and openness. A second assumption is that design instructors and their students, and designers and their clients' imagined images are at least similar if not identical. The third assumption is that everyone involved in reviewing a plan has a mutually agreed upon understanding of the character and content in and across the plan's spaces. Part of the problem, as identified by Arnold (1980), resides in the medium by which we develop, explore, and convey designs. Although stated earlier, this bears repeating:

*It seems curious that the essential tools used by the landscape architect and urban designer...nearly always give the impression that trees are visual barriers at eye level because of the way the tree crowns are shown superimposed solidly on the ground plane. The established drawing technique does not conform to the architectural convention of*

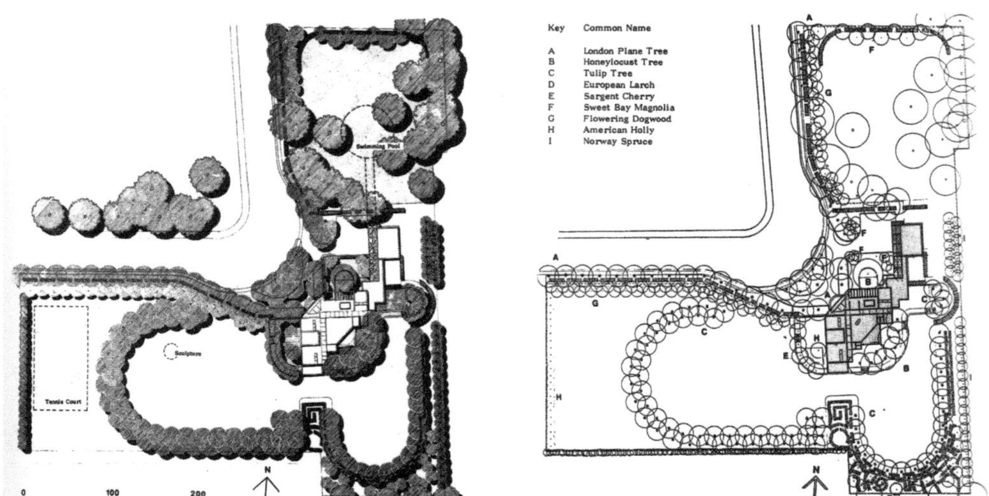


Figure 6.3 Trees as solid and see-through features in plan view (Arnold, 1991, 75). Courtesy John Wiley & Sons.

*showing overhead forms with a dotted line.*
*(Arnold, 1980, 74)*

Even with the various graphic techniques employed to convey the plan view character and content of landscapes they remain laden with assumptions as to participants' developing experiences across a site. Eye-level perspectives are a step toward conveying site participants' experiences but still lack the clarity that can be expressed through an accompanying first- or second-person narrative.

Arnold is not alone. Camillo Sitte (1945), writing in the late 1800s, notes in a complaining manner throughout his work that a T-square and triangle mentality robs urban spaces of their vitality. In his analysis of Italian and Swiss public spaces as people used them, and historically felt comfortable in them, he argued against the homogenizing of places by setting their object boundaries at 90-degree angles. Also, while the third dimension (height) and the fourth dimension (time/movement) remain in the background of many of Ashihara's (1970) statements about space, he frequently relates his thoughts to people's activities. In doing so a sense of volume is again unavoidably present. But it remains a sensed volume; a more explicit treatment is needed for beginning designers.

The arrival of video cameras and now the video capability of smartphones have brought a tremendous sense of depth and progression to the design process. Even so, both Arnold's and Jane Jacobs' maxims (1961) to get out and observe remain the best supplement to thinking of plan views and construction drawings as conveying four dimensions. This distinction between actual and abstract experiences of space, particularly designed space, is revisited in Chapter 10 when I discuss the ethical responsibilities of landscape architects' role in designing the living, work, and play environments for others.

*...two contrasting ways of handling time...monochronic is characteristic of low-involvement peoples, who compartmentalize time; they schedule one thing at a time and become disoriented if they have to deal with too many things at once. Polychronic people, possibly because they are so much involved with each other, tend to keep several operations going at once, like jugglers...the monochronic person often finds it easier to function if he can separate activities in space, whereas the polychronic person tends to collect activities. If, however, these two types are interacting with each other, much of the difficulty they experience can be overcome by the proper structuring of space.*
*(Hall, 1969, 173)*

Even with the advent of wire-frame models, then the likes of the computer program SketchUp, and now the ability to generate fly-throughs, the ability to study people's movement toward, into, around, and through spaces, designers still tend to consider pedestrian and vehicular movement in relation to the objects that define spaces in plan view. Our preoccupation with objects rather than volumetric spaces continues to fix our attention on what "looks good" (so often in plan view) and not on what the unfolding narration of a landscape design communicates at eye level. Now, with the ability to test or sample movement through spaces (fly-through) as they interrelate various parts of a design we do have the ability to assess volumetric spaces' defining features, evaluate the sequencing of experiences relative to those volumes, and more easily review and edit a design's landscape or spatial narrative. To do this calls for a new starting point in the development of a design. That starting point, once you become familiar with the potential users of your design, is your writing of a first- or second-person experiential narrative from the landscape participant's viewpoint before objects are sketched out and flown through.

*At every instance, there is more than the eye can see, more than the ear can hear, a setting or a view waiting to be explored. Nothing is experienced by itself, but always in relation to its surroundings, the sequences of events leading up to it, the memory of past experiences.*
*(Lynch, 1960, 1)*

*In his realization that people learn, or become socialized to the environmental cues they read in their surroundings, Lynch notes, "This image is the product both of immediate sensation and of the memory of past experiences, and it is used to interpret information and to guide action".*
*(Lynch, 1960, 4)*

In much the same way as when a two-dimensional plane moves through space to inscribe a volume (Wong, 1972, 7) written paragraphs build page upon page, eventually providing a three-dimensional volume as a complete narrative or story. The metaphor of a defined volume as the physical feature that contains, and presents, a written narrative opens a number of avenues in our understanding of design. Simply, the physical structure of a book is made up of chapters or three-dimensional sub-volumes in the same way a site is made of subordinate spaces transitioning from one to another. We also see that as dimensioned units the linear aspect of sentences, like paths, hallways, and roadways, carry us through the two-dimensional area of each paragraph: from topical sentence to transitioning sentence to next paragraph. With our movement, paragraph after paragraph, through the landscape user's mind, eyes, hands, and feet the volume, the three-dimensional space, comes into being. Any collection of subordinate units, chapters, is in turn set within a larger context of a bound volume. Remember Alexander's dictum, which he actually says in a variety of ways throughout *A Pattern Language*,

*no pattern is an isolated entity. Each pattern can exist in the world, only to the extent that is supported by other patterns: the larger patterns in which it is embedded, the patterns of the same size that surround it, and the smaller patterns which are embedded in it.*
*(1977, xiii)*

*In a narrative text, the reader is forced to make choices all the time. Indeed, this obligation to choose is found even at the level of the individual sentence – at least, every time a transitive verb occurs. Whenever the speaker is about to end a sentence, we as readers or listeners make a bet (albeit unconsciously): we predict his or her choice, or anxiously wonder what choice will be made…*

*(Eco, 1994, 6)*

Our culture has come to expect paragraphs to lead to chapters and chapters to culminate in a book, a volume, or a series of volumes each with a sense of closure. However, our expectations go beyond the physical characteristics that allow us to identify a book, or for that matter any of a number of electronic narrative forms. In a narrative, we expect a complete story. We expect closure: introduction, body, climax, anticlimax, and conclusion. Relative to the various narrative genre (e.g., mystery, humor, romance, technical), we expect to find what is in a book to present us with particular vocabularies through anticipated presentation strategies; on the book's outside we expect covers with particular visual representations, lines, colors, and fonts. The developmental structure of a narrative is intertwined with and to an extent controlled by the character of the physical structure of the book itself. We have expectations and good authors know how to play with our expectations.

For a moment, imagine yourself headed to a local bookstore. You're looking for a book that would be located in Humor. What are your expectations? Picture yourself standing just inside the doorway of the bookstore, with the ability to see all the shelves of books at once. There are no signs identifying the various sections as "Humor," "Fiction," or "Political Science." What clues would you look for to locate Humor? If you could see the shelves of Gothic Romances set next to shelves of books of silliness how would the book covers appear different from each other: color, images on the covers, types of lines, visual forms, and lettering making up the covers? In the same way culture contributes to your preconceptions about book covers, their range of colors, characteristic lines, forms, and fonts, you carry about – and rely upon – you also carry about preconceived ideas of particular landscapes. Even for many hard-core urbanites, the typical imagery associated with a university campus is that of large expanses of grass, stately brick buildings with ivy covered walls, and wide walkways. Sites, therefore, are not discrete entities disconnected from their greater geographical or historical contexts or without some expected organization structuring the smaller spaces within them and their connective transitions with their surroundings.

Figure 6.4 Washington State University, Spokane Campus. Photo by Campus Photographer Cori Kogan.

An anticipated organization of superordinate and subordinate spatial units applies to many sites' narratives. Like our approach to stories or technical reports, the readers of a landscape have certain expectations regarding a site's message format and syntax. The most elementary organization includes an easily read approach, arrival, entry, route(s) to a destination or destinations, way back or way out, and exit. Keep in mind that the users' expectations and the vocabularies read by bicyclists, motorists, and truck drivers vary. Imagine yourself traversing the same roadway on a bicycle, in a compact car, and in the cab of a tractor-trailer truck. Your expectations of what constitutes approaches and entryways vary in scale; and those expectations are what authors rely on. In much the same way authors know their audiences, landscape architects have a good idea of those people who will likely use their designs. Authoring a landscape narrative calls for designers to anticipate and rely on people's expectations. What says "Welcome," "Come this way," and what says "Stay away"? What communicates happiness, better yet gaiety, and what says "You are approaching a solemn place, behave accordingly."

An excellent self-reflective exercise is to take a few moments and consider the messages that make up three or four sites with which you are familiar. You can even think back over your first-time visit to a place such as a new campus or employer's office, a spa, or event center. Consider the template of images you anticipated fulfilling. Lay out the route you traveled and the story's parts along the way. Remember how your expectations were modified, refined, drastically changed, or ignored. Now consider a time and place where you got lost. What messages violated your anticipated actions? From your point of view, what wasn't in place? What didn't you find that you were consciously or unconsciously looking for? What signs or signals weren't you able to identify and understand? Designers need to anticipate why people come to a particular place and work with what people expect (either in support of or contrast to), what they anticipate finding as they move through a landscape or an interior space.

## SPACES WITHIN SPACES

Cullens' serial vision (1961, 17–20) can be a fairly strict, linear, additive, sequential narrative experience or it can be more like Jorge Luis Borges's *The Garden of the Forking Paths* (Borges, 1962). Borges's *Garden* plays with the multiple paths a person can take between two points and end up at the same conclusion (destination). Space and time provide alternative opportunities from which to choose one path at the expense of another, one narrative construct as opposed to another and yet arrive at the same end point. It is a lot like sailing. You are always plotting a course to the same destination while responding to changing tides, currents, and winds.

Conceived in this way, designs can be prepared to allow their readers to have a hand in selecting from a variety of narratives on a variety of occasions. Consider the variety of experiential narratives one can have in a Japanese garden, a national park, Costco, or a friend's backyard. Our movement away from believing a design capable of providing a single, sequential chronology of events, to one that provides multiple, potentially overlapping, narratives opens the door for us to review and apply other genres' formats such as mystery, romance, anxiety, and more that, in turn, employ flashbacks, foreshadowing, frame stories, jump cuts, and fade ins and fade outs. Whichever path is taken by the reader, there remains the design's need for that route to be readable and safe, to work with the appropriate sensory vocabulary. To move safely through landscape spaces requires a straightforward functional narrative. But they needn't be as simplistic as "place Tab A into slot B" but that's a good analogy to keep in mind. The overarching need to address public safety and welfare is what most distinguishes landscape narratives from that of the narratives found in the arts: paintings, music, storytelling.

## DIALECTIC STRUCTURE

A dialectic is a form of two-way or back-and-forth communication, a dialog, or discourse that, to a great extent, shapes itself as it moves along. Throughout any conversation you are constantly adjusting your stance, word choice, tone of voice in response to who you are talking with as they do the same with you.

Understanding what a designer needs to consider as the interaction between her landscape (or architecture, or interior) design is made clearer when we consider Jurgen Habermas's rules for "ideal speech acts" (Scarfo, 1989). In that as designer-authors we produce landscapes that express narratives, the goal of communicating those narratives has the expectation of a mutual level of understanding between designer and user. If in fact there are signs and symbols in landscape and built forms that communicate information, directions, emotionally and culturally laden messages, then for them to be understood and appreciated, an unspoken contract exists between the designer-author, the client, and the user-reader.

Figure 6.5 Part 1 (6.5.1–6.5.5): serial vision as narrative sequence. Photos by author.

The break in the vegetation and low stone wall along the edge of the sidewalk says "Come on in."

Then there are actually three possible routes to join the party: one through the arbor to the side yard, and two that call you off to the right to the front door.

Walking toward the arbor then opens a framed view of the side yard and deck with the stepping-stones telling you to "Come this way, and oh look, steps up on to the side deck."

The other possible approach, maybe for first-time visitors versus long-time friends, is to follow the opening in the planting beds and turn to the right, then follow the lawn which turns to the left and the front door.

What is interesting here, like Jorge Luis Borges's *The Garden of the Forking Paths* (Borges, 1962), there are (not shown here) multiple ways to arrive up at the front door or side deck. Some more enclosed with plantings and others more open.

Part 2 (6.5.6–6.5.8): serial vision as narrative sequence. Photos by author.

That designer-user contract says that what is expressed by a landscape calls for it to be:

1. Comprehensible – designed features introduced into an environment are perceived, identified, understood, and interpreted correctly; otherwise, we'd be walking into walls, off cliffs, or into traffic.
2. Sincere – expressed directions, invitations, warnings, and emotions are honestly stated/displayed. If landscape narrative statements were not true then we'd never believe what we were being asked to act out. Imagine this situation on an expressway, an urban walkway, a marina or urban park.
3. Appropriate – landscape features "are appropriate within their social [cultural] rules structuring the situation" (Scarfo, 1989, 19–20). Imagine the cultural rules employed in the design of an English landscape garden used to develop the layout of a military base.
4. True, factually correct – here I can't help but imagine landscapes designed by M. C. Escher, or those that employ trompe-l'oeil, or those prepared by the best of the sidewalk chalk artists.

If you've read this far, then you've likely bought into the idea of landscape narratives. As such, people approaching, entering, and enjoying the landscapes between their front doors and wherever they end up involve a dialectic, a dialog between that person and the environment. In much the same way two people meet and greet each other or have an extended conversation, there is a back and forth with what is perceived, identified, interpreted, and in a sense tested, in an effort to safely and enjoyably traverse a place.

From the recipient/user's viewpoint, the merger of a design's functional and artistic features is an achieved balance of identification, recognition, and interpretation. Our ability to identify the measurable characteristics of landscape spaces, to know that we can pass through, drive into, or walk over them, does little to foster the kinds of appreciation associated with the emotional pleasure we get from strictly functional messages gussied up with aesthetically pleasing modifications.

As the users of landscapes, we typically move about with little conscious thought to the explicit identification of this object, then those textures, then that pattern, and that sequence.

Figure 6.6 Visual illusion. Iceland uses optical illusion to slow traffic. Courtesy of Gautur Ivar Halldorsson, photographer and owner Vegamálun GIH.

Like grammar and syntax, we rarely think of how this and that rule is to be employed in composing a sentence. We just, in a sense, say what comes to mind. We hear, we listen, we interpret at multiple levels, and, as a result, we communicate. In this process, we also modify our ability to read and respond to the other as we go along. What we hear, and how it is expressed, causes us to modify what we say and how we express ourselves. That is the dialectic, the simultaneous forming and informing of the reader, the participant in the narrative, and the narrative itself. At any given place and time, what is significant to us as readers gives form to our decoding process, our experiences of moving through a space, which in turn reinforces or modifies what is significant to us as the reader. As a fundamental building block in our coming to understand a design, the interaction between what is expressed and what is interpreted resides in the realm of the social dimensions of communication (Hall, 1969; 1981; Sommer, 1969; Newman, 1972). If you would like to really get into the nuances of the expressive qualities of landscape, architecture, and interior design take some time to associate various forms of nonverbal communication – body movement, posture, facial expressions, gestures, eye contact, comfortable social spaces and social distances – in environments you experience every day.

## TEXTURAL ORGANIZATION IN SPACE AND TIME

Space and time are the fundamental qualities with which we work. Geography is one discipline spatial designers should reference for data in support of many of their arguments. Geographical studies run the spectrum from the social and spatial analysis of rooms in a house (Wood and Beck, 1994) and residential streets (Schmidt, 1975), to towns, villages, and cities (Ley, 1983; Smith, 1984; Relph, 1987; Yamin and Metheny, 1996) and regions and nation states (Hartshorne, 1939; Hagerstrand, 1970). In a similar fashion, these geographical realms of life are reflected in the temporal works of historians (Harington, 1986; Cronon, 1991; 1983). For designers working in particular countries, a wonderful blend of geographical and historical contextual analysis is found in what is often referred to as American Studies, Canadian Studies, Latin American Studies, and the study of specific nation states, or cultural regions. Their works describe and explain the manifestation of cultural values in such things as literature, poetry, painting, household items, neighborhoods, landscapes, and more. Their works identify various sensual languages (visual, audio, taste) and their interpretation, or the ways in which they can be read. One wonderful example that covers much of what was mentioned above is John Conron's *American Picturesque* (2000).

Space and time, or as geographers say "time-space geography," exist in each of The Analogy's three realms of narrative expression: functional, social cultural, and collective unconscious. How space and time contribute to the organization of written, basic design, or landscape texts resides in both the physical and social realms in which they are set. Space and its construct provide the designer with features that are both constraints and opportunities. You can't imprint a 12″ square object on an 8.5″ ×11″ piece of paper, but you can draw people's attention with only one word on an otherwise blank sheet of paper. There are only a finite number of ways you can employ an ascending slope, descending slope, and horizon in a landscape design. The organization of the physical aspects of a design – location, orientation, and sequencing – is socially and culturally structured and each brings with it its own set of opportunities and constraints. Design, as a process, does have structure. It occurs not in a vacuum but within the larger physical, social, and expressive contexts in which it is being designed and eventually built.

In their visual form as recorded text, letters and words are placed, or located on a page or within a framal reference. That location immediately sets up or denies the addition of certain other words. Excluding, for now, Avant-garde forms of communication, if the location of the first word of a text is in the middle of a page, then additional text typically moves across and down the page from that point. Written words in other languages, Chinese and Hebrew, move across and up and down pages according to their cultural parameters. Nevertheless, the starting point of the text, within the greater social and cultural contexts, greatly determines the direction and quantity of what follows on that page, and what the reader expects to find on subsequent pages. Given various written texts, all have a starting point, middle, and end as should a design and its narrative.

> *Art is a social product (1)... artistic practice is situated practice, the mediation of aesthetic codes...and ideological, social, and material processes and institutions. At the same time [the sociology of art] insists that we do not lose sight of the artist as the locus of this mediation and the facilitator of its expression ... the artist as producer...[and] the nature of artistic work as located production.*
> *(Wolff, 1981, 136)*

Even the process of editing a written text responds to both physical and social parameters. The placement of an editor's notes, even an author's or reader's editorial remarks, is typically added outside the bounds of a text. Such notes are not only physically located outside the text, in the margins, but suggested improvements are also located outside the context of what was originally written, the narrative. An editor's remarks and reader's thoughts are added within the page's borders as critique or reaction to the text's contents and outside the text's boundaries. Examples of this are often found whenever a designer thinks graphically. The heart of a design, whether a concept diagram or isometric or perspective sketch, is typically surrounded by critical thoughts, variations on a theme, or whole new views.

The placement of words on pages is also directed by the physical context within which the message is being produced: page layout in portrait or landscape format "the part of graphic design that deals in the arrangement of visual elements on a page. It generally involves organizational principles of composition to achieve specific communication objectives" (Wikipedia, https://en.wikipedia.org/wiki/Page_layout). The use of the physical space, the area of a page, is sectioned or formatted differently for advertisements, technical reports, grant applications, cartoons, essays, and the like. Even photographs are generally composed using the "rule of thirds." Imagine a sheet of paper used to communicate a National Endowment of the Arts or National Science Foundation grant proposal formatted in a manner similar to the pages of *Mad Magazine* or Dell Comics. If you still doubt there are rules surrounding the location and placement of marks on a text, watch the reaction of design students when an instructor pulls out a felt-tipped pen and marks up their drawings. The normative behavior, the typical behavior in that situation, is what James Adams, in *Conceptual Blockbusting* (1976), refers to as "conceptual blocks." Our normative thinking is to recognize a problem and "pick the first solution that comes to mind and run with it" (xi). Adams helps us shift from that normative approach of convergent thinking to a more divergent approach to problem solving

Figures 6.7.1–6.7.3 Three social spaces, similar directional messages. Photos by author.

that provides for a number of appropriate and complete solutions from which to work.

We often think of a bounded area, or framal reference, or the boundaries of a perspective sketch, as the physical context within which things may be placed or activities carried out within varying degrees of social approval (social/cultural context). As we move from one location to another, one country to another country, how similar types of spaces are defined change along with the social mores that guide the activities that occur within and around those spaces.

Time, the period during which an activity occurs, or doesn't occur, is another bounded context across which the values associated with places and things change. Holiday meals at the grandparents: how many of us remember eating at the "children's" table during family holiday gatherings? How many of us now, as adults, often wish at holiday gatherings that we were again off in the kitchen with the kids? Toys and play places fade from places of importance to nostalgic remembrances. What was once nouveau is now retro. What was once "a happening" is now a "New York Minute." As with oral history, the words, characters, and places may change over the years, but the messages may remain the same (Rubin, 1995). While composing a landscape narrative may find you unconsciously considering time, it may be worth the effort to consciously consider your designs' planned-for activities in terms of their temporal characteristics: fast, slow, rhythmic, sporadic: paving patterns; tree canopy shadows; spacing of plants, planters, benches. How does the landscape narrative transition into and then come out and away from an activity? Are the landscape features positioned? You can see how they associate with adjacent activities within the site and to the adjacent sites. And what landscape features communicate the changes, the transitions linking the activity spaces?

Good design is contextual on so many levels. By now you are probably thinking that good design works with local social and larger cultural contexts. It is guided by the potential interpretations of those who will be left to experience the design within the social, cultural, and ecological contexts of its location. Good design fits into the social and natural dynamic equilibrium of a place. The better a designer knows the eventual users of a design, the kinds of spatial volumes and sequences of spatial volumes with which they are most comfortable, the more the likely the users will, as Norberg-Schulz notes, identify with, find personal and collective meaning in, and feel comfortable with where they are, their sense of orientation within their own life and their social and cultural contexts.

## DESIGN PRINCIPLES AS APPLIED TO VOLUMETRIC SPACE AND NARRATIVE

A study of Christopher Alexander's *Pattern Language* (1977) provides us with two interesting points. First, it would be easy to argue that almost all 253 patterns are nouns, things. A few can be considered verbs, actions or behaviors, like #63 Dancing in the Street, #138 Sleeping to the East, # 165 Opening to the Street, and #168 Connection to the Earth. Still, their descriptions are all about objects, their placement, and their relative positions. The unspoken thought is that the objects represent, call for, and support particular behaviors in particular spaces. The assumption being the delineated space accommodates the activities. In other words, the verbs of people's actions are somehow set in among the nouns making up the physical spaces. The second point of interest, and more startling, even though it does make sense, is that the word "space" is almost nonexistent in *A Pattern Language.* I said it makes sense because all the noun patterns, and their associated subordinate and superordinate noun patterns, take up space and in doing so give shape, content, and character to spaces and sequences of spaces all calling for the given activities. One strong exception is Pattern #106, Positive Outdoor Space. Here Alexander delves into positive and negative space, how to identify it, figure/ground, degree of enclosure, and people's feelings of comfort and discomfort. A second exception is found in #121 Path Shape. Space is mentioned but not to the extent it is explored in #106. A few others include #132 Short Passages, #146 Flexible Office Space, # 152 Workspace Enclosure, and #191 Shape of Outdoor Space. Even with these, the spaces are typically referred to as what they are, passages, rooms, halls, corridors, and the architectural features that define them. Still, "space" is not as explicitly prevalent as we would expect.

In stating that a pattern can only exist within larger patterns and simultaneously be composed of smaller patterns Alexander speaks to the serial vision of volumetric spaces. He

> *...says that when you build a thing you cannot merely build that thing in isolation, but must also repair the world around it, and within it, so that the larger world at that one place becomes more coherent, and more whole; and the thing which you make takes its place in the web of nature, as you made it.*
>
> *(1977, xiii)*

In how he speaks to *A Pattern Language*, especially in its counterpart, *The Timeless Way of Building* (1979), I can't help but feel Alexander is a product of his generation, both chronologically and professionally. His "quality without a name" is so Taoist. "There is a central quality which is the root criterion of life and spirit in a man, a town, a building, or in a wilderness. This quality is objective and precise, but it cannot be named" (Alexander, 1977, 19). In an effort to express "this quality when it is one with itself...true to its own inner forces...[and] at peace with itself" (Alexander, 1979, 26), Alexander shares words which he feels are commonly used in an effort to identify the unidentifiable quality: alive, whole, comfortable, free, exact, egoless, and eternal. He concludes, "...in spite of every effort to give this quality a name, there is no single name which captures it" (Alexander, 1977, 39). Yet, there is a seamless coming together of parts, a flow between nouns to provide an ongoing comfortable appreciation. Done well, we could consider this blend of patterns and paths as the building blocks of an esthetic experience.

Much like the Lao Tzu's *Tao Te Ching*, once *The Way* is named, is given a single label to identify it, it is lost.

> *The Way that can be spoken of*
> *Is not the constant way;*
> *The name than can be named*
> *Is not the constant name.*
> *The nameless was the beginning*
> *of heaven and earth;*
> *The named was the mother of*
> *the myriad creatures.*
> *(Lao Tse, 1963, 57)*

Or, more simply stated,

> *The name that can be named is*
> *not the real Name.*
> *(Lao Tse, 1955, 37)*

Alexander (1977) reveals two criteria regarding the patterns, or "elements of this language" (x). The first is connectivity or "connections between patterns" (xii). He notes that as the means by which larger and smaller patterns are brought together each is tied "to all those smaller patterns in a language, which are needed to complete [the larger] pattern, to embellish it, to fill it out" (xi). Without using the word, Alexander's realization speaks to syntax as applied to activities and the volumetric spaces and sequential spaces of which they are comprised. What narrative brings to *A Pattern Language* is the aspect of motion, time, change, growth, and becoming. With this realization a wonderful interplay can be explored by bringing the nouns making up Alexander's *A Pattern Language* together with the verbs making up the basis of Bacon's *Design of Cities* (1974) and Aminzadeh et al.'s scenario writing (2016). Bacon brings "movement" to Alexander's patterns. Blended, they bring setting and surroundings to the actions and activities expressed in a narrative. Combined, they enhance the expression of Alexander's settings and Bacon's syntax and Aminzadeh et al.'s scenario writing as used in the production of films.

> *Up to this point architecture has been discussed as a series of linked spaces, each possessing a particular quality and each related to the other... a continuous, unbroken, flow of impressions that assault their senses as they move through it. For a design to be a work of art, the impressions it produces in the participator must be not only continuous, but harmonious at every instant and from every viewpoint.*
> *(Bacon, 1974, 20)*

If there is a spectrum connecting nouns and verbs to be drawn from Simonds' realization that by designing experiences, we come to identify the spaces, places, and things making up our designs, then Alexander is at the spaces, places, and things end while Bacon is at the experiences end.

I agree with Alexander, there is a quality to the spaces in which we live that can never quite be captured in a word, in a name. I am sure most people have a place that lives in their heart. It is a feeling that pervades their soul but can't be simply expressed to another person. At the same time there is a quality that when recognized and brought into *A Pattern Language* brings life with it. That quality is motion, movement. Life as motion and movement is what brings the things and the experiences of growth, learning, and fulfillment to a landscape participant. "Cinematic technique, with its sequencing of spaces and events makes users aware of the importance of time and motion" (Aminzadeh et al., 2016, 176). In a well-written narrative, moving from word to word is how the narrative takes on shape and substance. While Alexander doesn't speak directly to it, realizing the interplay of movement in *A Pattern Language* does make it more complete.

Bacon's recognition of movement calls for the "involvement of the whole self" (Bacon, 1974, 15). In fact, when you read *Design of Cities* (Bacon, 1974) with an eye to motion you quickly realize that space as experience is fundamental to Bacon's whole thesis. And with this realization it is impossible not to again announce the critical importance of Simonds' call for us to recognize our contribution to the design of people's experiences. Given all The Analogy has provided you thus far, you have a strong foundation for the four-dimensional, sensual language of landscape, architecture, and interior design as it guides and is guided by the authoring of narratives.

## LEADING TO NARRATIVE

The communication of messages within each domain of people's lives responds to the rules of grammar and syntax and their various interpretations (Alexander, 1977; Bacon, 1974; Benzel, 1998; Forman and Godron, 1986; Smith, 1977; Watts, 1975). When a composition of physical objects and identifiable spaces comes together as an orchestration of sensations, the principles of design are brought into play in conjunction with the structural aspects of story and discourse: the welcoming greetings (front stoops, gateways, entries) expressed before a formal entry and subsequent area for the removal of jackets, hats and shoes, and so on (Alexander, 1977; Wood and Beck, 1994). These sequential volumetric arrival areas are what Alexander (1977) refers to as the "elements" of his pattern language.

The structural flow of Alexander's pattern elements is further substantiated by what we can refer to as a dynamic conformity. This structuring of conformity (not sameness) across an area, a site, is found in Cullen's serial vision and Sitte's concern that

> *the ideal street must form a completely enclosed unit... where the gaze cannot be lost in infinity...the winding character of the ancient streets kept sealing off perspective views in them while offering the eye a new aspect at each succeeding turn.*
>
> *(Collins and Collins, 1986, 199)*

For other examples of this less-is-more approach to experiential places made up of enclosed units see Bacon (1974) and Ardalan and Bakhtiar (1973) and as you do, remember Alexander's tenet that

> *no pattern is an isolated entity. Each pattern can exist in the world, only to the extent that is supported by other patterns: the larger patterns in which it is embedded, the patterns of the same size that surround it, and the smaller patterns which are embedded in it.*
>
> *(1977, xiii)*

Stated differently, in regard to movement into and through an area, what was the larger pattern of an area's context becomes the pattern being experienced. As such, as you continue to move forward, the experienced pattern becomes one of the smaller patterns now in support of the next larger one. All together they contribute to sequences of called for behaviors and experiences.

I hope you've begun to realize we are characters in our own stories. As both designer-as-author and user-as-reader we experience landscape narratives and with a little self-reflection we learn from those experiences whether walking or riding, bicycling, or using public transit, enjoying a public garden, or a private space on an urban roof top. Life is serial vision, sequences of interconnected or at times juxtaposed spaces and emerging experiences. They are our daily stories, our daily dance. They are the narratives that connect us with our environments, regions, social groups, and nation states. Paraphrasing the close of each episode of the TV series "The Naked City," "There are a million stories in the naked city." Each day yours is just one of them. No, they weren't referring to landscape, architecture, or interior spaces but like today's detective and adventure stories you do so often see the action sequences unfolding in settings you've come to expect.

*As used in this book, 'spatial form' in its simplest sense designates the techniques by which novelists subvert the chronological sequence inherent in narrative. We read narratives one word after another, and in this sense all narratives are chronological sequences. But the novelist's arrangement of events within this linear flow of words often departs in varying degrees from strict chronological order.*
*(Smitten and Daghistany, 1981, 13)*

An odd thought comes to mind. Is it possible to arrive at a space where there is an absence of activity: an activity-less space? What kinds of feelings would you have? What expectations would not be being met? In answering those questions it begins to become clear, to me at least, there is no place where there is no narrative. Like the pauses between notes help produce music, so to an absence of information, of incomplete messages, is part of a larger narrative. Accepting that, we must concern ourselves with context, theme, connectivity, and interconnectivity of narrative, and place, and users' purposes. In doing so we accept the sequencing of selected nouns and verbs moving across and down a piece of paper with the expectation of character development, connectivity from word to word, action to action, and a coherency between, we could say, Alexander's noun/things and Bacon's verb/actions.

*To read a novel expecting the satisfaction of closure and the receipt of a complete message is what most people find enough to do [when reading]; they are easier with this message because it resembles the one that works for ordinary acts of communication. In this way the gap is closed between what is sent and what is received... [as a result] it is not uncommon for large parts of a novel to go virtually unread.*
*(Mitchell, 1981, 84)*

While Mitchell refers to narrative, specifically in a poem, I can't help but apply what he says to landscape narrative. "Of course the poem encourages us to do these things [make connections and draw conclusions] by ending [the poem] with the sort of message that seems possible and proper" (81). Among the "things" Mitchell says we seek is closure even when it isn't what the author intended, but "in the end, [we have] made it say something that suits us" (81), that speaks to a landscape participant's expectations. We expect completed actions. We expect closure, anticipate life trajectories, character development. Given the sequential nature of a developing landscape narrative, we anticipate between departure from one place and arrival at another place a sequence of serial completions: I crossed the street; I arrived at the small park; I found a coffee stand; and so on.

In his analysis of narrative, Mitchell presents us with an interesting quality. Narratives have an immediate dimension and a deeper, less obvious, "secret" dimension. For us, as landscape designer-authors, our "secret" dimension beyond the immediate functional messages include the socio-cultural aspects of period and place and the deeper associations touched in our collective unconscious.

> *...Eliot and Pound, for example – undermine the inherent consecutiveness of language, forcing the reader to perceive elements of the poem not as unrolling in time but as juxtaposed in space [across the page]...The suppression of these connectives alters the whole character of the literary work and forces the reader to perceive it in a new, unconventional way.*
> *(Smitten and Daghistany, 1981, 17)*

New York's High Line is a good example of immediate and deeper more secret messages. Most immediately the 1.45 mile elevated linear park, is just that a linear environment that connects the Meatpacking District to the Javis Center and any number of places along the way. Slightly less immediate messages revolve around the fact that

> *Built on a historic, elevated rail line, the High Line was always intended to be more than a park. You can walk through gardens, view art, experience a performance, savor delicious food, or connect with friends and neighbors-all while enjoying a unique perspective of New York City.*
> *(https://www.thehighline.org/about/)*

Even deeper, the park symbolically tells people "New York cares about you; New York cares about its history and its neighborhoods." "We are building on the inventiveness of our past." So, with landscape architect James Corner's comment that "the High Line's design 'is characterized by an intimate choreography of movement'" we delve a little deeper still into the narrative. Regarding the levels of narrative expressed, Mitchell says that "by various signs" other matters than the most immediate storyline are detectable "for the sake of clearness and effect" (85). With that, let's explore narrative in general, and landscape narrative in particular, a little more deeply.

The Analogy's first six chapters laid a foundation on which to build the upcoming narrative-storyboard-design process. A number of times to this point I've wanted to say I realized how complicated and at times seemingly convoluted the three-tiered analogy is. I didn't. In playing with The Analogy I'm still recognizing new lessons that improve my understanding of landscape architectural design. There remains so much to be discovered. HRH Prince Charles, in his effort to "support public participation in planning," sums up my thinking. Although he is speaking of architecture and urban design, his thinking is applicable to landscape architecture and interior design.

*Buildings should reflect these [Classical] harmonies, for architecture is like a language. You cannot construct pleasing sentences in English unless you have a thorough knowledge of the grammatical ground rules. If you abandon these basic principles of grammar the result is discordant and inharmonious.*

*(Moughtin, 2003, 11)*

With the beginnings of "grammatical ground rules" now in hand, we turn to their application of the rules to landscape narratives.

# 7

# Narrative

Landscapes, as with all built environments, whether purposefully designed or the outcome of decades or centuries of vernacular development, express narratives. Their stories may be experienced as the spiritual redemption of a cathedral's forecourt and entryway or as simple, and yet meaningful, as the unavoidable communal sharing at a farmers' market. Alexander captures this breadth of expression when he says, "A pattern language gives each person who uses it the power to create an infinite variety of new and unique buildings, just as his ordinary language gives him the power to create an infinite variety of sentences" (Alexander, 1979, xi). In how these environments express themselves, I agree with Robert Vischer who sensed

> *an almost mythic quality in empathy and spoke of a person forming an emotional union with an external object. Observing that feelings may be aroused by experiencing totally abstract objects (as well storms, sunsets, trees), he surmises that we may empathize with objects by projecting our personal emotions on them.*
> *(Bloomer and Moore, 1977, 29)*

Figure 7.1 Horseshoe Lake, Deer Park, WA. Photo by author.

Figure 7.2 Grand Teton National Park, WY. Photo by author.

Complimentary, personally expanding, or humbling, those projections are recognitions of the existential qualities of our environments. Social geographer Marwyn Samuels sees such existential environments as

> *always an authored space and, in methodological terms,*

DOI: 10.4324/9781003286981-8

*always has a biography as well as an intellectual history. The method of an existential geography is to 'begin with the subjective,' which is to say with some author rooted in some situation.*

*(Samuels, 1978, 33)*

The situation in which I ask the readers of *Storytelling for Landscape Architecture* to be rooted are the expectations of those people most likely to experience their designs. Exploring narrative as employed in landscapes requires we look beyond design and into language, literature, and a bit of philosophy.

There is more written on landscape as narrative and on aspects of landscape narrative outside landscape architecture than within. With that comes good and bad news. The good news is that many non-design professional authors mine the landscape-narrative metaphor much more deeply than designers do and we benefit from their work. Social and cultural geographers and philosophers of art, in particular, refine the metaphor and provide designers with philosophical rationales for practical applications of narrative. If there is bad news, it is that many art historians and philosophers can't seem to see beyond architecture as their object of study and admit to the existence and role of landscapes.

Figure 7.3 Wave Action, Plymouth Beach, MA. Photo by author.

Light and Smith's (2005) compilation of works in *The Aesthetics of Everyday Life* has a lot to offer designers' understanding of environmental and everyday aesthetics. At the same time a number of the contributing authors' works exemplify a shortcoming in art and environmental philosophers, aestheticians, and art historians' approaches to aesthetics. Simply, they are primarily object oriented. Their thinking grows out of a historic, passive appreciation of singular objects such as paintings, sculptures, or pieces of music. So conditioned are they to viewing "an object" that they fail to see, and appreciate the volumetric spaces around, preceding and leading up to those objects: in other words, landscape. Landscape is just something connected to or growing out of a building's influences. Yet, my first project out of school had a multimillion dollar complex of buildings responding to the landscape, and not the other way around.

*One can think of landscapes as scenes, or carefully designed and choreographed spaces controlling human actions or movements: for example, public plazas, park meadows, rooftop gardens, or parking garages, with various thresholds, vistas, and niches working cumulatively both to encourage and arrest spatial experience.*

*(Berger, 1998, 8)*

With that realization, be cautious as you explore the aesthetics of architecture. I know from experience, how easy it is to get drawn into and accept their incomplete arguments. They are incomplete in two senses. First, in that their aesthetics are all too often those of passive observers, not active participants. Their approaches to environmental aesthetics are hindered by an attitude that architecture is to be appreciated much like you would a sculpture or painting and not participated in as you would a landscape, or a dance, or a stream when fly fishing, or a rock climb. Second, as stated above, landscape is all too often not recognized as landscape but as an extension of architecture. Philosophy Professor, Arnold Berleant, gives a good example of that thinking.

> *When we turn to architecture, we find this enlarged space extended inward as well as outward. Mass opens within to enclose space, and this interior volume is designed for people… At the same time, architecture reaches far beyond the aura of sculpture into the surrounding space of its site, incorporating that space into an ensemble with its physical structure. Such an enlarged space may extend beyond the site and adjoining grounds to contiguous or nearby structures.*
> *(Berleant, 2005, 25)*

Berleant continues,

> *To regard architecture merely as the art of building distorts the way in which it actually organizes an entire environment. The design of a building determines not just its own features, it also affects the site on which the building rests and our perception of the structures that stand nearby. Sometimes, in fact, a building casts its character over an entire neighborhood, for buildings are not self-sufficient objects but places for human activity, determining the patterns of movement toward, into, and out of them, as well as within them. This fact transforms our understanding of architecture from an art of physical structures into an art of complex social and environmental organization.*
> *(Berleant, 2005, 29–30)*

And still no direct mention or attention given landscape, or the influence of the landscape in which the architecture is set.

I can't help but wonder how a landscape architect would have written Berleant's two previous quotes. Would the building "actually organize the entire environment"? Would the building, alone, effect the site on which it rests? And would the aesthetic experience be derived from a thing, an object, or from the progression through spaces, over time, starting somewhere early in the day or week with a person's expectations, then first impressions, then actual arrival and entry into the building? To what extent was the building's design influenced by the topography, soil mechanics, bedrock, neighborhood, and surrounding landscapes? When I conducted study sessions for architects preparing for their registration exams I learned from them that the most difficult section of their exam was the section dealing with topography and site design. It was obvious they had learned to think from the building out and not the landscape in.

Figure 7.4 Sedona's Chapel of the Holy Cross. Shutterstock ID: 2094954961 Contributor: Angel DiBilio.

Berleant further helps us understand the lack of attention given landscape when he says

*The distinctive pleasure we associate with the arts involves an aesthetic of objects – an aesthetic of art objects and sometimes of objects in nature. These stand at the theoretical center, and most discussion concerns the distinctiveness of such objects, their properties, qualities, and other features, their forms and order, and how they transfigure the world beyond in ways that render it aesthetically enjoyable. Contemplating this aesthetic of objects is a distinctive way of appreciating them.*

*(25)*

He reinforces the singular view of architecture as an art object comparable to a painting, sculpture, piece of music, or theatrical or dance production. I find it inconceivable that anyone could look at, could passively contemplate Sedona's Chapel of the Holy Cross only as an aesthetic object without they're thoughts being influenced by and growing out of the landscape in which it is situated. Unlike Berleant I see "how the world beyond" transfigured the chapel "in ways that render it aesthetically enjoyable."

Unlike the "passive and contemplative" appreciations associated with art objects, most architecture, landscape, and interior aesthetic experiences are whole-body experiences that extend from the awe inspiring to the mundane. It was an "aha" moment the day I realized my butt had immediate, mundane aesthetic experiences with a comfortable chair, a well-engineered car or motorcycle seat, or an ergonomically designed park bench. Aesthetics can be passively derived: I will stand here and contemplate that painting, the painter's brush strokes, the lighting across the canvas. Given that standing still is an experience, how did the fact you'd been standing in front of this or that paining for almost six hours influence your contemplation?

Aesthetics are experienced as interactive encounters across a plaza, beach boardwalk, approach to and arrival at a 20-story atrium. Aesthetic pleasures are every day, mundane experiences and not just those derived from the high arts. Also, when you think of it, your experiences filling 15 to 20 to 30 minutes leading up to your arrival at a building put you in a mood that likely colored your aesthetic state of mind which is either enhanced by the building once arrived at, or not. But not the other way around. You can't have a building's interior spaces you have yet to experience, color the landscape experience on the way to the building. Well, maybe you can if you read a critique of the architect's interior or saw a documentary on television. But you still won't escape the surrounding landscape's tug as you approach the structure.

Professor of Aesthetics Arto Haapala recognizes life as involvement and the aesthetics of everyday life.

*The first step in the process of interpretation is making the*

*strange surroundings familiar. As I said, this is something we start to do as soon as we enter a new environment. Perhaps it is more appropriate to say that it starts to happen; first on a purely sensuous level – what we see and hear is not as striking as it first was; we get used to the visual and auditory features of the surroundings. We also start to recognize the functions and meanings of different buildings and sites: there is a shop, there is a school, kindergarten, police station, church, these are residential buildings, the market is there, the harbor further away, etc.*

*(Haapala, 2005, 47)*

Haapala does mention "sites." Yet, while he still gives architecture most of his attention, especially as related to other architecture, he does recognize the influences and pleasures associated with nature, fields, the sky, and birds (49). Even so, his recognition of landscape remains limited. All the "buildings" to which he refers are seen from a distance and preceded by and set in a landscape the character of which he admits is read and interpreted in ways that the individual's state of mind (attitude, mood) is preparing them for their arrival at the respective structure or place.

However, a wonderful compilation of articles in New Zealand's *Landscape Review* between 1995 and 1998 finds landscape architects delving into landscape as text and landscape as narrative. Throughout the issues of *Landscape Review* there is a mutual understanding that cultural landscapes express themselves, at least express information that we often turn into stories. The difficulty is that the authors do not consider the progressive sequencing of statements that is narrative or storytelling. They focus on objects, scenes, and locations (places). What is "read" is read in, or as emanating from, an object much like visiting Plymouth Rock in Plymouth, Massachusetts and recalling the stories of the Pilgrims' landing, sickness, death, barely surviving the first winter, interacting with the indigenous people. These again are stories associated with the object, Plymouth Rock. They are not the landscape narratives that come into play as people come to Plymouth to visit the rock; not the narratives people experience while planning their trip, arriving in downtown Plymouth or on the waterfront, walking through a residential or commercial area to the site as sounds of traffic compete with the ocean's waves and fishing boats. The stories associated with objects or places do provide one with narratives but they are not the narratives that comprise The Analogy, the kind associated with eye-level serial vision seen by Bacon (1974) and Cullen (1961), the kind of sequential continuum experienced while getting to a destination.

While social and cultural geographers consider landscape texts at the pedestrian level, those in the landscape community tend to deal with landscape primarily as an academic subject. In the late 1960s and early 1970s with rising attention to literary criticism and literary theory, landscape architects gave more attention to the meta-theories than the substance of everyday life. There remains, however, an extensive body of related literature. The various authors may not refer to the landscape as a text, they likely don't even mention it as a medium to be read, but they clearly deal with texts in the interpretations and arguments they construct of what exists as built and inhabited environments.

## MESSAGE MEDIA

The fundamental qualities that allow basic design elements to be recognized are those that also provide contrast and give character to a landscape text. The characteristics of the individual media used to express parts of a design are the phonemes and morphemes of the landscape-as-text vocabulary. These are the smallest units of expression that cannot be further subdivided. In basic design, what is it about graphite that distinguishes it as a unique medium from, say, charcoal? The same question can be applied to landscape materials: pea gravel vs. coarse gravel vs. sand, teak vs. oak vs. pine, types of concrete vs. their particular applications and finishes.

*Phonemes: any of the perceptually distinct units of sound in a specified language that distinguish one word from another, for example p, b, d, and t in the English words pad, pat, bad, and bat. (see https://www.dictionary.com/browse/phonemes, viewed March 2022)*

*Morphemes: any of the minimal grammatical units of a language, each constituting a word or meaningful part of a word, that cannot be divided into smaller independent grammatical parts, as the, write, or the -ed of waited. (see https://www.dictionary.com/browse/morphemes, viewed March 2022)*

Whether paint or vegetation, stone or milled lumber, brick or fabric, the discernible qualities of a media, and by that I mean discernible by our five senses, enable the communication (expression) of functional messages, personal identity, regional character, and national ideology. Media selection and manipulation allow artists and designers to express statements, talk with their work, test ideas with other artisans, and express themselves to the public. How materials are shaped and their surfaces finished express cultural values and myths, they are also more typically expressions of such idiosyncrasies as local nuances in personal taste and personal character. It is like peeling back the wallpaper in the living room of a 200-year-old house. With a little research you can identify the approximate years each layer was applied and wonder about the homeowners' taste.

Within the logic with which messages are initiated through the selection of basic materials, we quickly realize the link between intended messages and a narrowing in the number and kinds of materials from which we can chose. The shapes, colors, and textures associated with gaiety would not support such activity spaces as a contemplative garden, a cemetery or mausoleum, or a Shakespearean theater. Well, depending upon the play. This realization also works to dispel the notion that good design is haphazard. Of course, a corollary is that when design is haphazard the narrative is weak, people are likely confused and potentially disoriented. I'd argue even further that neither professional nor nonprofessional work recognized as good design is haphazard. Luck and intuition do play a role in the design process, but by no means the primary role. Designs that are both appropriate to their settings and cohesive across their activity settings are rarely if ever derived solely from random ideas or insights or personal opinion instead of decisions made on observed facts.

*It is well known that smoothness is always an attribute of perfection because its opposite reveals a technical and typically human operation of assembling. (Barthes, 1972, 88)*

As stated earlier, at the detailed end of the elementary-landscape scale the purity of mathematics does not set well with designers. Someone could have a lot of fun rewriting the entire theory of design if they were to apply mathematical definitions to it: points have no dimensions, lines no width only length, and planes no thickness. When aspects of size and scale are considered, any effort to come to strict definitions of points, lines, planes, and volumes takes on a Zen-like aura. The questions raised, however, while complicated in the ambiguity of their answers, provide for important and valuable discussions to improve our understanding of design. The value of exploring such questions as, "When is a sidewalk an edge and when is it a linear area? When is a hedge row a vertical plane or linear volume?" How do we come to understand what makes up design and how to apply it to the narrative form called landscape design? In what follows, the answers to such questions help illuminate a myriad of applications and interpretations of each basic design unit making up landscapes as narratives.

Figure 7.5 Stonehenge, a prehistoric monument in the fields of Wiltshire, England. Photo by author.

## NARRATIVE BASICS

In the multiplicity of messages associated with any medium over time, there are fundamental forms associated with fundamental messages: gateway, positions of power associated with speaker's height above a crowd, and clan, family, and tribal gathering sites. The more basic the media in which they are worked the more collective are people's interpretative responses to those forms: understanding the work's existence or reason for being. The previous thought reflects the recognition in oral traditions that the more generic a story's message the greater the variation in characters and locations' settings expressed over time while maintaining the original overall story/message.

Figure 7.6 Machu Picchu is a 15th-century Inca citadel. Photo by Karen Shields.

Although fiction, I take James Michener's *The Source* (1965) as an excellent example of the power of a place to which people gathered time and again for millennia: Stonehenge, Mesa Verde, and Machu Pichu are other examples of places of power. Similarities in people's reactions, across cultures, evoked in specific landscapes and architecture give credence to the existence of archetypes: the "'first form' or 'original model' as it exists as a basis for all later variations and combinations" (Thiis—Evensen, 1987, 17).

*Archetypes are not inborn ideas, but 'typical forms of behavior which, once they become conscious, naturally present themselves as ideas*

*and images, like everything else that becomes a content of consciousness.' Archetypes have an organizing influence (my emphasis) on images and ideas.*
*(Storr, 1983, 16)*

"Jung came to think of archetypes...as organizers...[that] 'manifest themselves only through their ability to *organize* images and ideas'" (Storr, 1983, 25). This organizing, or gathering, power is the quality echoed in Christopher Alexander's elementary patterns. Elementary patterns are such because they "occur over and over again in our environment." However, the singular nature of their core forms can be used a "million times over, without ever doing it the same way twice" (Alexander, 1977, x). Elementary patterns are themselves identified by the even more basic characteristics of frequency and repetitiveness over time and space. In addition, each time we use a landscape feature without changing it we in fact reinforce its existence, its message(s), and the authority (longstanding cultural belief) behind them.

*In reference to the Villa Lante's axial water feature, Moore says: The main glory of the garden is its axis of water...the journey is a curious one, since it starts most of the way down the axis, and the goal is ambiguous – either the source up the hill or the magnificent fountain below, toward which the water drops.*
*(Moore et al., 1988, 147)*

In the gathering power of points is our recognition that they can't stand alone. Most usually

Figure 7.7 The Water Chain at the Villa Lante, Bagnaia, Viterbo, Italy. Photo by author.

points as intersections or landmarks are tied to circulation patterns if not lines of sight. Axes not only link points of origin and destination visually but usually allow for people to move either alongside the axis or in the axis itself. The former is an interesting point in the case of the Kashmir garden Shalamar Bagh. "The main pathways are placed on either side of the canal. This means of course, that you cannot walk on the axis but are always slightly offset from it...Only the ducks cross the axis without formality" (Moore et al., 1988, 171). Of interest is that nature, in the form of the ducks, has a freedom to roam that we do not feel we have in our confinement to linear walkways.

The ability of points to collect people, or provide for activities, calls for a progression of people-people and people-environment interactions. The type of interactions, and certain of the procedures associated with the interactions, is usually expressed along the route leading to the point. At human scale that information may be read by a person with both arms and hands free, as opposed to someone carrying two bags of groceries, pushing a shopping cart or a carriage, or on crutches or in a wheelchair. Each, given their stature and abilities, looks at the landscape ahead and asks "Can I get to where I wish to go?" "How should I prepare for what I will encounter along the way?" "Once I get there, will I be able to carry out my wishes?" even as questions, these are expectations looking to be met. At the vehicular scale the information says whether the vehicle can enter, turn, backup. Imagine riding your bicycle to work. You get to the fire house and the bell goes off. You climb up into the back end of the ladder truck and off you go. You're the tillerman. As you leave the station, do you read the environment the same way you did while riding your bicycle? Both realms of information are communicated along a progression, a linear progression of spatial information further defined by features containing the volumetric spaces, or those areas delineated by edges.

## READING THE LANDSCAPE

The Analogy builds upon our tacit understanding of how we organize stories into recognizable narrative forms such as technical or comedic, meditative, or recreational. By reflecting on how we regularly read landscapes we bring our unspoken understanding of design into the open. Once made conscious, we can then explore a design as a new and unknown subject from the comfortably familiar surroundings of what we do day in and day out. With the change in perspective comes a shift from "what do I do to make this place look good?" to "what does this place need to communicate its messages well?" both here and now, and within its larger story (surrounding context). Answering that question immediately calls for an understanding of "what" is to be communicated. The answer to that question brings to the beginning of the design process what is usually found toward the end. Knowing what your design needs to communicate guides your design decisions in such a way that the overall design gains in continuity, coherency, and consistency.

> *…became more deliberate and watchful. He often stopped to examine the trees; nor did he cross a rivulet without attentively considering the quantity, the velocity, and the color of the waters. Distrusting his own judgment, his appeals to the opinion of Chingachgook were frequent and earnest.*
> *(Cooper, 1980, 252)*

The fact that we read our environments should be inescapable by now. If we fail to interpret the signs, signals, and symbols then we are unable to recognize gateways, walkways, doorways, overlooks, and how to get there. If this wasn't true, then we would find piles of bodies, bicycles, and cars where paths and roads intersect, we'd have students and faculty standing around loading docks rather than the main entrances waiting for libraries to open, and we'd have parking lots where the parks should be, and God knows where we'd find the parks.

> *Their advance was rapid; and made with as much confidence as a traveler would proceed along a wide highway. If a rock, or a rivulet, or a bit of earth harder than common, severed the links of the clue they followed, the true eye of the scout recovered them at a distance, and seldom rendered the delay of a single moment necessary.*
> *(Cooper, 1980, 253)*

At the completion of an architecture student's presentation of his design of a new library structure on the University of Kentucky campus, I asked, "Imagine coming around the corner of this building," a building 60–70 yards away from his proposed library. I pointed to it and located him at a point that caused him to see two sides of his building. One side included the main entrance. I continued,

*And, imagine you wanted to use the library and it was your first time going there. From this location, as you scan the building's two facades, where would you likely walk to if you were searching out the library's main entrance?*

Looking over his elevations, his classmates began to chuckle. He even admitted he would walk toward the large darkened rectangular areas delineated by the sunlit, highlighted fenestration: the library's loading dock. While the activity spaces of his structure were well organized and interrelated, he had not considered how people would read the signals – lines, outlines, and darkened areas of his design – as they approached the building.

*People make attachments to places that are critical to their well-being or distress. An individual's sense of place is both a biological response to the surrounding physical environment and a cultural creation, as geographer Yi-Fu Tuan has argued. From childhood, humans come to know places through engaging all five senses, sight as well as sound, smell, touch, and taste… simultaneous engagement of several senses in orientation and wayfinding. (Hayden, 1995, 16)*

At our most pragmatic level we read both built and natural landscapes. While the former is authored and the latter is (in a sense) not, natural environments can still read at various levels of detail. Any how-to books dealing with woodland survival or living off the land provide ready examples of how to read the woods (Nesbitt et al., 1959). Satellite images and the forms, tones, and textures in aerial photographs are read, information is organized, data are analyzed, and conclusions are reached. Doug Way, in *Terrain Analysis* (1973), uses the opportunity to identify particular drainage patterns as seen in aerial photographs. He shows how each drainage pattern is the signature of a particular subsurface parent material and the specific soil composition to which it gives rise. In conjunction with the identified vegetation Way determines the land uses for which a particular locale are best suited with regard to particular soil structure, soil load-bearing capacity, and subsurface parent materials.

At a more down-to-earth level, May Theilgaard Watts, *Reading the Landscape of America* (1975),

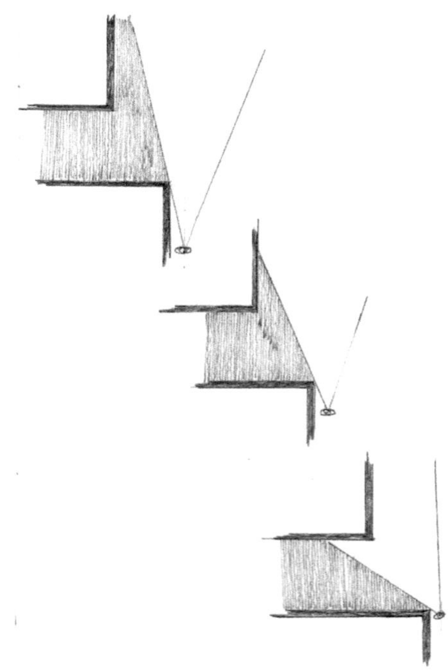

Figure 7.8 Message delivery sequence. Drawn by author.

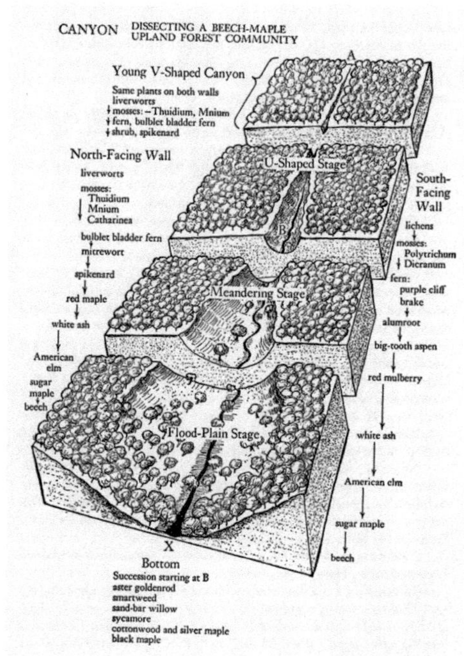


Figure 7.9 Canyon Story in *Reading the Landscape of America* (Watts, 1975, 50). Courtesy of the Estate of May Theilgaard Watts.

provides "readings" of a wonderful spectrum of landscapes, from the natural settings of "Canyon Story" and "Picnic in a Gritty Wind" to the humanly altered landscapes of "Readin', Ritin', and Recess" and "Looking Down on Improved Property." With similar powers of observation, a civil engineer I once worked with amazed me (something I shouldn't, as a landscape architect, admit to) with how he could, from a mile or more distant, identify the vegetation on a site. Once identified, he would go on to describe the area's soil type and moisture level, birds and animals inhabiting the place, their culture, and the advantages and disadvantages we'd have with the construction processes we were preparing to propose.

At an even more personal level, both fiction and non-fiction are replete with examples of environments being read by the stories' characters. Authors rely not only on a reader's ability to visually construct the environments in which their characters are set, remember Ichabod Crane's galloping ahead of the headless horseman through Sleepy Hollow, but also on the readers' ability to enhance their images by appealing to memories held by their other senses. Edgar Allen Poe and Sarah Orne Jewett are only two of many writers who orchestrate their readers' sensual memories to the benefit of their imaginations. As you read the following, consider which of your senses are being (re)called into Jewett's story, *The Country of the Pointed Firs.*

> *It was a queer little garden and puzzling to a stranger, the few flowers being put at a disadvantage by so much greenery; but the discovery was soon made that Mrs. Todd was an ardent lover of herbs…the sea breezes blew into the low end-window of the house laden with not only sweet-brier and sweet-mary, but balm and sage and borage and mint, wormwood and southernwood. If Mrs. Todd had occasion to step into the far corner of her herb plot, she trod heavily upon thyme, and made its fragrant presence known with all the rest…You could always tell when she was stepping out there, even when you were half awake in the morning, and learned to know, in the course of a few weeks' experience, in exactly which corner of the garden she might be.*
>
> *(1956, 14)*

Jewett provides us with an olfactory garden, or what could, just for fun, be referred to as a geography of the nose. Jewett introduces the potential of an olfactory text and its use in locating people within a landscape. Landscape designers could easily transpose Jewett's geography of the nose into a powerfully subliminal landscape. I can easily imagine a choreography across a landscape based on aromas.

What with so many people's limited attention to the world around them, even more now with their attention to their smartphones and social media, residential sites and even urban streets and plazas could be made into subtle olfactory landscapes. Walls where people sit, areas through which short cuts are taken, and intersections where people gather, once planted with scented herbs, would become locations, or landmarks, of memorable scents and aromas. As I leave for work each morning and return each evening, brushing against the lemon-scented herbs planted alongside where I park my van, the aromas enhance the planting's visual farewells and welcomes home with wonderful scents.

Poe, however, uses our five senses to help his readers locate the depths of the fears to which they allow themselves to be brought as they read "The Pit and the Pendulum." Identify which of your five senses he brings into this paragraph:

> *In the confusion attending my fall…and while I still lay prostrate…my chin rested upon the floor of the prison, but my lips, and the upper portion of my head…touched nothing. At the same time, my forehead seemed bathed in a clammy vapor, and the peculiar smell of decayed fungus arose to my nostrils. I put forward my arm…I had fallen at the very brink of a circular pit, whose extent, of course, I had no means of ascertaining…I succeeded in dislodging a small fragment [of masonry wall], and let it fall into the abyss. For many seconds I harkened to its reverberations as it dashed against the sides of the chasm in its descent; at length, there was a sullen plunge into water, succeeded by loud echoes. At the same moment, sound resembling a quick opening and as rapid closing of a door overhead, while a faint gleam of light flashed suddenly through the gloom, and as suddenly faded away.*
>
> *(Poe, 1904, 13)*

While I periodically relate aspects of landscape as text to theories of narration and communication, The Analogy primarily focuses on the texts through which people's encounters with their environments occur and are made possible in the landscapes of everyday life. The seduction of entering into academic explorations of the importance of structuration theory (Giddens, 1984), deconstruction (Derrida, 1978), and communication theories (Giddens, 1984; Habermas, 1981) as they inform landscape as text and narration is great. But at this time such explorations must be left to others with the hope that they jump into what I see as both a necessary exploration and debate valuable to the design professions and those they serve.

At this point we have recognized the interdependent links tying the character of designed features with the character of the messages they express. Integral to this interdependence is the association that binds the object world with both our meaning-full and value-laden social and personal worlds. That association is both temporal and spatial. Learned over time, and through numerous encounters, the interdependence of particular shapes, colors, textures, objects, and their respective meanings is temporal. In a sense, temporary to a place and a time. Europeans' rapidly changing associations regarding the New World between the late 1500s and early 1600s exemplify the temporality of meanings and place (Marx, 1981, 73–145). A landscape made up of parts of different older landscapes, parts of which are seen as making up portions of the newer one, is referred to as a palimpsest. I believe it was originally a painterly term referring to a painter adding bits of new layers on top of the original layer which shows through the newer ones.

*pal·imp·sest [ˈpaləm(p)ˌsest] NOUN, Geography Word of the Week: Palimpsest – While palimpsest often refers to a writing material on which the original script has been erased (though not completely) and written over again, in geography, the word means a place or landscape in which something new is superimposed over traces of something preceding it [parts of which show through].*

*(Canadian Geographic, 2015)*

The recent fury in the United States over what constitutes art and whether it is of sufficient value to be supported with Federal funds is another example of the almost fickle nature of people's object-meaning associations. Fickle or not there are still rules guiding actions and related outcomes.

*Images of America over a couple hundred years went from merciless, howling, cannibal-infested darkness to "the master symbol of the garden…expressing fecundity, growth, increase, and blissful labor."*

*(Marx, 1981, 142)*

Syntax and grammar guide the quality of the interrelationship between individual words and the sentences in which their messages are developed. Each of the arts has its own design principles, all which rest upon their socio-cultural psychology of perception (Arnheim, 1969; Gombrich, 1972; Wolff, 1981). As our culture changed so have we as a society. What early colonists and settlers valued and what they saw centuries ago in their landscapes that were and were not there is quite different now (Marx, 1981). The principles guiding acceptable dance, eating behaviors and food preparation, music, and landscape design are always evolving. We are the seeds of our future and the fruits of our past.

*Apparently, an unverified statement, but it has been going around for a long time: And in some parts of New England, serving lobster to prison inmates more than once a week was forbidden by law, as doing so was considered cruel and unusual punishment.*

*(https://www.reddit.com/r/AskHistorians/comments/11hakg/lobster_for_prisoners_cruel_and_unusual/)*

The good news is that the guiding principles are tied to society and with even a limited degree of accuracy we can interpret artifacts and identify a society's valued structures, appreciations, and the forms with which society identifies (Yamin and Metheny, 1996). This culturally based approach to landscape-as-text is what makes Potteiger and Purinton's (1998) *Landscape Narratives*. What they present as readings of images, signs, and symbols requires the "reader" have a background in the history and cultural evolution of the society responsible for the landscape texts being read. This is not The Analogy. The Analogy is about reading everyday landscapes by everyday participants such as tourists and local residents. Walking or driving along the route of Paul Revere's ride does not require you know the story. But avoiding being hit by a car or truck does require you be able to read the roadway-traffic narrative.

Locally available materials, period styles and production, and social practices are themselves mediated by cultural conventions. Personal expressions are formed through social and aesthetic codes (Rapoport, 1969; 1982; Wolff, 1981). The daily transformation of cultural conventions forms and informs more than the world of art (Wolff, 1981); it is found in cities (Domosh, 1996) and the fiction embedded in the cities (Sharpe and Wallock, 1987). Normative codes formulate the character and context of our myths about Martians (Barthes, 1979, 27–29) and other cultures (Lowenthal and Bowden, 1976). Each of these environmental manifestations of values, beliefs, and understandings contributes to the production of texts read and reread, whether urban form (Alexander, 1979; Smith, 1977) or mythic image (Bacon, 1974; Campbell, 1974).

Manipulating a medium is itself an act of reading, or more to the point, learning how to read it. We are not born with this capability. We learn the grammar of the media with which we work. As for selecting our parents, we have little choice. We are born into our culture and our language and the signs and symbols through which they are expressed. Each medium comes with its basic characteristics, they are the resources that await our arrival and our experimentation in an effort to learn their potentials. A child learns how to recognize, identify, and interpret audio and visual qualities that make up particular types of objects and language units. Babble becomes language, erratic behavior becomes dance, crayon scribbles on floors and walls give way to painting, and for some PB&J transforms into culinary delights. Not that PB&J isn't a delight. A sculptor interprets the potentials held in the crystalline structure of a stone, a scout follows a trail (Cooper, 1980), or an interpreter of aerial photos or satellite images recognizes what is referred to as a signature on the ground (Way, 1973). In each instance signature patterns become identifiable through a repetition of encounters over time (Dondis, 1973, 2–12).

> *[Henry Moore remembers] Then I had to think in the same way that Michelangelo might have done, so that one had to wait until the idea came that fitted the shape of the stone and that was seen, the idea, in that block.*
>
> *(Bronowski, 1973, 115)*

When Henry Moore, like Michelangelo, sought his next sculpture, it is said he would look for a particular piece of granite that held within it what he envisioned as the final sculpture. In theory, all he would then need to do is remove the excess. Actually, his work displayed his knowledge of the crystalline structure of a stone, how it would and would not fracture from chisel blows from this or that angle, and the surface finishes it would allow him to produce. Moore expressed the idea of reading what a medium brought him, not imposing his thoughts on it through his tools. In the stone's crystalline grain he recognized the sculpture's existence and the ways he would direct his tools to bring (by subtraction) the sculpted form to the surface.

*Of course, it cannot be literally true that what the sculptor imagines and carves out is already there, hidden in the block. And yet the metaphor tells the truth about the relation of discovery that exists between man and nature... Discovery is a double relation of analysis and synthesis together. As an analysis, it probes for what is there; but then, as a synthesis, it puts the parts together in a form by which the creative mind transcends the bare limits, the bare skeleton, that nature provides.*

*(Bronowski, 1973, 115)*

The ideal driving planning and design is that the solution is massaged out of rather than imposed upon a given situation. The end product appears not to have been designed, e.g., the forests designed by Frederick Law Olmsted surrounding the Biltmore Estate in Ashville, North Carolina.

People and place, heritage and habitat, are the resources a designer's creative energies weave into a design. Carried into the realm of ecology, an approach to design and planning that works within a given situation supports the concept of sustainable design: working with the natural and social dynamics of a situation, rather than against them. This is the basis of the permaculture movement (Mollison, 1988; Morgan, 2006) and the rise of ecological concepts as applied to production and commerce (Hawken, 1993).

Carried to its logical conclusion, such designed environments, ones that work within the existing ecological dynamics of a site's location, minimize the costs and time spent in ongoing maintenance. Good designs are like conduits. Through them pass the existing energy flows of the place and the people the designs influence. Done well, an introduced design does not interrupt the existing energy flows. Much of the natural and social dynamics of a place are sustained and perpetuated. This is where the creativity of landscape architectural design resides, along that balance beam where things change but relationships don't, in that realm of dynamic equilibrium. This means that designers, as part of their typical project inventory and analysis, need to become familiar with the sensate vocabularies of the place in which their project is located.

*When I (Bob Scarfo) asked for a definition of dynamic equilibrium, a Reardon, WA 6th grader told me how he saw it in he and his brother's bedroom. "We make piles of dirty clothes that our mom takes away, washes, folds, and puts on our shelves which we then take, wear, and make more piles of dirty clothes. That's dynamic equilibrium, the room keeps changing but always looks the same." Yes, this definition came from a 6th grade student.*

## NARRATIVE

Like the opportunities brought forth through growth, and the opportunities associated with the cycling of seasons, communication's dynamic quality allows us the potential to manipulate narrative materials to fit a variety of situations over a variety of times. This aspect of change over time is found in folklore and wonderfully suits landscape architecture since it deals with plants that change over time. Over time we may tell the same story in many different ways, or we may tell different stories in the same way. This cyclical growth, of maturing, is found in each of us. In so many ways we are each the same person we were years ago, while in many other ways we are different. Note Joseph Campbell's identification of the similar structures, but different settings and characters, in various cultures' myths, in *Hero with a Thousand Faces* (1949). Rubin also helps us recognize commonalities of expressive themes across the epics found in oral traditions (1995), and Bettina Knapp's *Archetype, Architecture, and the Writer*, in which she analyzes archetypes used by architects to fit their needs (1986).

*Poseidon's trident (which in India is Śiva's) became thus the Devil's popular pitchfork; Poseidon's great bull, sire of the Minotaur (In India, Shiva's bull Nandi) gave the Devil his cloven foot and horns; the very name, Hades, of the god of the underworld became a designation of that inferno which Heinrich Zimmer once described wittily as 'Mr. Lucifer's luxury skyscraper apartment-hotel for lifers...*

*(Campbell, 1968, 20–21)*

The flexibility of grammar, in combination with our imagination, allows us to draw on the finite resource of our vocabulary and create a seemingly endless variety of messages and types of messages: poetic, technical, romantic, and financial, to name a few. In the same way that life is not a single photograph, messages are not singular expressions. Accents and exclamations are not solitary, fixed points in time or space. They are part of a larger continuum, a narrative. The highlight of flowering dogwoods in the Virginia woodlands of spring is fleeting, but they are part of a greater seasonal and, in turn, annual narrative. Over time their color accents even move, through succession, moving into and across once plowed fields. Sun angles, where the sun rises and sets, what is illuminated and what isn't in the course of a day, a month, a year, also vary greatly. Taking advantage of the dynamic qualities of these natural exclamations requires knowing both the grammar and the syntax of a landscape's resources, cycles, and successions.

In the retelling, or rereading, of a built environment, similar messages can change from one location to another and from one time to another. Our ability to read their messages, even when they vary from place to place, is as much a factor of the organization of the place's basic vocabulary as it is a factor of what preceded and succeeded them. Common examples include the apron areas in front of doors, entries to outdoor public gathering places, and sitting areas. Doorway aprons are typically wider and of a different material than the walk that led up to them. Even with universal design, they are often raised and have an overhead canopy to provide protection from the weather and a sense of arrival at a private place aside from the more public walkway.

Outdoor public gathering places, again, are usually preceded by a paved area larger than the entryway it precedes. The area is signaled by and distinguished from adjacent areas by changes in paving patterns, if not also the materials used. People mill about, get their bearings, find friends, and then move toward a smaller entry area flanked by posts, towers, or some such vertical features. The towers may house ticket booths. They echo the columns often seen flanking a gated entryway. Sitting areas are also often found physically separated. A seating area may be adjacent to but outside of a walkway's stream of traffic. It may be completely separated from the walkway with plantings between the bench and the walk. Other benches may be located right on the walkway, but again a different paving material or pattern often

Figures 7.10.1–7.10.5 A succession of arrival points providing for a holistic experiential flow. Hallgrimskirkja Lutheran (Church of Iceland), Reykjavik, Iceland. Photos by author.

visually separates it from the circulation portion of the walkway. The stability of the planes, verticals, points, textures over time and between places are what allows us to read similar, yet new, environments with a degree of personal safety and security. By stability I mean they have survived the test of time. People have for years successfully read the messages. The messages expressed an understandable story and the appropriateness of what they expressed didn't call for any user editing (making changes to the landscape features or what we sometimes refer to as vandalism).

*Expectations can be used to, in a phrase, trip you up:*

*....small design detail of the stairway – the so-called false step. The riser of the false step was one inch higher than the other steps so that it would trip up anybody unfamiliar*

*with it and serve as a primitive burglar alarm.*
*(Berendt, 1994, 63)*

The aspect of stability, or consistency, in form and structure on which many of our built environments are based is also fundamental to our oral traditions. The counting-out rhyme, Eenie, meenie, miney, mo, "remained stable for over a century. There have been only two systematic changes" (Rubin, 1995, 4). The word "hollers" is not a British word and has, in the United Kingdom, been changed to "squeals." The word "tiger" replaced an admittedly derogatory social connotation. In epic poetry, many poems are not as stable, in the strict sense. "The plot, the rhythm, the general cast of characters, the formulaic descriptions of characters and places…remain stable, but the exact words sung do not" (Rubin, 1995, 4). The epic poem's basic organization, as with many of the spaces that comprise our built environments, remains the same at the practical level. The repetition over time, from one generation to the next, of the makeup of our most cherished spaces and collective experiences, is the basis of much of what comes to be our heritage and personal and familial histories. From the routine of shared stories, and shared physical features, we acquire a sense of orientation, of belonging. When a place's vocabulary, and the structural organization of that vocabulary varies, we often sense something as being "out of place" or, to quote Dorothy, "Toto, I've a feeling we're not in Kansas anymore."

*Eenie, meenie, miney, mo,*
*Catch a tiger by the toe,*
*If he hollers, let him go,*
*Eenie, meenie, miney, mo.*
*(Rubin, 1995, 227)*

*Now I lay me down to sleep,*
*I pray the Lord my soul to keep;*
*If I should die before I wake,*
*I pray the Lord my soul to take.*
*(Anonymous, 12th century)*

*SPOKANE, Wash. – A new stop sign at the intersection of High Dr. and Grand Blvd. has been confusing for many drivers on Spokane's South Hill.*

*Where drivers used to be able [for decades] to turn onto High Dr. from Grand Blvd. without stopping, a "traffic revision ahead" sign warns them about the added road sign.*

*But this sign doesn't seem to be enough to remind some drivers to stop before continuing through the intersection.*
*(Waltower, 2019)*

In design, as in communication, there are basic narrative structures that when followed contribute to the survival of oral histories across thousands of years (Rubin, 1995). Contrary to most students', and many practitioners' reactions to the idea that design involves constraints in no way diminishes the creativity called upon for designers to narrate place messages of intense personal, social, and cultural significance. In realizing the sensual language (the role of the five senses in reading a landscape) that is a users' social context, in combination with the expressive capabilities of vegetation and building materials to be used in a design, a designer acquires a sense of freedom. That freedom comes with knowing what media are available and their capabilities and knowing the limits of what will and will likely not be understood, appreciated, and sustained by the users of a design. By working with the social character of a people and a place, and this can also mean consciously working against the character of the design's contexts – the operative word being "consciously" – artisans are able to interpret and record emotions into narratives that can be identified, read, and appreciated by the landscape users.

Marc Treib notes the downfall of taking things out of context to create a design. He specifically refers to designers reproducing early earth sculptures in modern environments.

> *In borrowing primitive forms without their systems of significance, by wrenching them from their settings and their societies, we risk confusing syntax for semantics... in the process, we can acquire only a representation of a representation, compounding the entropic process and the dissolution of meaning.*
> *(1991, 107)*

> *The built forms of late nineteenth-century New York City and Boston differed dramatically. New York's landscape exemplified the ultimate material expression of a capitalist city... unlimited expansion, both horizontal and vertical... a never-ending search for outlets of conspicuous consumption, using their residences and commercial buildings as the most evident forms of display... In contrast, Boston's bourgeois class fashioned their city and its landscape into a cultural capital and center of intellectual society. The expansion of commercial districts was limited by certain civic values that were translated into a ban on skyscraper construction and regulations that prevented the [commercial] development of the Boston Common.*
> *(Domosh, 1996, 1)*

As social products, interior and landscape spaces form and inform both their designers and their users. Like words in a narrative, landscape spaces are storage bins of thoughts, values, and the ideologies and local biographies of particular people, places, and periods of time. Landscapes and their spaces are expressions of power (Louis XIV's Versailles' axis cut to the horizon, and the Vanderbilt estate near Ashville, NC), of subtlety (the park lands of Olmsted or Jens Jensen), of sensitivity (many ecotourism resorts), and of obscenity (grid—like clear-cuts, canals cut for oil barges through Texas swamplands, altered ecologies in the name of recreation). When people read spatial messages as appropriate to their personal, social, and ecological contexts, the dialogue between people and place is one that sustains the place. A design that respects and works with those systems basically cares for itself. Dialogue in this sense is the interaction of people with their surroundings. When the messages are inappropriate, the dialogue is one of negative critique. It takes the form of people making changes to their surroundings. People altering environments do so to suit their sense of identity. Two well-documented examples

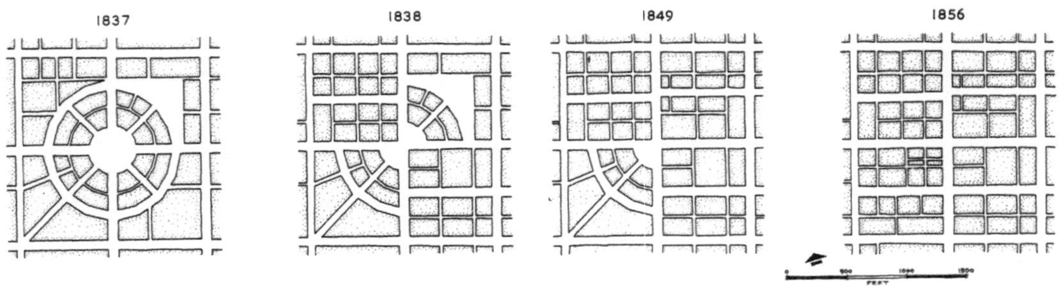


Figure 7.11 The Redevelopment of Circleville, Ohio 1837-1856 (Reps, 1965, 302). Courtesy John Reps' Estate.

are found in Mona Domosh's, *Inventing Cities: The Creation of Landscape in Nineteenth-Century New York and Boston* (1996) and in John Reps' account of the squaring of Circleville (1969, 300–305).

We modify our environments to suit our personal and collective sense of *Self* to the point that such places – front yard, backyard, patio, office – reflect who we are, our personality. As designers, we do not work in a vacuum. But as designers, and to the benefit of our clients and eventual design users, we must be able to step outside our personal heritage and put it aside as best we can. Design is an act and a product derived from social-cultural and ecological contexts which at times are not our own. Languages are set within cultures, including professions, religions, street gangs, and the like. Some read left to right, some top to bottom. Our very appreciation of form and flow is likely influenced by the graphic symbols that comprise our alphabets: Arabic, Chinese, Japanese, and Indo-Iranian. We also grow up within the language boundaries of place, time, occupation, and activity: technical, scientific, religious, romantic, economic, artistic, and more. As we go back further and further in time, the distinctions that brought us our specialized vocabularies blur. Art and science were not the severely distinguished areas of study, kinds of research methodologies, and specialized vocabularies they are today.

*The rules of each street were set by different gangs. If I wore certain colors in the wrong neighborhood, I could get shot. If I wore my baseball cap slightly to the left or to the right in the wrong neighborhood, I could get shot. If I scratched my head and it looked like my fingers were making a certain gesture in the wrong neighborhood, I could get shot.*
*(Cooper, 2020, 1)*

Erasmus Darwin, Charles's father and a medical doctor, wrote up the results of his scientific experiments as poems. In this form the dissemination of his work educated the greater public to the emerging sciences (Jordanova, 1986, 159–203). A more contemporary example is found with Loren Eiseley. His anthropologist's wonder with life had him negatively labeled by some as a mystic; his sense of awe expressed in his prosaic style was considered highly unscientific (Heuer, 1987, 161). Even fiction can be seen to convey new social and scientific knowledge and shifts in art appreciation and landscape design. In 1849, E. A. Poe, in his trilogy that opened with The Landscape Garden, educated his American readers to arguments in Britain and Europe regarding the Picturesque, Sublime, and Beautiful. Without them realizing it, he prepared his American readers to understand the forms of art coming to America.

*Those who wrote about nature in the eighteenth and nineteenth centuries negotiated through their writings not only their relationships with their readers, but also their relationships with God, and with nature itself.*
*(Jordanova, 1986, 21)*

*Loren [Eiseley] was criticized by his scientific colleagues for…venturing into the field of literature and appealing to the general reader. Other anthropologists had their careers hurt for similar writings: Margaret Mead for Coming of Age in Samoa and, more recently, Colin Turnbull for People of the Forest.*
*(Heuer, 1987, 160)*

The particular ways in which cultures shape their landscape, architectural, and interior vocabularies become identified with styles such as Romanesque, Baroque, Elizabethan, Victorian, Modern, Post Modern, Zen, and West African Swept Yards (Westmacott, 1992; Raver, 1993). Crafted pieces and landscape spaces, like most written works, are specific to an individual people, region, and time. As designers, our boundaries of understanding expand through observation and interpretation; our experiences over time allow us to become familiar with a place's vocabulary, syntax, and grammar. What we know about a place, our ability to accurately read constructed places, has the potential of expanding well beyond the period in which, and for whom, they were produced (Deetz, 1977). An archeology of kitchen dumps has enlarged our knowledge of the working spaces of historic Williamsburg, Virginia and the Black neighborhoods of Annapolis, Maryland (Leone, 2015). Archeology and anthropology are two ways in which we learn to read what landscapes can tell us (Yamin and Metheny, 1996).

As designers the narrative structure we carry about with us is both our resource and our baggage. Our ability to recognize it and hold it at bay while coming to know a people and a place is a skill rarely taught in design schools. Philosophers call this technique "bracketing," that is to identify through self-reflection personal values and attitudes and then put them aside. A designer may be called upon to produce a design he or she would not want to live in but its users would be excited to live with and care for. Developing this skill of holding one's value-laden landscape texts at bay is helped not only by anthropologists and archaeologists who delve into the past, but also by those who do more contemporary work in applied anthropology. Major differences exist in vocabulary, dialects, and diction between and within cultures in spoken and landscape languages. During my last visit to Germany, many older Germans were bemoaning the fact that their regional dialects were disappearing. The reason given for the homogenization of their language was the radio, television, film, and music industries' use of High German. This raises the same question asked by Sharpe and Wallock in their *Visions of the Modern City* about interpreting modern urban landscapes.

*There is, moreover, an important temporal dimension to the problem of seeing the city clearly. How can the contemporary critic, a product of his own time and of the very forces he seeks to objectify, manage to cross the hermeneutic horizon separating him from the historical moment of his subject? (Sharpe and Wallock, 1987, 9)*

Many say it is impossible for us to read historic landscape texts.

When my University of Guelph students worked with an Ojibwa Cree band in Northern Ontario we often heard, "How can you design for them, you can never become one of them." My response was that the critique was valid, but by working with the Cree, by reading their lore, asking questions, and observing their behaviors in nature we could bring our design talents more closely into line with their personal, tribal, and cultural values (Ndubisi, 1988). The results of our efforts would be a lot closer to an expression of the other's structures, landscapes, and interior vocabularies than if we had not taken the time to be open to their environmental vocabularies and done our best to verify their spatial vocabularies through workshops and charrettes with the Cree.

## NARRATIVE STRUCTURE

The idea of narrative structure is not unique. Structure is found in the literature and practice of the sciences, humanities, arts, and everyday life. Whether seen in science (Capra, 1982), rules of written and oral composition (Guth, 1961), creative problem solving (Gordon, 1961; Prince, 1970), or the daily routines of life in a house (Wood and Beck, 1994), the structures are characterized by wholeness, transformation, and self-regulation (Chatman, 1990). In fact, once you start looking at books on urbanism, town planning, cinematography, writing, culture, consumerism, and just about anything you care to name you begin to realize that many of the works deal with structure in one way or another. Sure there are critiques questioning the structure discussed in each of the writings – Marxist, non-Marxist, Skinner, Jung, Freud, Duany, Calthorpe, Olmsted. Still, each is a way of understanding, and that is as much the key to learning to become a good designer as it is to practice good design. I no longer believe there is one right way to do design. Our goal is to find and become comfortable with a structure, maybe a mixture of approaches, that allows us to analyze, interpret, apply, and substantiate our expertise, one that works from within and does not impose outside values upon a people or a place. The Analogy is just one more structure to sample and hopefully adopt – at least parts of it.

Recognizing the possibility that landscape narratives have a structure opens up the potential of identifying a logical approach to doing landscape architectural design. A landscape's narrative structure is shaped by a place's theme, the activities it provides, and the user-reader's expectations. In developing The Analogy I learned that a design process based upon narration calls for a proposed landscape's narrative to lead the design process. After getting acquainted with the proposed design's user group(s), client's wishes, and local ecology and history, the designer-as-author writes a first- or second-person narrative based upon a landscape user's expectations. The narrative doesn't describe the expectations but allows us to follow the landscape user as he or she carries out activities that fulfill their expectations. Such a narrative leads the designer to select and orchestrate the activity spaces best suited to the site design's theme and at least satisfy, if not excite, the user. Like writing a letter or short story, the question (design problem) you need to answer directs your character's actions and, in so doing, gives you what needs to be said. Then you select and organize the particular grammar and syntax (word choice and message structure). There is a logic to this narrative-storyboard-design process that becomes evident in the built product and, in turn, in the narrative read by its users. When I say "logical approach" I infer two things. First, we are a step closer to that elusive formula I've been looking for: the writing of a first- or second-person narrative + its interpretation into a storyboard = an experientially based design. Second, and more importantly, this is an approach to doing design that can be replicated and in doing so culminate in good designs time and again.

Well-planned narratives begin with a first word, an opening scene, an initial musical note. Selected carefully, the first word sets an impression, a mood, a situation, and, we can say, a direction. This is both the power and the responsibility associated with starting points. A critical question that needs to be answered is, where, as one approaches a designed space, does she encounter the first message? Does it echo the mindset with which she approaches the site: determination, introspection, gaiety? Where does she get her first impression of the design that sets the stage for so much that follows? How does it develop from there on? Having the last word, the closing experience at the terminus of a series of spaces acts in a reciprocal manner. Their power and reciprocity between first and last experiences rely on the landscape user's ability to draw together the site's exit point with what came before. In film terms the user is looking for closure and in environmental design terminology, a continuity across the site. In terms of a lasting impression, I often asked a student struggling with a design, "It is three days after someone has visited your site. What would be the title of the article they write in a local paper about your design?"

What is being called for here is "Programming," specifically a programming of the landscape users' experiences that, through a first- or second-person narrative, help you give form, character, and content to the proposed physical environment. Developing a strong program that addresses people's experiences strengthens your understanding of the users, the activities they'll likely perform, and their expectations. Instead of numbers of people to be accommodated and the quantities of spaces and materials that need to accommodate them in particular activities, we consider a design program that anticipates people's experiences as an expression of their needs. Considering the quantities and areas that will be needed is dealt with after you have seen through your user's eyes what the site looks like.

This is the kind of background research needed to write a good narrative. It provides for "the definition and analysis of human needs" (Motloch, 1991, 296). A narrative's building blocks, from the smallest unit to the final chapter in a volume, interrelated in ways that provide a wholeness, a completeness, a continuity of flow, and a sense of closure. With this background you are prepared, as Rae says, to partner with your characters, to empathize with them. "See the world through their eyes, touch it with their hands, smell it with their noses…to partner with our storymaker" (1996, 5).

The idea of partnering "with your characters" is critical to good design. Good design is not haphazard, it is not an I-like-it, I-want-it, kind of thing. Design narratives, like storytelling, draw their structure from (1) the functional practicalities of people's movement; (2) their social norms, or what is deemed acceptable to a people's laws, religions, or collective values; and (3) their expressiveness, or what they find aesthetically pleasing. Guided by your character's movement, social norms, and aesthetic appreciations, you, as designer-author, make, with your narrative characters' help, decisions as to what messages to communicate. With each decision you begin to limit the range of possible spaces, spatial dimensions, design elements, and organizational patterns that may be used. Rubin refers to this as a combining of constraints (1995, 90–121). The recognition of a design's message-delivery domains, and the selection of a thematic category across those domains, immediately helps guide a designer's selection of the first word – whether spatial or material – as an entry into the eventual narrative.

This process of becoming familiar with your users is helped by Habermas's Domains of Knowledge (1979). Considering the domains below helps us understand if we are being honest in how we are expressing our design thinking. I can't talk about the social implications of landscapes if I use the language and thinking of technology. Using the language of geometry does not allow me to express landscape users' sensations and appreciations. Understanding the Knowledge Realm in which you are working makes the interpretation of landscape narrative into a storyboard easier.

When I first became familiar with Habermas's Domains of Knowledge, the triad that brought me the most clarity was The World – Our World – My

| Technology | Society | Expression |
|---|---|---|
| Objective | Normative | Subjective |
| Fact | Consensus | Experiential |
| The World | Our World | My World |
| Theory | Practicality | Self in culture |
| Technical Interests | Practical Interests | Natural Interests |
| Technical Discourse | Practical Discourse | Expressive Discourse |
| World- as- it- is | Institutions | Reflective- self |
| Physical World | Social World | Inner World |
| Value- less | General Values | Essential Values |
| Instrumental Action | Communicative Action | Evaluative Action |
| Science/Technical | Legal/Moral | Aesthetics/Expression |
| Means of Production | Ends | Self- reflection |

Figure 7.12 Three domains of knowledge.

World. Do take time to go through the 12 triads below. It is worth your time, energy, and discussions that emerge.

## A PERSONAL EVOLUTION

*The Dutch [paintings'] scenes require a gradual and complete reading; we must begin at one edge and finish at the other, audit the painting like an accountant, not forgetting this corner, that margin, that background in which is inscribed yet another perfectly rendered object adding its unit to this patient weighing of property or of merchandise.*
*(Barthes, 1982, 67)*

Harry Porter's question to me back at the University of Virginia years ago of how The Analogy could be applied to design stayed with me for years. Looking back, I came to realize that as I continued working to answer his question I became more and more aware of the social, cultural, and personal meanings and values that are simultaneously overlaid and embedded in all environments' physical features. What I came to understand as a landscape's potential to communicate messages, to simultaneously build and bind places and people, took on a dynamic, interactive character. I began to question the structural dynamics influencing the selection and organization of the objects to be included in my designs – a process that had previously remained unquestioned, unexplored, subtly unconscious. In my search to find what would define the act of "doing" design, I found guidance in writings on narrative form and analysis of film (Chatman, 1980), poetry (Conron, 1974), and painting (Novack, 1980). My eventual introduction to communication theory (Giddens, 1984; Habermas, 1979; 1981) and critical theory (Habermas, 1970) squarely set my interpretation of designed products into their socio-cultural contexts. Whether landscape or architectural, personal, or cultural, space is a social product (Lefebvre, 1976; 1979; 1994). Understanding the built environment as a social product was the first of a number of my realizations of the daily interplay between people and their physical surroundings and the languages and abstractions designers use to record and recreate those same surroundings. Lefebvre's statement that space is a social product finally allowed me to comprehend John Simonds' realization that as designers, we "plan experiences. The places, spaces, and things take their form from the planned experience" (Simonds, 1961, 225). The structure embedded in the idea that as designers we author experiential narratives was becoming clearer in my mind.

For a start, as the authors of designed narratives we design in two realms. Those realms both deal with the transmission of information, but one does so through the expression of the landscape narrative, while the other does so through the manifest form of that which is being expressed. Chatman (1980, 23) recognizes this distinction as story (the message(s) being expressed) and discourse (the things expressing the messages). This distinction is echoed by Conron respectively as tenor and vehicle (Conron, 1974, 3). Another distinction is between what the designer/author intends and what the user/reader interprets, or as you read earlier, the closing of the gap between what is sent (authored) and what is received (read).

On one hand, those new to design tend to "willy-nilly" claim they know what a design should include and with that belief seemingly gather together what they consider as necessary landscape features. Not that they don't argue (vehemently) for their design decisions; they do. All too often each decision is grounded in its own immediate and unique circumstances thereby bringing a discontinuity across the array of parts making up the whole. On the other hand, writing a first- or second-person narrative depicting an individual or group's experiences in and around a design brings an inescapable continuity to the landscape features and spaces. By continuity I mean a smooth flow in your narrative character's experiences as he moves toward, into, and through your design. Unless your character is in an episode of Star Trek he won't be beaming in and out of disconnected settings. To a great extent, writing a narrative brings with it a high degree of continuity, which is reinforced with follow-up discussions with those you asked to critique your narrative.

The act of narrating a story or stories that run throughout a design brings a high degree of continuity to the design process and to the resulting designed spaces and spatial definers. A well-authored landscape narrative provides a logical progression of experiences that direct the selection and location of objects you will apply to the creation of spaces.

> *"Narrate" is from the Latin narrare ("to tell") which is akin to the Latin gnarus ("knowing," "acquainted with," "expert in") both derivative from the Indo-European root gna ("to know") whence the vast family of words deriving from the Latin cognoscere, including "cognition" itself... whence the Modern English "know." Narrative is, it would seem, rather an appropriate term for a reflexive activity which seeks to "know"... antecedent events and the meaning of those events. Drama [is derived] from the Greek dran ("to ado or act"); hence narrative is knowledge (and/or gnosis) emerging from action, that is, experiential knowledge.*
>
> *(Turner, 1980, 167)*

In developing that continuity through the selection of things that constitute a designed landscape, structure, or interior there are five thematic categories that may be used to both analyze an existing site and synthesize a new site. Four of the ways to interpret inhabited space are derived from a course on American Landscapes taught by John Conron in 1985 at Clark University in Worcester, Massachusetts. They include the categories of (1) function, (2) history, (3) style, and (4) symbolism. The fifth category is what I call the affective or emotional. Used to guide the application of narrative in landscape design, in conjunction with a unifying viewpoint (the user or users' expectations), these thematic categories provide design decisions with a coherence, a harmony in the developing message content of an interior, architectural, or landscape architectural design. While these ways of interpreting landscapes worked for Professor Conron (1974; 2000) and Meinig's "Ten Versions of the Same Scene" (1979) worked for him, my three categories of narrative expression – function, social/cultural, and archetypical – situated under the umbrella of emotion work for me. It would be worthwhile to review those works and decide what categories or combination of categories work best for you.

> *Narrative always says less than it knows, but it often makes known more than it says.*
> *(Genette, 1980, 198)*

While both the story (the message(s) being expressed) and the discourse (the things expressing the message(s)) communicate meaning, the information expressed through the meaningful objects (discourse) is the more explicit of the two. The selection, organization, and detailing of objects, and the spaces they define and populate are read as explicit expressions of what can and cannot be carried out as an activity. A curb with a six-foot radius displays the arc bicyclists, rollerbladers, motorcyclists, and automobile drivers have to work with and at what speeds they can and cannot make a turn. We read function, and immediately determine whether we can execute the intended behavior/skill we need or want to. Although I admit young kids can play pick-up games of street hockey or stickball just about anywhere, there are still limits to the minimal space and shape of space needed for a descent game.

Three similarly shaped and dimensioned walkways may allow the same number of people to use each one. But, comprised of different materials each walkway expresses not only its use but also its social character and cultural setting. Is it a casual walkway, a thoughtful artsy walkway, or a highly efficient military walkway? Parking lots and parking experiences are part of funerals, carnivals, government centers, grocery stores, universities, and more. Each supports the mechanics of driving and parking while calling for its socially appropriate behavior on the part of the participants. One may say "prepare to have fun" another may say "respect those who came before you." With this in mind, the communication of explicit functional messages is dealt with shortly, followed with a look into the deeper meanings conveyed through a narrative, the meanings that evolve through the experiences one has in a place as reflected against a blend of their cultural, familial, and personal heritages.

Before each topic is explored, the delivery of both explicit and implicit meanings through a landscape narrative is approached in two ways. First, the messages' clarity, as determined by the designer-as-author, is considered with regard to the public's health, safety, and welfare. Second, the role of the design user-as-reader is considered. Both means of understanding a potential design's appropriateness, or fit, recognize that a landscape, a building or buildings, and interior spaces introduce a relationship that binds the designer not just with the client but more importantly with the design's users. The relationship is seen as the production, transmission, and receipt of readable messages. However, one experiences a narrative, he or she responds with an interpretation: "they cannot avoid participating in the transaction" (Chatman, 1980, 28). People come to designed spaces, for that matter any spaces, for a reason. First, and foremost, that reason is to use the space in a particular way for particular reasons, to be given direction in how to use it, and to feel comfortable and safe while using it.

That particularly interactive, interpretative process of the user-readers gains clarity with the help of an understanding of oral history (Rubin, 1995) and the narrative structure employed in fiction and film (Chatman, 1980; 1990). Bringing that understanding of story and discourse into the lived-in world of designed products and processes is helped even more with Giddens's structuration theory (Giddens, 1979; 184) and the analyses of spatial readings by such geographers as Duncan (1990), Barnes and Duncan (1992), Duncan and Ley (1993), and Wood and Beck (1994). These are a few authors who recognize that spaces are read and that while there may be multiple interpretations there are limits to the interpretations a reader may make of a given text. They also recognize that by determining a theme design boundaries are introduced within which their design (narrative) decisions are to be made. The act of doing design, of dealing with what seems at times to be an infinite number of decisions to be made, is really not as hopeless as many students believe as they exclaim, "I don't know what to do." "Oh, my God, how can I make all those decisions?" The act of doing design is not the wide open anything goes world that so many design students believe it to be. If a design is to be readable, coherent, and cohesive a designers' creativity must work within certain contexts and, as a consequence, within certain parameters. Creativity is not a license to do whatever we please. In authoring a text, within a given theme, what is to be communicated, how it is communicated, and the spatial and temporal sequences used each gain their organization from the "recursive character of social life" (Giddens, 1979, 41). In other words, contained in the repetitive nature of our daily activities and expectations is the determination of what preceded it. This understanding is found in Rubin's (1995) "constraint,"" Chatman's (1980) "reading out,"" and Rapoport's (1982) "cues," all of which you will shortly be asked to use in your development of a narrative.

## WHOLENESS

Whether met or left unmet, what people anticipate as they move through a landscape, what they find, and the experiences they have contribute to a sense that a design is whole or complete. Whether moving from room to room, building to building, or city to city, within the realm of anticipation people envision typical types of spaces and spatial cues that will make up their destination, and, to some extent, those spaces they encounter along the route to their destination. How do you "know the way" or know when you've arrived? People anticipate particular landscape, architectural, and interior spaces that comprise their approach, arrival, entry into, progression through, and exit from a medical center, a university campus, or an urban or natural park. If we don't find the anticipated signs and signals, if the expected messages are not there to be read, we often feel lost, or at least disoriented, or find ourselves walking toward a building's loading dock rather than its main entrance. Yes, the GPS on our phone can get us there. But I'm sure most of us have looked up from the phone's screen to where we are and thought, "This can't be right." The surroundings are what we expected.

Within our landscape narrative experiences, movement through spaces, reading of a design's features and organization of the features that make up institutional, personal, and wilderness places are not easily interchanged. Their relevance to a particular place or activity setting is an aspect of their wholeness. To substitute the narrative features read as an approach to and arrival at an Atlantic City, New Jersey or a Reno, Nevada casino for those of a medical center or church would likely confuse a lot of people. While the spirituality of hallowed grounds may include the same tree and shrub species and paving materials as a complex of government offices, their narrative organization would be quite different. Their composition as related to the story to be told and the context in which it is being told is the heart of the recursive character of social life, that what we expect to be able to read (identify and find meaning in) is determined by what has come before. We know what we know, and anticipate what we anticipate, because of our past experiences. Our daily routines, our repeated experiences, refine both our understanding of a place and our expectations of related or seemingly similar places (even in foreign countries). Fulfilling our expectations as we move about new places can give rise to comfortable encounters with those new places. Driving the suburban neighborhoods of Montreal is unnerving until you come to expect stop signs located in the middle of a residential block. Even though the red octagon sign is out of place, for many first-time visitors, their learned reading of the shape's message from previous experiences, and the deeper symbolic meaning of authority related to the object, gives rise to quick glances in the rearview mirror as you drive through the first few mid-block stop signs you encounter. Yes, I am speaking from experience.

The delivery of information, in a storytelling form, provides a designer with an opportunity to orchestrate or choreograph narratives on three levels. It is quite helpful, prior to writing a landscape narrative, to include the following in your research: functional program, sense of place, and social attitudes and values. Most would say design starts with an outline or definition of the design problem. Typically, the listing of a client's wishes and the site's needs is referred to as the Program. For an excellent overview of what makes up a design program, see Lynch (1971, 28–31; 260–262). Predominantly quantitative, the program notes the required needs to be fulfilled by a design. In doing so, it introduces the first domain of constraints along with the first of the three primary structural domains of narrative. The domain of functional narrative notes the human and mechanical functions, and the quantities of people and things, to be accommodated by a design. Determining things like numbers of vehicles and their sizes, turning radius, number of people, tables and chairs or benches needed to accommodate the activities for which a design is being done brings with them certain constraints. Qualitative aspects like the character of the movement (fast, slow, smooth, broken; comfortable; exciting), the nature of the gathering spaces (business group, intimate, solo), and the personality of the place (introverted, extroverted) also introduce constraints to the size, openness, shape, orientation, and sequencing of spaces. These aspects begin to introduce a palette of materials and finishes from which detailed design decisions are made later in the design process. For now, they set the foundation on which your narrative's character will move about their sequence of anticipated experiences.

The second structural domain, the place-making aspects of social norms, deals with the communication of socially accepted, and socially appropriate, messages. This and the third structural domain, that of deeply held collective beliefs, is dealt with below. Consider adding a new dimension to your program, one that includes the users' anticipated experiences and their sequencing. Again, this will help you when you let your narrative's characters lose to enjoy the design.

In the course of your lifetime as a designer you may work to provide for two or more design programs that are the same. It wouldn't be unusual for two elementary schools, in two different parts of the country, to have the same requirements for interior and exterior activities, and the same functional requirements for teaching, administrative organization and oversight, and recreational activities, to say nothing of natural and artificial lighting. Yet, set in a New England town, the resulting design would (I hope) appear materially different from one located in Taos, New Mexico or Moose Jaw, Saskatchewan. The domain of social norms, what is considered, accepted, and promoted by local residents as standard or typical spatial character, overlays the functional domain with another set of constraints.

In that space is a social product, a volumetric product defined by people's bodies in motion, those constraints primarily give shape, form, and character to the functional activity areas in your design. How do educators approach teaching and learning? How many students, at what level, are to be accommodated at one time? What are the parents' attitudes toward classroom learning? What would the kids like to experience? What style of education is promoted by the State Board of Education? What is legally permitted regarding science, health, recreation, and social events? These constraints again introduce both observable and measurable criteria that would influence spatial organization. And finally, the way in which the residents of a design express their sense of identity overlays a third domain of constraints. Knowing how teachers wish to adorn their rooms, express their character, or how clients typically relate to organized clutter or extreme neatness also influences a design's spatial and material narrative. The ways people personally and collectively express themselves call into play the finishing touches to the building's original structural organization. What kinds of spaces prepare one for a chemistry class, social studies, math, or physics? And what kinds of experiences leading up to particular parts of the building or surrounding landscape would prepare students to be learners?

Identifying the likely social groups that will experience your design is important to preparing your narrative to be authentic, true to the users. Willis argues that

> *the particular meaning and scope of the role of institutions in the reproduction [of social attitudes and values] may be less to do with their formal nature and manifest communications than with the unintended and often unseen results of their relationships and habituated patterns of interaction.*
>
> *(1977, 176)*

This view is reinforced by Sieber's recognition that "informal processes contribute to the pupils' learning of formal social behavior" (1979, 274). Sieber's study shows how many of us experienced early education as both explicit/formal and the more subtle and informal learnings. While Sieber focuses on kids' socialization through informal interactions with teachers and parents, N. B. Johnson (1982) argues for how the spatial character of preschool to sixth grade classrooms acts as a "hidden curriculum" socializing kids to sociocultural norms.

Think back to how your early learning environments (preschool) went from pretty much unstructured space, with seats around the perimeter of the room, to kindergarten to first grade's partially unstructured spaces with a few chairs/desks facing "the front" (teacher's area and blackboard), to no unstructured space and chairs and desks aligned and "facing forward." Recognizing the differences between what is taught and what is learned necessitates giving attention to the communicative dynamics of the educational arena – the informal aspects of formal education. You face front toward the authority. You tend not to answer questions unless you know the answer. Speculation and creativity are not rewarded. I always enjoy the story of a friend and his cat. Every time the cat would scratch, or worse climb, the living room drapes my friend would put the cat outside. One day I asked, "Have you ever considered that maybe when your cat wants to go outside it scratches the curtains?" And we think we train them.

An excellent example of how the three realms – physical sciences, social sciences, arts and expression – of message delivery structure our "doing design" is found in landscape architects' rainfall runoff calculations. Consider the interplay of the three areas of knowledge in the application of Manning's formulas to a drainage problem. Scientific data, derived from hydrology, soil mechanics, botany, and climatology, provide ecologically sound solutions for the removal of surface water. Manning's formulas are universal; they can be successfully applied anywhere on the surface of the earth. The result of their application is a technically adequate drainage swale that safely collects and removes specified amounts of surface water. The application of a technically efficient drainage swale to a cornfield in Iowa, the grounds of a French chateau, or a Hopi community introduces design concerns outside the mathematical formulas' capabilities. Once technically correct, the swale is then required to satisfy the second domain's criteria of social appropriateness. Social context introduces an array of norms associated with local politics, attitudes, and special interests to be recognized in the design: the appropriateness of the swale's location, course, and configuration. Neither Manning's formulas nor Manning's charts can be used to validate decisions made relative to local social norms, values, and history.

Within people's everyday lives, the acceptance of a technically adequate, socially acceptable drainage swale greatly depends on local residents' subjective expressions or aesthetic appreciations. Here Chatman would distinguish between viewpoint and point of view. Local approval of a swale's surface materials, crosss—ectional shape, edge as it meets existing grade, and rate of transition to adjacent land uses and landscape character are expressions of local people's social or group sensitivities. Such appreciations of the immediate setting are an experience of the solution's "character" and "fit." Although part of a larger flood control project, the reworking of Englewood, Colorado's Little Dry Creek provides excellent before and after examples of aesthetic expressions of a public land use (Grove, 1990). Other examples are found in Brian Howard's National Geographic article "Buried Streams See the Light of Day" (2014) and Pinkham's "Daylighting: new life for buried streams" (2000).

*Nearly a hundred years ago a stream was covered over and diverted into an underground culvert in Northwest Washington, D.C. Now environmentalists are taking steps to restore the stream to a more natural flow and revive the native flora and fauna. (https://www.youtube.com/watch?v=0cpl4Oii95U, last viewed March 2022)*

A further narrowing of the boundaries within which a design's decisions are made occurs with the introduction of a theme or themes for a design. Themes may express period, place, or style: Colonial Pennsylvania, modern Dallas, Texas, Arcadian, Nova Scotia. Each has historical precedents that may be drawn on and replicated with some ease through the retention of a place's historical character. This approach, the identification and implementation of a place's local architectural and landscape character, contributed to what is referred to as "Performance Zoning" (Kendig, 1980). Another set of themes introduces the affective or emotional experiences of joy, contemplation, strictly business, and the like. These again have some standards, but typically require the designer to be alert to the character of the clients' current environment, how he or she expresses ideas, how they move about, and what their resultant spaces look like. There will likely be differences in density of plantings and the plants' visual permeability for a happy extrovert versus a happy introvert.

A designer's ability to identify the constraints introduced within each of the domains of knowledge, and the thematic realms, reflects on the breadth, depth, and character of the designer's formal and informal education. A working knowledge of the sciences, both physical and social, and an understanding of the arts as a communicative medium make for a wonderful foundation. At this crossroad is the heart of the design professions' frequent claims to being "the art and science of design." This, however, is a whole other debate, one best taken on by educators in conjunction with practitioners, and, I would suggest, futurists. A starting point is the Carnegie Foundation study, "Building Community" (Boyer and Mitgang, 1996). Additional fundamental information is found in the works of Forester (1989; 1985), Cuff (1991), Baum (1983), and Scarfo (1989).

## RELATING NARRATIVE STRUCTURE TO THE DESIGN PROCESS

Authoring a narrative across a landscape, determining the objects, spaces, colors, and so on from within the rubric of what users expect to find contributes to a design's wholeness. When people meet they share stories. I've asked hundreds of newly arrived freshmen at different universities how they got to know each other during their first days together. They quickly conclude that storytelling was the major medium. They recognized: how they expressed their individual personalities through shared personal stories; how they discovered commonalties among themselves; and how they used the stories they heard as a means to select the people with whom they would be most comfortable being around.

When we initiate the telling of one (personal) story over another our selection immediately brings into play a palette (or family) of words to choose from, displayed facial expressions, and body language. Taken individually and out of the context of the story being told, the facial expressions and behaviors may appear to be those of someone experiencing a seizure or epileptic fit. Try watching America's Got Talent or The Voice with the sound muted. But, organized through the overlay of the story the person is telling, their body language, facial expressions, and more become coherent and meaningful. In a like manner, a story told in a neutral tone without facial expressions or audio inflections would seem odd and difficult to understand. This interplay or orchestration of multiple senses is equally important to the authoring and reading of a landscape narrative.

Difficult as it may be to imagine design is a narrowing process. Taken as just so many dimensioned and measured individual items in the thousand-plus pages of *Architectural Graphic Standard* (AGS), the construction details and design specifications acquire a coherence when filtered through a design's identified program of needs. The particularity of the activities selected for a site design not only introduces a helpful narrowing of constraints, a limiting of the landscape features to be included in the design, but also brings with a continuity to the organization of things selected for inclusion in the design. In turn, the locations, sequences, and interrelationships give the organization of the details a coherency across space and through time. As a landscape narrative is "written" for a majestic overview, a forecourt to a photographer's studio, or an entryway to a historic district, the story being "written" acts to filter the multitude of diagrams found in AGS down to a particular few. The same filtering process occurs as we move our design ideas from concept diagram through a schematic to a final plan.

Without the guidance provided by imagining a design's participants and writing a narrative of their behaviors and experiences many design students take the wrong fork in the road. Too many students design by selecting objects because "they look good" and then argue for their goodness based upon poorly informed opinions. Even the design of a parking lot can often be heard to be argued from the position of "I liked," and "I wanted," and punctuated with the dimensions for a parking stall and the turning radius of a car. Designers' faulty reliance on technological and geometrical reasoning in their substantiation of a design for people is echoed by Baum (1983, 89–90,127–130), Forester (1985, 206–220), Cuff (1991, 44–45, 66), and Scarfo (1989, 176–182).

> *What architects want us to hear about design practice often tells us more about beliefs and ideals than about the principles that guide action, or theories-in-use.*
>
> *(Cuff, 1991, 20)*

> *...school [design] problems and those of practice... The school disembodies the primary professional activity of design from its context.*
>
> *(Cuff, 1991, 66)*

One approach to taking the correct fork in the road comes with young designers becoming part of their design through their narrative's characters' thoughts and actions, and becoming a participant in their design as a walker, bicyclist, driver, or passenger. Then, after writing that first- or second-person experiential narrative, the technical criteria and dimensioned layout of chosen construction details can be organized into a coherent progression appropriate to its setting. Please reread the previous two sentences. Think about it. Your design process answers broad, general questions first and later the more specific ones. The functions to be accommodated by a design within the domain of the activities sought as the users' expectations are where the wholeness of a design resides. The wholeness of a design's narrative is found in the appropriateness of what is selected for inclusion in a design. Given a potential user's perspective through a first-person narrative, the selected items are organized and articulated to reflect the continuity of a design across and through its spaces in a storyboard.

> *What if all drawing were taught as diagramming, as merely a sequence of increasingly elaborate graphic memoranda? …graphic communication can be broken down into levels parallel to the normal design evolution/communication process: conceptual diagramming, pure diagramming, generalized diagramming, and specific diagramming.*
> *(Wester, 1990, 28)*

As a side note, help in exercising our visual and graphic thinking in the three structural realms and selected thematic categories is found in a variety of graphic communication and visual thinking books. Two approaches to a clearer understanding of how to test and turn original ideas into functional diagrams and then into final designs are found in Lari Wester's promotion of diagramming (1990, 28–29), thinking conceptually, and Paul Laseau's use of a graphic language (1980, 51–63). Wester's approach helps overcome the belief that either you are born with good graphics or you are not. Laseau complements Wester's levels of diagramming by echoing the fundamentals of The Analogy: "the graphic language proposed here has grammatical rules comparable to those of verbal language" (Laseau, 1980, 51). In being able to graphically express (storyboard) the written narrative of a design's users the potential of authoring the design's landscape narrative comes a step closer to reality. This help in authoring a landscape narrative becomes apparent as you reflect your written narrative off your storyboard. That back-and-forth process between word narrative and visual narrative makes clear the value of being able to express yourself well as a means of seeing your design clearly.

## TRANSFORMATION AND SELF-REGULATION

Two other characteristics of narrative, in addition to wholeness, are found in Piaget's concepts of transformation and self-regulation. They refer to aspects of cohesion in narrative. In Piaget's words the "transformations inherent in a structure never lead beyond the system but always engender elements that belong to it and preserve its laws…" (Chatman, 1980, 21). This means that when you decide to vacation in Florida during Spring Break you'll bring a wardrobe and accessories for Spring Break in Florida and not skiing in Aspen. Yeah, sounds pretty silly. When you think about it, it all sounds pretty simple too. It is simple. Design, as practiced through The Analogy, calls on storytelling which we all do every day. Make a decision to vacation in Florida. Look in your closet, go through your dresser, and select the appropriate attire and accessories. Having chosen Florida the content of your decisions never leads beyond what your Florida trip calls for. The gloves, woolen socks, down parka, and thermal undies stay home. What you've done isn't much different from outlining a design's story, using it to lead you or your character to the contents and character of a physical design, and then use those decisions to filter the appropriate construction details offered in *Architectural Graphics Standards* or a construction manual. The transformational part is the progression from idea to substantive product. But again, those are detailed decisions made further along the narrative-storyboard-design process. When a design's narrative is laid out and is appropriate to a place, its activities, and the users' expectations, the parts of the design's messages are most likely transformed into shapes, forms, colors, textures, and more that all fit the place, provide for a story, provide for a cohesive sequence of experiences, and maintain a stability throughout. Users' expectations are fulfilled. There is a sense of wholeness.

Within the idea of traveling to Florida, your decision helped you select items appropriate to the trip and your stay. The same selection rationale applies to those decisions you'll make at the more aesthetic end of your travel planning: decisions like what is "in" this year on the beaches and not "should I dress in my parents' swimsuits and party wear?" A decision to do one thing over another directs you to opportunities, and here again it is worth remembering Piaget's thought that "the transformations inherent in a structure never lead beyond the system but always engender elements that belong to [the system] and preserve its laws" (Chatman, 1980, 21).

*This holistic view has the landscape as a system consisting of interdependent subsystems. Elements of the landscape are seen as expressions of, and cues for understanding, underlying processes...[the meanings of which] are to be accrued not primarily from elements, but from the interrelationship of those elements with one another, with their behavior as systems, and with the processes that have created them. Landscape as system also holds that elements holistically express the various systems of which they are part.*
*(Motloch, 1991, 14)*

As for preserving the laws of "a system," first determine which system and which laws you will work with during your design decision-making process. Remember the three realms of knowledge to which we constantly respond: science/technical, social/normative, and arts/expressive. Each has its own database and employs its own related problem-solving methodologies (Habermas, 1981, 157–185; Scarfo, 1989a). In a sense, each has its own rules of the game. The rules are like those that we call etiquette, or that Wood and Beck recognize as the codes that come into being through our "consciousness of what does and does not satisfy [our wish to be comfortable]" (1994, 24). As you move beyond your narrative to the greater detail found in your storyboard, a blend of Maslow's hierarchy (1970) and Manning's formulas become easier to apply. The systems we strive to satisfy, and I argue that we read as the primary landscape texts, are related to our application of Manning's formula (Lynch, 1971, 170–171). First, deal with the basic laws of the world. The physics and chemistry of climate, topography, and even our sustenance and physical structure set the parameters within which we select types (not styles) of footwear, clothing, and playthings. With our biological and physiological comforts cared for we then deal with and read the environment in search of the applicable social or normative laws that will ensure our personal safety and acceptance: "what's socially acceptable and expected" for this time and place.

*...in practice ecologists look for a relative homogeneity enabling them to characterize and understand the ecosystem with a reasonable number of measurements.*
*(Forman and Godron, 1986, 14)*

The civil laws of Jacksonville or Daytona Beach, Florida (remember when you used to be able to park a car and sleep on the beach?) influence both our selection of accessories and the activity settings in which we will use them (ten people in a two-person motel room). But there are also the social laws of the pack. We live in peer groups. We strive for our peers' respect while acting out our beach front, boardwalk behaviors. We react to our peers while testing just how much the local residents (versus the merchants) will tolerate during the very lucrative Spring-Break week. In the summer of 1993, Ocean City, Maryland reacted to the difference between the local residents' intolerance for loud "noise" and the vacationing students' craving for loud music. The legal response was the introduction of noise ordinances. People were arrested and parties broken up when the music could be heard more than 50′ outside an apartment or house. Seriously, 50′? What self-respecting student on Spring Break doesn't want to let the world know what's contained in his or her playlist?

Further along the narrative-storyboard-design process, as your decisions call for greater detail, you find yourself increasingly clarifying your design's context and contents. A design's context, both physical and social, and the design program to which the context gives rise provide the parameters within which we work to transform ideas into built products. That transformation, of turning our narrative into a storyboard and then a landscape design, typically occurs in three stages: program, theme, and expression. First, the design program's transformation of ideas into physical forms is determined: number of people to be accommodated, square footage for each activity setting, quantities of materials, and related acceptable costs. Second, the organization of those physical forms is laid out: passive to active, personal to public to institutional, and so on. To remain in context, to both support and be supported by its surroundings, and best ensure a design's readability, the organization of the parts of a design follows the rules, or codes (social norms), associated with the design's institutional and public contexts. Institutional codes organize activities, how they are arranged across a site, and their defining elements: e.g., approaches to churches, entry into them, and movement through them as a ceremony (Wood and Beck, 1994). Japanese gardens often provide previews of distant destinations upon entry then take them from your view so you can remain more focused on the immediate experiences but with the knowledge of where you are going. The institutional formats, socially derived through the cultural heritage of a place and its historical background being designed for, are much like the formats associated with particular types of stories: mystery versus romance, documentary versus humor. Consider such places as Salem, Massachusetts, especially around Halloween, Savannah, Georgia's Historic District, Washington, DC' National Cathedral, or any one of hundreds of amusement parks.

I find it easier to think of the system as a family of things that come together and work together within a given topic. Botanical garden, riverside dining, wooded picnic area, or lakeside beach, they all have their own particular words, paraphernalia, and procedures associated with their activities that can't be interchanged without violating their respective system's wholeness. The words, paraphernalia, and procedures are the narratives into which those systems are transformed.

The third structural domain giving form to a landscape narrative is that of expression. Habermas identifies that domain of knowledge and experience with aesthetics and the arts. How we come to express our environments draws on our personal and our collective idiosyncrasies, our tastes. Everyone may wear their hat backward this year but which hat to wear, our school's hat, that of a sports team, a favorite bar or association we support? How people interpret local ordinances and rules, whether in church, school, or work, is how they chose to express themselves as individuals within the greater norm. Again, remember, there is plenty of latitude for personal and cultural expression within what at first glance seems to be a totally restrictive number of constraints. This personalizing of social norms, through the transformation of physical and social laws into desirable behaviors, introduces the aesthetic expressiveness of the arts.

As a reminder, you need to do research prior to writing a first- or second-person narrative, a story of someone or group approaching and experiencing your design. As noted in Chapter 8, Authoring a Landscape Narrative, all your background work will prepare your narrative's characters to experience your design.

## PARTS OF A NARRATIVE

Whether an everyday routine that has you getting from point A to point B safely, or entering into a more personally moving, meditative or contemplative place, narratives are comprised of and communicated through a blend of form and content. The interplay between form and content that gives rise to narratives that are cohesive and coherent, whether at the purely functional end of the spectrum or the sublimely aesthetic end, rests upon the existence of an underlying structure that holds together "the elements of storytelling, their combination and articulation" (Chatman, 1980, 15).

Chatman separates storytelling into the following four aspects of narrative and discourse. Remember Chatman defines story as the message(s) being expressed and discourse as the things expressing the messages (1980, 24). I'm sure you can think of at least ten things that say "Enter here" or "Don't enter here, go around." There may be a given message but various ways, appropriate to the place, to express it.

An analysis of the narrative structure's ability to communicate meaning is laid out in what Chatman calls a "quadripartite array." Understanding the array is helped with Chatman explaining how the "Story is the content of the narrative expression, while discourse is the form of that expression" (1980, 24). Think about it this way. You are asked to design an entryway to a park or football stadium. Both ask you to express "arrival." How you achieve that in terms of content, substance, and form in which substances are arranged across an area will vary.

Juggling all this in the production of a landscape narrative is much easier when you think about your final design as narrative of an individual or group approaching, arriving at, entering, moving about, and leaving it. In other words, you write a first- or second-person narrative. Rae's *Movies of the Mind* (1996) does an excellent job interpreting Chatman into lay terms, common

| | Expression | Content |
|---|---|---|
| **Substance** | **Substance of Expression:** the material nature of language elements - those things being read as meaning-full.<br>Visual objects, size, color, texture<br>Sounds, tinkling/light, resonating/heavy<br>Tactile, pressure, heat/cold, wet/dry<br>Smell, sweet, rancid, light, oppressive,<br>Taste, sweet, sour, salty, bitter | **Substance of Content:** the palette of thoughts and emotions common to people "independently of the languages they speak" (Chatman, 1980, 23; John Lyons, *Introduction to Theoretical Linguistics*, Cambridge, 1969)<br>-mental experiences<br>-people, things etc. as preprocessed by the author's cultural codes. |
| **Form** | **Form of Expression:** the particular arrangement (selection and order) of the material expressions that provide for narrative discourse - bringing the system together in particular ways - the structure of narrative transmission.<br>The ordering of material expression for a humorous setting vs. solemn vs. market. | **Form of Content:** the story line or "structure of relationships", the narrative story components as organized within the story line - humor, passion, caution...<br>Events- actions, happenings.<br>Existents - characters, settings |

Figure 7.13 The composition of narrative structure (Chatman, 1980, 22-25). As paraphrased by the author.

speak as it were. Rae (1996) supports my belief that wholeness and coherency are natural byproducts of the authoring of a narrative. Think about how many times you've met someone and started telling a story about yourself, one that expresses values in which you do or do not believe. Your stories have an introduction, middle, and end. They are not a disjointed sequence of seemingly unrelated thoughts. In much the same way, when it comes to writing it, your landscape narrative's character(s) cannot approach, participate in, or leave any activity setting without experiencing the appropriate objects and spaces and their appropriate sequence, scale, and orientation. Remember Piaget's thought that a character transitioning from one location to another can "never lead beyond the system but always engender elements that belong to it and preserve its laws" (Chatman, 1980, 21). Telling a story of someone approaching, arriving at, and enjoying a skatepark will, with very little effort, "engender" all the features making up a skatepark and not a corporate plaza or, for that matter, an elementary school playground. This is where your creativity is called into play; there are the shapes and spaces that make up a skatepark, but their sequencing can make up an okay park or a great one. As Rubin (1995) notes, the stories making a society's oral history may be told at different times with different characters, in different settings, but , let's say movie or theater of the mind, walk-away messages are the same.

Chatman's diagram of what constitutes narrative structure, the components shown below, introduces an approach to bringing narrative into the more traditional design process. In this way, a design process dominated by a view to the production of landscape spaces, their content, and spatial sequence becomes a process now balanced through the authoring of cohesive and coherent designs. This continuity across space and time is explored further toward the end of this section. For now, the following diagram summarizes Chatman's elements of narrative theory (1980, 26).

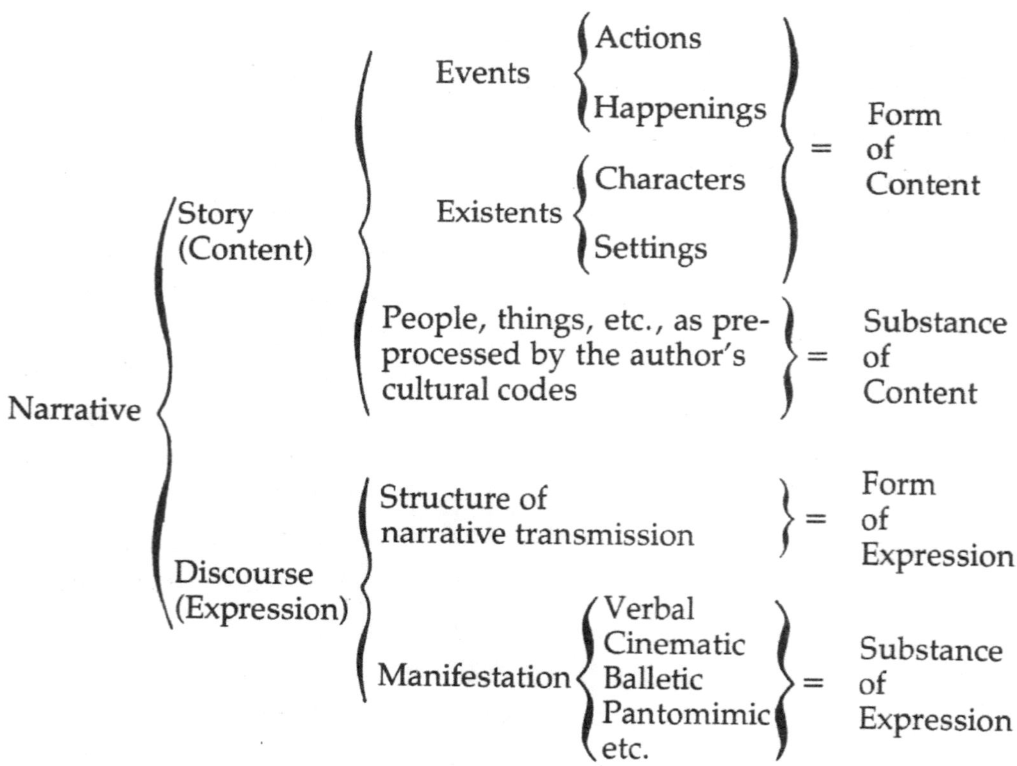


Figure 7.14 Elements of narrative theory (Chatman, 1980, 26). Courtesy Cornell University Press.

Just for fun, take a few minutes and interpret Chatman's diagram to represent the elements of a landscape narrative. Use the above diagram to help you better see a narrative's application to design.

## PLACE MAKING AS NARRATIVE

As a product of social processes, space by necessity is derived from the interaction within and between groups of people (Lefebrvre, 1994). Remember Habermas's taken-for-granted rules or what he calls "validity claims" (1979) back in Chapter 2. His validity claims help us recognize that people are daily formed and informed through their interactions with the land and with others. While his theory is almost exclusive to the social interaction afforded by words and sentences, or speech acts, his basic rules of communication do inform us, as designers, of four fundamental trusts that bind people with the narrative character of most landscape, architectural, and interior design spaces. Transposing his four validity claims into the realm of designed spaces helps us recognize that the users of our designs to be able to comprehend what they encounter. Second, that our design is honest in that it doesn't intentionally mislead the users, that is unless the design is one in which people expect to be played with. Third, the nature of the design is factual. It works with the local ecology and meets the program set by the designer and agreed upon by the client. And last, the design is sincere. This may seem a stretch, because when you think about sincerity you recognize that it can only be proven over time. I take sincerity in design to mean that a design will have a longevity, it will contribute to its own sustainability. It will become a place in people's hearts and minds. It will meet people's expectations to such an extent that their daily routines work with the local ecology to sustain the site and the design. For this reason, designs are prepared with the aim that they work within the natural and social dynamics of their location. We won't play semantic games and call five-acre zoning a resource preservation area, 20-acre zoning an agricultural preserve, and we won't lie to ourselves that we can create wetlands where none previously existed. Or, like Spirn's Mill Creek, we wouldn't be, as she notes, "illiterate" regarding the creek's existence and force (1998). Taken together, these four trusts require that as designers we know whom we are designing for, what resources are most available and how they may be utilized, where the site is situated in terms of its ecological and social contexts, and why anything on the site is to be altered. These trusts are easily transposed into the ethical responsibilities accepted by environmental designers as noted in Chapter 10, Ethics in Landscape Architectural Design.

To those trusts Anthony Giddens's structuration theory brings a way to interpret language not only in terms of space and time but also in terms of the planning and design fundamentals of position, locale, and regionalism (Giddens, 1984). These geographical concepts, so frequently found in designers' descriptive and prescriptive thinking, are for Giddens also the medium of people's individual and collective memories: the history people recount on a daily basis as their heritage. The geography of the places we call garden, home, bedroom, and neighborhood that we regularly recreate through our daily routines in turn reinforces who we are. As individuals, our daily routines sustain our material sense of identity. We daily reinforce the location and organizational character of those things and spaces in our rooms, homes, front, side, and backyards, and workspaces. We daily reinforce our existence as to who we are and the culture and subcultures of which we are a part. In admitting to the fact that we daily recreate the content and character of our personal environments we must also recognize that we do so in conjunction with others. Thus, we have rush hour. Our daily routines are a mesh that includes those routines of our companions, peers, grocery clerks and postal workers, teachers and administrators, parking attendants, and others. Edward T. Hall refers to this mesh of routines as *The Dance of Life* (Hall, 1983). Giddens along with the semiologists (Barthes, 1979; Parsons, 1964) and certain of the deconstructionists (Derrida, 1978) provide us with a bridge between language and landscape-as-text, designer-as-author, and user-as-reader.

Giddens's critique of functionalism, particularly the predictability of what is to come in terms of shape and form, in a written or spoken narrative and along routes we travel between places speaks to our expectations as we travel with intent. He introduces us to a building block of The Analogy. Functionalism provides the aspects of locale, or points in which transactions occur. To that Giddens introduces our routine practices, the aspect of routine as related to everyday life. To these "situated practices" he speaks to a person's active involvement in the transformation of codes or rules. As he calls it, this duality of structure notes that "the structural properties of social systems are both [the] medium and [the] outcome of the practices they recursively organize. Structure is not 'external' to individuals: as memory traces…" (Giddens, 1984, 25) but more internal to who we are as a society and a culture.

Recursive actions as speech acts are simultaneously generated from previous actions. As you and I talk with each other we are not only listening to each other, we are listening to ourselves and adjusting our respective tone of voice, word choice, body stance, and manners. Like sailing, we tack and trim our way from point A to point B, but not necessarily along the course or line originally plotted. Our relations are socially structured, sustained in, and reproduced by our actions in "situated contexts." While they may vary from time to time they remain contextual. My activities at a BBQ pit are not those I carry out at the credit union. We've all watched the movie scene where someone, more often than not an off-duty policeman, walks into a diner and no one is talking, they are not paying attention to their meals, to their food, no one is counting out change. You know there is a bad guy (or woman) nearby and the action is about to erupt.

> *But more fundamentally, we mean that, insofar as these arrangements emerge from the pasts of those living the room, the room is a memory of those pasts.*
> *(Wood and Beck, 1994, xv)*

Wood's and Beck's definition of mnemonic, their view of a room as containing memory aids, supports Giddens's idea of situated contexts. A living room, or any room for that matter, "is a mnemonic, certainly: *it stores in the arrangement of its parts* (authors' italics) how, for instance, we will sit with each other so that we do not have to figure it out every time anew" (Wood and Beck, 1994, xv). They identify inanimate objects as having the ability to trigger routine behaviors. They recognize situated contexts as transmitting behavioral regularities from generation to generation: living room, Grandma's kitchen, bus stop, plaza refreshment stand. An important realization by Wood and Beck is that the spatial and sequential spatial character of each place is a critical part of what they call a mnemonic and Giddens calls a situated context: memory triggers are refined by the volumetric spaces defined by the accompanying objects, the spaces' arrangements, and the overall character of the spaces and their definers. So, when it comes to writing a landscape narrative, you will introduce motion on the part of your character. Follow-up discussions within yourself and with others will add objects of a given spatial character, and their placement that will slowly refine the bases of your character's experiences and their expectations. Your path or walkway will become unique as it becomes, in the storyboard, a particular part of a particular place.

I am going to repeat, in a somewhat different manner, points you've already read about the landscape as text, and as a narrative that expresses the character of and activities in an environment. I do this first as an important reminder of the existence of and foundation for landscape narratives before getting into functional narratives. Second, to introduce the connection between personal and local texts and those of larger political and historical regions, e.g., the New Englander and New England, Cascadian residents in the Pacific Northwest or the New Jersey Pine Barons region, and Western versus Central versus Eastern Provincial Canadian peoples and places.

Consider what happens whenever you leave your house or apartment in someone else's care while away on vacation. What do you find upon your return? In public and professional workshops and seminars on design, I've asked hundreds of people this question. Other than comments on the toilet paper being turned "the wrong way" and the new smells in the kitchen, a phrase commonly heard is that "Things are out of place." People talk about feeling strange, their living room isn't quite comfortable, and it lacks a sense of identity, their identity. They are not just talking about specific objects, they are expressing feelings derived from a reading of an environment's physical attributes or what we typically refer to as basic design elements: line, plane, volumetric form, spatial sequence, and texture. Shortly after returning home, and finding things "out of place," you bring the objects around your house back into place. This act of realigning your world, of recreating YOUR spaces, is not unusual, it occurs daily. Whether arranging pencils on a desktop or sofas and chairs in a living room, our routine restructuring (and reaffirming) of our environment is the idea of maintenance found in both the definition of economics and ecology. For me, this is a fundamental basis of sustainability. Our daily routines not only maintain but also recreate our sense of identity and in doing so regularly reinforce who we are. Through what we see, smell, touch, hear, and taste we sustain aspects of ourselves that cannot be seen, and yet are taken for granted as they provide for our comforts.

> *The Greek word oikos, meaning house, is the root of both "economics" and "ecology." Over time the significance shifted from the house itself to what it contained: a living community, the household. Economics, the older of the two concepts, concerns the study of how the community manages its time, labor, and material resources in order to maintain itself.*
>
> *(Nash, 1989, 55)*

Space is a meaning-full social product. With this realization, as I reread my dust-covered undergraduate sociology textbooks, I began, for the first time, to see Berger (1963), Berger and Luckmann (1967), and Goffman's (1959; 1963) interpretations of people in terms of the spaces they produced and sustained over time. As I became aware of geographers and historian's respective views of space and time, my understanding of the bases on which landscape-as-text is written and read was further broadened. Heritage as evidenced in built and modified environments (farm fields, harvested woodlands, realigned waterways) communicates to people their sense of being, belonging, and what they are becoming. People, as active agents of change, whether professional producers of landscapes or the users, reflect social and cultural values and meanings through the products of their activities. Set on the land the products of people's daily lives become what designers, architects, and planners call local and regional landscape character (Jackson, 1994; Zube & Zube, 1977). This collective social production of a place's character, its deeper meanings, and its heritage are in a constant state of becoming. The protracted climate change debate, the consternation expressed by existing neighbors to a proposed new development, and the extended discussions and deaths associated with oil interests taking natural resources and cultural heritage sites are all pronounced states of becoming: as a person, a society, and a culture. As a species.

## MEANING CONTENT

In speaking of the capacity of language to allow us to interpret meanings, as from a dream, and integrate those meanings into everyday life, "such transposition belong, in a sense, to both spheres of reality. They are 'located' in one reality, but 'refer' to another" (40). The emotional devastation of the Twin Towers on September 11, 2001 rivals the physical devastation, loss of lives, and symbolic trauma.

Burger and Luckmann talk of "semantic fields or zones of meaning."

> *Vocabulary, grammar, and syntax are geared to the organization of these semantic fields. Thus language builds up classification schemes to different objects [and relationships] by 'gender' (a quite different matter from sex, of course) or by number; forms to make statements of action as against statements of being; modes of indicating degrees of social intimacy, and so on.*
> *(Burger and Luckmann, 1967, 41)*

Semantic fields are also recognized as related to space and the relationships that bring into being and sustain those spaces. Mention your vegetable garden. Immediately there are practical questions and personal questions about what you are growing, how you prepare the soil, what you will make with your harvested vegetables, and how good you are at growing things. A red, ripe tomato embodies so much more meaning than the salad it will become a part of: sensitivity to nature, learning about one's self, providing for family, self-sufficiency.

Semantic fields, or our social stocks of knowledge, are that accumulated knowledge that we build over time and that give order to what we do. The objects and spaces that are the semantic fields act as triggers, or readable outlines of behaviors in the workplace, the home, the street in particular parts of town, and the park, that are acceptable, somewhat acceptable, and to be spurned. This could be considered the realm of not-so-deep meanings, those symbolic realms not so far below the surface of the pragmatic world, that they are still so much a part of our commonsense world. This is the realm of accumulated knowledge, what is forgotten and what is retained by the individual and society. When in a group

> *I know that others share at least part of this [common-sense] knowledge, and they know that I know this. My interaction with others, in everyday life is, therefore, constantly affected by our common participation in the available social stock of knowledge.*
> *(Burger and Luckmann, 1967, 41)*

Regarding driving we refer to the "Rules of the Road." As North Americans, standing in a line brings with it a whole set of expectations that are, in a sense, thrown out the window when visiting many foreign countries.

## AUTHORING A LANDSCAPE NARRATIVE

Authoring a landscape narrative is a juggling act, but a fun one. I hesitate to say that there is a beginning. Saying that sets the stage for us to imagine a linear process. Writing a landscape, architectural, or interior narrative is not a linear process. Deciding what is to be expressed, and read, is an admixture, a blending of the reading and writing skills associated with storytelling and the design process. The narrative process simultaneously gives substance to and takes guidance from the design process. While the design process, with which we are most familiar, is typically driven by a quantitative program, and the measurable criteria associated with amounts, areas, and rates, the narrative process is driven by a more ambiguous, subjective program of theme, messages and message sequencing, meanings to be expressed, and remembrances. Blending the object world with the type and character of social and personal messages to be expressed is found in the process of production and expression that makes Giddens's structuration process (Giddens, 1984).

## TYPICAL DESIGN PROCESS

A sampling of design processes provides us with an outline of a typical design process. Even during the 1960s, when design was more of a check-list exercise – do I have one of these, four of those, did I do this or that – we were still instructed to go through a generic problem-solving process of problem definition, inventory, analysis, ideation, synthesis, evaluation, and final production. This process readily tends to impose a design on a site, and subsequently on the users. Since then our awareness of ecology, such concepts as dynamic equilibrium, restoration, and natural systems have introduced design processes that are more fluid and context driven than earlier processes. Today's designs tend to be more responsive to their settings, their uses and users, and their future. Today, designers and educators talk of the typical design process as arising out of the site and the users' history and their needs. They talk about it; however, my experiences with senior landscape architectural, architectural, and interior design students indicate we haven't come very far since the check-list approach to design.

When design students are asked to describe the process they are going through with a particular design project – fire interpretation center, motorcycle showroom and repair shop, wetlands restoration, low-income housing, refurbished elementary school, neighborhood park – they regurgitate the answer: "first I define the problem, then I do the inventory." And onward they go. Rarely, very rarely, does a student express their design process steps or phases in terms of the specifics of the project on which they are working. Rarely have I heard students express their project in terms of its physical and social contexts: location, orientation, and client and user character. Rarely does the student express an understanding (let alone a comprehension) of the inventory of their structure or site within the context of their project.

| | | | | | |
|---|---|---|---|---|---|
| Birth of a Problem<br>Acceptance<br>Defining the Problem<br>Ideation<br>Implementation<br>Evaluation<br>*The Universal Traveler*<br>Koberg & Bagnall, 1973 | Speculate<br>Fluency in thoughts<br>Flexibility in approaches<br>Making connections<br>Testing the Patterns<br>Congruency between mind's image and drawing<br>*Drawing: a creative process,*<br>Ching, 1990 | Resources<br>Score<br>Performance<br>Valuation<br>*Taking Part,*<br>Halprin & Burns, 1974 | Definition of project<br>Analysis of possibilities<br>Definition of specific problem<br>Examine alternatives<br>Select alternative<br>Act of production<br>Clarification<br>*Discovery Through Design*<br>Bevlin, 1977 | Training<br>Analytical Phase with Experiential feedback loops<br>Programming<br>Data Collection<br>Analysis<br>Creative Phase w/feedback loops<br>Synthesis<br>Development<br>Executive Phase<br>Communication<br>*Systematic Methods of Design*<br>Archer, 1963-1964 | Identify and understand the problem.<br>Collect relevant information.<br>Mull it over.<br>Speculate.<br>Develop ideas.<br>Select best idea.<br>Implement it.<br>*The Practice of Creativity,*<br>Prince, 1970 |

Figure 7.15 Design Process Sampler. Prepared by author.

A semester and a half into his project, the redesign of a neighborhood park under the rubric of crime prevention through environmental design (CPTED), a senior landscape architecture student could only express his inventory in the most generic terms. The terms were from the design textbook: solar orientation, precipitation, vegetation, topo, views into and out of the site, and so on. That's fine, but what value are each of these typical site inventory categories when applied to crime prevention, to a person feeling safe and secure in the park? What I heard him and so many other students express as their inventory (analysis, or whatever) was so generic that it could be taken and pinned up with any other of their classmates' projects. In fact I did just that. On the day our seniors were to present their projects I interchanged the sheets displaying their design processes and they all fit in their new locations as part of completely different projects.

So, the question emerges, is it possible to have a design process that contextualizes the usual design process questions? Answering that question automatically finds a designer replacing all the science, technology, and geometry with contextual human behavior, regional values, and more. Consider a scientific/technical approach to an application of CPTED versus a narrative of someone apprehensively walking through a dimly lit park afraid for their safety. How might this latter approach, this shift in your thinking, contribute to your first-person narrative, storyboard, and final design?

Specific to crime prevention through environmental design the existing vegetation of a park site could be inventoried as to height of shrub mass, height at which tree canopies begin above the ground, leaf and branching density along a transparency-translucent-opaque scale. The criteria could be considered sight lines for safety. Could someone hide behind the leaf mass? Tree trunk density at eye level changes in elevation as enhanced by the presence of existing and, later, mature ground cover, shrub and tree growth. With an inventory of this nature, one specific to CPTED, the designer concerned with public safety could conduct an analysis of the park's vegetation in a manner that identifies the existing vegetative assets and liabilities specific to public safety and welfare during the day, during the night. A generic site inventory would not open that door and reflect and inform the design's public safety theme. Thoughts of public safety would remain assumptions.

This approach to design, the interpretation of a design process into the framework of a particular project's needs, works to the benefit of the designer authoring a narrative. First, and foremost, defining the words "problem (or project) statement" in terms of the project's particular experiential goals works to introduce the theme of the design to be narrated. Second, replacing the ambiguous statements of problem definition, inventory, and so on, with the physical, social, and expressiveness particular to the project we begin to adopt the sensory vocabulary that we will use in our narration. A police officer and a thief should read the environment in the same way, as should housing authorities and the poor, and bike path designers and bicyclists.

## NARRATIVE DESIGN PROCESS

The introduction of narrative into the design process draws our attention away from a tendency to focus on and to deal with just the objects in our designs. The designer, as story builder, now gives attention to those aspects of the designed space based in users' anticipated experiences as they call for the spaces' activities and character. The inclusion of particular messages requires us as designers to interrelate the immediate, or local, settings' qualities with those of previously experienced spaces. Even when a space is designed to provide a dramatic contrast that drama can only occur in relation to what came before, whether what came before is actually within the site boundaries or the surrounding landscape. The aspects of cohesion and continuity that link past and present also work to have a design's activities and settings reinforced by what follows them. A good narrative brings the reader into a future that grows out of and reflects back on the present. The future, those spaces ahead of you in an environment, reinforces the present. The objects that fill and define space in the past, present, and future moments of a designed area are tied together by the interrelationships that grow out of a good narrative that makes sense.

When a narrative's events and activities make sense, as part of a larger story, the quality of cohesion in design is inescapable. Narratives are not a random gathering of discrete statements; the storyline at any point is verified and reinforced further along in a design's narrative or the users likely become disoriented. But even in films that depend on the disorientation of the viewer, like *Chinatown*, *Inception*, *Gotham*, and *Bladerunner*, and such works of fiction as those by Edgar Allen Poe, there exist patterns to be discovered. Cohesion and continuity still exist in narratives of mystery and intrigue. As the antagonist in many films and fictional works is so often told, "You don't know what you're getting into." Meanwhile we work to find the ties that link the parts together, eventually identifying the glue that binds the events, activities, and views.

Narrative design is an admixture of a typical environmental design process and the authoring of a story. Identifying what constitutes a typical design process led me to Kevin Lynch's classic work, *Site Planning* (1971). The value of his work made up portions of landscape architects' Uniform National Exam (UNE) through the 1980s and more recently the Landscape Architecture Registration Exam (LARE). For this reason I start with a review of pertinent phases of what I feel comfortable referring to as the typical design process.

While greatly enhancing the design process Lynch is aware of the typical process's shortcoming. More accurately, the shortcomings of the designers who applied it. Deep into his description of his enhanced process, he notes, "The preceding discussion takes it for granted that we know why we are manipulating the sensuous form [of our environments]. Our recommendations imply the reasons, and most designers are content to leave them implicit" (1971, 224). He continues, in the next paragraph, to say that we need to be "explicit' in what we call for but doesn't clarify what "the more precise statements of these [sensuous] criteria" are. In character with much of the rest of *Site Planning*, those "more precise statements" are likely interpreted as related to the design of discrete objects, views, and spaces. Again, there is a call for a mechanism that brings together the parts, the statements, into a cohesive, coherent progression of spatial experiences leading to, into, and through a site. A narrative approach to design development is that mechanism.

Program development is part of "specifying for whom the [yet to be designed] settings are intended and how predicted conflicts are handled" (29). "Since qualities of the setting are now being determined, design has already begun when a program is made [including required access between features]" (Lynch, 1971, 28). "By describing [landscape] units in terms of behavior settings, it describes the intended outcome" (28). What becomes clear with Lynch's view of what constitutes a program is that he sees it as a blending of quantities and qualities. The identified activity settings, the circulation routes linking them and within them, and the forms they take on are all scaled to meet certain numbers of people carrying out particular activities. The activity settings are given observable character with regard to their materials' finishes, sense of openness and enclosure, softness and loudness of sounds, and so on.

"The fully developed program will include desired character, intensity of use…[it will be] rich and detailed, always expressing environment, management, and behavior" (29). His wording specifies the character of the activity yet remains ambiguous. I know this sounds like an oxymoron. I guess it is but it is one that works. At this stage of the design process bench seating for two is better expressed as intimate seating. A barbecue pit is a food preparation area. Words, as do the graphics employed early on in the design process, deal in concepts. They identify the relationships sought in the final design. More detailed questions are left until later in the process, in that way the designer minimizes the confusion of simultaneously working to determine the materials and finishes along with the general kinds of spaces and their contents. Development of a program, even a simple one, through discussions with clients and users, provides a good start to identifying your narrative's theme. As you begin to draft your narrative and get into reviewing the connections between landscape sentences (the mental images seen and experienced through the eyes of a user) increasing degrees of detailed design thinking arise.

As for a design's hierarchical settings (Alexander's larger and smaller patterns), Lynch's approach to preparing a program gives them more definition. Informal dining of 4–8 people implies a different character to the space than formal dining of 30–40 people, or 15 foot candles of light on the walking surface provide quite a different late-night ambiance than 80–100 foot candles. Taking a primarily quantitative approach early in a design process leaves the final character or thematic nature of a designed place to chance. To begin with a definition of a place's theme (your user's state of mind and expectations) provides a palette of potential characteristic spaces, patterns of spatial organization, materials and finishes. Identifying activities to be included in your design begins to illuminate location, orientation, and sequencing requirements across a site (Lynch, 1971, 30). In making these decisions you are eliminating a lot of what shouldn't be included in your design.

> *But the basic measure of value [that comes out of a program] is the degree to which the behavior settings allow people to feel competent in doing what they want to do and the degree to which they favor personal development of competence and purpose.*
> *(Lynch, 1971, 30)*

Realize, as true as the above statement is, it leaves implicit and therefore relies on each designer to determine just "how" people experience the site so as to feel competent doing what they want to do. In this approach, there remains a call for a socially agreed upon mechanism that accommodates people feeling competent. His is a valid approach but in stopping short of recognizing that materials, shapes, magnitudes, and more speak to the users he seems to assume the designer knows best. If this is the case, then designs are imposed upon rather than drawn out of those most likely to use the site.

> *The site plan deals in its essence with three fundamental patterns of location in space and time: the patterns of activity, the patterns of circulation, and the physical forms. These are the subject matter of the [designer's] first sketches and remain dominant themes throughout the work.*
> *(246)*

Within The Analogy, the dominant theme throughout a work is what the main character(s) is experiencing walking along the "patterns of circulation." Before providing any storyboard sketches, the narrative introduces rates and rhythms of movement, views and sensual stimulations, all colored with the proposed landscape users' experiences. It makes conscious what so much design literature leaves, at best, implied and, at worse, unconsidered: that is people, their experiences, emotions, and expectations.

The summary Lynch provides of his approach to doing a design (1971, 246) seems a bit haphazard in its lack of a theme to unify the decisions made each step along the way. He seems to rely on the assumed skill and goodness of the designer's sensitivity. In so doing Lynch ignores so much of the sensitivity he earlier called for as relying on and referenced to contexts. This assumed intuitive nature of a designer leaves to chance a continuity across the site design. There is a need for a developmental logic between activity settings, something a designer's day-to-day changing moods are not likely to provide. People do not just beam in and beam out of points throughout a site. A conscious application of a narrative logic to that of the setting's "ecological" logic would strengthen the adequacy, appropriateness, and suitability a person would sense throughout the entire design-as-narrative text. With those environmental, material qualities noted in your character's movement along his or her circulation route in hand the designer-author is then prepared to do the sketches that comprise a storyboard.

*A designer's "studies proceed in the schizophrenic manner common to all design…relaxed and uncritical, allowing his subconscious mind to suggest new forms and conditions… at other times he looks sharply at these suggestions, consciously testing them for the way in which they apply to his purposes…Part of the skill of the designer lies in managing these two states of mind so that his critical powers do not inhibit creative suggestion, while his irrational processes do not prevent adequate analysis. (Lynch, 1971, 247–248)*

Lynch's thinking, provided above, adds to my wish for the exclusion of independent statements throughout a design, expressions of messages and meanings that are not part of the developing storyline. A developing storyline, a first- or second-person narrative, brings with it a coherent progression across spaces, views, and activities. Given Lynch's description, the process sounds egotistical, leaving the designers with so much ambiguous authority. Again, Lynch relies too heavily on the assumed goodness of the designer's abilities which, in turn, rests on education, practice in an office, and licensure. How, on what basis, does the designer "look sharply" at "new forms and connections"? On what basis does the designer to whom Lynch refers know "good design" when he or she sees it or "test them for the way in which they apply to his purposes"? "His purposes"? How can we be confident that the designer's purposes echo those of the eventual user of the design?

Lynch talks about the design alternating between part and whole, part and parts, section and plan, elevation and perspective, and "managing" these dual states of mind "so that [the designer's] critical powers do not inhibit creative suggestion…" (248). These thoughts are describing skills, leaving us to question what directs the selection and organization of content, the affective substance of the design, its program. Okay, but even though Lynch recognizes what he calls a sensuous program, his process still calls for a designer to be able to orchestrate a cohesive continuum of messages. One that can be read to others, with others, and make perfect sense. Lynch's approach leaves the emergence of a landscape narrative to chance whereas The Analogy starts with a narrative. In doing so, the narrative-storyboard-design process provides in a more logically progressive way much of what Lynch calls for.

I don't agree with what follows but it is telling of what Lynch recognizes:

*The normal program simply gives the quantities of space to be provided by broad categories of standard function…The quality of those spaces, the behavior expected…and the purposes and attitudes of their users get little attention. (260–261)*

Lynch does state that much of this shortcoming can be overcome "if a program is properly made" (261) and that such a program calls for the designer beginning "by specifying what actions are expected to take place [throughout the final site design]" (261). Rather than relying on the designer to specify the expected actions, Rae, in *Movies of the Mind* (1996), would say to let your character(s) take over. Let them take you into and through the emerging design. In their use of the site they will help you see so much of what Lynch leaves to the designer.

What is important to note is that writing a narrative, as a starting point in a design's development, is only limited by the people for whom it is written. The narrative and therefore its storyboard and final design respond to the likely users' expectations. Identifying the users' expectations is something, as designer-author, you come to know after interviews and research, and through discussions with others and before you start writing your narrative. In writing a landscape narrative by empathizing with the likely users' expectations, experiences, and wishes, you bring a more user-friendly approach to the design process than Lynch's designer's sense of what makes for a good design.

Lynch provides parameters of what could be interpreted as a landscape narrative but doesn't refer to them as such. He outlines what is needed but not how to provide for them. Needless to say, the designer-as-author and landscape-as-text metaphors answer the "How" question. In fact, taking Lynch's *Site Planning* as a whole we could use it to guide the design authoring process. Please note: the following is located on pages 224–227, only in the second edition of *Site Planning* (1971):

1. Sensations should be within the range of comfort.
2. An adequate diversity of sensations and environments should be available.
3. The sensuous environment should support the action people want to engage in there.
4. Places should have a clear perceptual identity: recognizable, memorable, vivid, engaging of attention, differentiated from other locations.
5. The identifiable parts should be so arranged that a normal observer can relate them to each other and can understand their pattern in time and space.
6. The environment should be perceived as meaningful, its visible parts not only related to each other in time and space but related to other aspects of life: functional activity, social structure, economic and political patterns, human values and aspirations, even individual idiosyncrasies and character.
7. Finally, the environment has a part to play in the intellectual, emotional, and physical development of the individual, particularly in childhood, but perhaps also in later years.

Lynch's seven-step design process can be used more as a test of completeness than a guide. By that I mean, once you understand your probable users and their reasons for coming to your design write a first draft of your narrative. Review it. Then use Lynch's seven steps to identify parts of your narrative that can be made more complete and provide for a smooth flow across the site. In applying Lynch's seven steps to your written narrative, each step is clarified in terms of your narrative's theme.

> *The environment is an enormous communications device – people read it, they seek practical information, they are curious, they are moved by what they see. Congruence of the visible and social world will facilitate action and make both those worlds more comprehensible....Formal legibility is at least the common visible base on which all groups can erect their own meaningful structures.*
>
> *(Lynch, 1971, 226)*

As grand an idea as this is in support of considering landscape-as-text, Lynch goes on to refer to the production of discrete statements with the assumption they contribute to a cohesive landscape narrative that flows.

What is interesting in Lynch's design process is his laying out of activity areas early in the process, his early recognition of the need for a design to deal with people's physiological needs. This is in line with Maslow's hierarchy. The second phase of Lynch's design process, the laying out of circulation, reflects Maslow's second level of actualization, that people are able to access locations where their needs can be satisfied, be it information (wayfinding), physiological (food, shelter, exercise), or an appreciation of nature or culture (views, aromas, birds or butterflies).

> *Thus it is that architectural genius has been able to realize spaces dedicated to voluptuousness (the Alhambra or Granada), to contemplation and wisdom (cloisters), to power (castles and chateaux) or to heightened perception (Japanese gardens)…the animating principle of such a body…[is] within those who use the space in question, within their lived experience.*
> *(Lefebvre, 1994, 137)*

How circulation routes approach and meet activity areas provides for and reinforces a person's sense of orientation. In working to provide for and satisfy people's basic needs, these first phases of the design process call for images that express the place's functional narrative. As a narrative, the messages' flow supports people's sense of comfort in terms of their health, safety, and welfare as participants in and users of the environments they traverse. Beyond these needs, as more detail is added to a design, that product – through the designer's sensitivities – responds to the locale's social norms and finally people's expressions of self. Again, my experience is that this comes about fairly easily with your writing of a first- or second-person narrative from your design's participants' point of view.

What you'll find, once you write a narrative and begin to refine it, is that the parts making up the landscape are increasingly responsive to the value-laden meanings held by the inhabitants of a place. As the details in a design are refined, they increasingly support the users achieving self-actualization. Landscape designs are interactive, sustained over time through the ways in which the spaces are used. Designs that do not allow the users to interact with their form and content do not allow people to become self-actualized. Consider the times and ways you've felt put upon or ignored by a landscape setting. What did you do? Did you behave in ways that sustained the place and its parts? Or, did you consciously or unconsciously behave in ways that modified the place or its features? Did you cut across that lawn? As Lefebvre notes, people who are unable to self-actualize by identifying themselves in a setting tend to incite a revolution, or at least change the spaces. Given what they find in a landscape, people act on refining their personal and collective senses of self and the spaces reflect that process of becoming. This can occur when a place allows people to "build" or in some way realize their own narratives: the selection of ancillary routes, various seating types and scales of

Figure 7.16 Green area by Hirschhorn Museum, Washington, DC. Courtesy of James Urban, the Office of James Urban, FASLA.

related space and elevation, views and suggested paths, an orchestration of multiple senses called upon, and so on.

I laughed when touring James Urban's design for the Hirshhorn Museum in Washington, D.C. with Jim. He pointed to two benches set on a small plot of green lawn on the west side of the museum. In this highly urban setting, especially in the summer heat and humidity of Washington, D.C., the bench seats invited you to walk across the lawn to get to them and take a break. The twist was that museum security staff saw the lawn as inviolable and yet would themselves walk across it to tell you you couldn't sit there. Apparently, like the time we visited, Jim would on occasion go to the museum's security office and remind them the seats were in fact there to be used.

The questions remain, what do we as designers narrate? How do we identify and select the story to be authored? Where do we find the answers that provide us with a sense that the evolving larger message is appropriate to the people and the place and that it will be identifiable and readable?

The real creativity of an emerging design is the designer's ability to author a narrative that at least holds all the critical points together, if not, at least holds the users' attention because the evolving story touches them personally. The reward may be as simple as getting between two points without feeling lost, or as deeply moving as sensing a closeness with the magnificence of technology (I am part of this city), spirituality (This place brought me closer to God; or the Earth), or self (Here I found new meaning in my life).

Identifying deeper meanings is helped through a case study provided in the next chapter. In the next chapter you are introduced to a project that was a collaboration between a sculptor, a landscape architect, the Providence Sisters, and Sacred Heart Hospital and the Center for Faith and Healing staff. The client was Sacred Heart Hospital (now Providence Sacred Heart Medical Center) in Spokane, Washington. The theme to be expressed was "Finding Solace."

For now, this chapter has dealt with the breadth and depth of narrative and landscape-as-narrative. It is not all here. There is so much more you can explore and, I hope, contribute to the idea and application of landscape-as-narrative; for that matter, designer-as-author and user-as-reader. What you have at this point is a basis for understanding what to look for in your clients and what to present to your clients as substantiation for your using the narrative-storyboard-design approach.

Now that you have a sensitive understanding of your client's wishes, the site to be designed, and the likely background of the landscape design's users, we can outline an approach to writing your narrative: where it came from; how to get started; and ways to perpetuate its opening up to you in increasing detail.

# 8

# Authoring a Landscape Narrative

Writing is a slower, more focused, process of self-expression than speaking. I have long believed that the process of selecting each word making up a written sentence and each sentence making up a paragraph is directly related to how well we see our designs' features, details, and contexts. If you can't express it, then you can't see it. And the "I'll know it when I see it" ploy is just a stalling tactic. This understanding goes hand in hand with how Betty Edwards (1999) helps us draw better by learning how to focus on and see more clearly what we want to draw. Our ability to express ourselves well in writing directly relates to how detailed and complete our drawings are.

In preparing to write your narrative, the following chapter introduces you to three writers – fiction writer Colleen Mariah Rae, novelist and non-fiction writer Anne Lamont, and poet Ted Kooser. These are three practical, down-to-earth writers whose love of expression, of clarity of imagination, helps you prepare to write a landscape narrative, cum storyboard and final design. These writers' thinking is summed up in the following three quotes and further unpacked throughout the chapter to help you become more comfortable with writing.

Rae's approach to writing is cinematic. She allows her characters to live out their lives in her mind as she watches and records their experiences and emotions. Rae helps us imagine that

*On that 'screen' [referred to as your mind's eye] you don't just see and hear as you would with a regular movie, you taste and touch and smell and slip into the bodies of your characters and see the world through their eyes, touch it with their hands, smell it with their noses.*
*(Rae, 1996, 5)*

However, Anne Lamott admits to the difficulties that, for many of us, characterize how we think of and deal with writing. Yet, she also speaks to how to deal with those difficulties and reap the rewards that come with perseverance.

*[In the midst of all the distractions] you clear a space for the writing voice…you begin to string words together like beads to tell a story. You are desperate to communicate, to edify or entertain, to preserve moments of grace or joy or transcendence, to make real or imagined events come alive. But you cannot will this to happen. It is a matter of persistence and faith and hard work. So you might as well just go ahead and get started.*
*(Lamott, 1994, 7)*

And then poet Ted Kooser helps us understand not only the critical importance of each word we select but also the importance of knowing who you are

DOI: 10.4324/9781003286981-9

writing for and also that you are writing (designing) for individuals who need to be able to read what you've written. In filtering his thoughts through The Analogy, you begin to recognize your narrative as the foundation of an environment in which people find their expected activities and fulfill their emotions. As with Kooser's poetry, your design is a form of communication, and for him

> *Poetry's purpose is to reach other people and to touch their hearts. If a poem [read landscape design] doesn't make sense to anybody but its author, nobody but the author will care a wit about it. That doesn't mean that your poems can't be cryptic, or elusive, or ambiguous if that is how you want to write, as long as you keep in mind that there is somebody on the other end of the communication.*
>
> *(Kooser, 2005, xi)*

Kooser's closing comment, above, captures the driving force behind the narrative-storyboard-design approach to landscape, architectural, and interior design. The beauty of the narrative approach is that it is almost impossible to not "keep in mind that there is somebody on the other end of the communication."

On one hand, describing a design, as is so often done, in geometric terms (focal point, ground plane, verticals, axis, etc.), budgeting, and efficient scheduling brings with it no responsibility to consider people as more than related assumptions. On the other hand, by starting your design process through the senses of your site's potential users, as a first- or second-person experiential narrative, you cannot avoid having somebody at the other end of your landscape narrative's communication. Given background research on a site, client, and probable users, your design emerges as seen through your characters' eyes, touched with their hands, smelled with their noses, heard through their ears, and maybe even tasted. Your written narrative is almost like a cerebral fly-through, one that leads you into and through your design and is so much more than a summary of objects and spaces in the completed design.

> *Fortunately, chronicling the movies in the mind is not only an easy job, it's one that is great fun. The most magical part about writing fiction – the thing impossible to understand until we experience it – is that stories play themselves out on the screen in our mind with a volition all their own. I've literally fallen out of my chair laughing at the antics of my characters. And that's what's fun about writing fiction – the pure entertainment value of watching the movies in the mind once we learn to get out of our own way and partner with our story maker.*
>
> *(Rae, 1996, 5)*

Writing your narrative is where the wonder of The Analogy's designer-as-author, landscape-as-text, and user-as-reader design progression pays big dividends. As a designer-author you design your site through your character's wishes, anticipations, and expectations. The design materializes as your character progresses from first impression, to movement toward the site, into it, and through it. Yes, this is a highly subjective approach but not without a foundation in normative behaviors. Since we are developing, to use Lynch's wording, a sensuous design, then doing it through the senses of those who will experience it is a great place to start.

I'll never forget one student's narrative. The design was the revitalization of a neighborhood's historic business district. A family made up of mom, dad, their young daughter, and their dog were on their way to enjoy a parade and street festival. The interplay between adults, child and dog, yes and the dog, as they approached the parade was priceless: animal, child, and adult movements, views at different eye levels, each pulled in different directions by various sights, sounds, and smells. The historic business district's design was, in a sense, coming into view.

As a starting point to your design, writing a first- or second-person narrative makes it simple. Your design will gain in depth and detail as you progress. Even a simple narrative, especially since you are not used to doing this, is an important way to introduce yourself to your characters before you let them go and allow them to help you get to better know your eventual landscape design. Yes, time is critical. You're probably working on a project for school or for a client and your likely, to some degree, on the clock, facing a deadline. Given what time you have, do as much background research as possible before you start writing. Start by familiarizing yourself with your clients, they know who they wish to attract, to entertain or please so they will spread the word and come back with friends. When you feel your client provided sufficient insight, do some good old-fashioned book and article research. If you're lucky enough to be near a university, or a public library, use it. Consult the librarian. Often there is a "research librarian" on staff and they are good at what they do. If you're not near a library, fire up Google, Bing, Yahoo, or whichever search engine you find most comfortable.

Don't confine your research to just the demographics and behaviors of your site's expected users. Get into their heads. Expand your thinking. Why will they be coming to your site? Get into their emotional state, age-related issues, social development, and likely outlook on life. Don't hesitate to call specialists and explain what you're working on. I've tracked down specialists (as far as Finland and Australia), called or emailed, and discussed questions I had. Remember John Simonds' epiphany,

> *The soul stirring secret... lay, not in its plan, or its forms, or its spaces, but in what one experienced there... One plans not places, or spaces or things – one plans experiences. The places, spaces, and things take their form from the planned experience.*
>
> *(Simonds, 1961, 225)*

Designers often address their potential users, in terms of what they will do, where they will do it, how they will go about doing it. Rarely do they, do we, consider the users' experiences, let alone use those experiences as the lead into a design. Hopefully, Kooser's call to keep in mind that someone is receiving your design is beginning to dovetail with Simonds' realization that your design participants are the recipients of the experiences your design is going to provide, in a sense, what they are asking you to design.

When you are done with background research, gestate, ruminate, contemplate – ideate. Let your subconscious do its thing for a day or two. Always carry a recorder, pencil and pad of paper, or notebook, iPad, or smartphone with you. As ideas come to mind regarding your user group and their likely sought-after experiences, record their points of view.

## PREPARING TO WRITE YOUR NARRATIVE

> *The result of this painstaking work was that Thoreau gradually elucidated most of the dominant patterns of seed dispersal and tree seedling recruitment within an*

*afternoon's walk of his home.*
*(Nabham, 1993, xv)*

What Rae, in *Movies of the Mind*, calls "the well" I referred to over the years as personal experiences. I've assured many first-semester design students that whether they grew up in a metropolitan city or a rural village, by the time they came to university they had experienced all they needed to become a great designer. Flannery O'Connor does me one better. He "said that enough has happened to each of us by the time we're eight years old for us to write about it forever" (Kooser, 2005, 74). For me, the caveat is that, no matter the age of my students, they had to be willing to ask deeply self-reflective questions of their personal experiences and be willing to dive equally deeply into those experiences, by way of their subconscious, to find the answers. Intuition isn't Shazam magic, it is work, but like learning to drive a car or ride a bicycle, once you got it down it literally becomes second nature and you can call on it when you need it. "For each of us the way to the well is different. That's why every artist must become aware of his or her own process and become the mapmaker, naming the terrain so that the way back is known" (Rae, 1996, 6).

Rae goes on to say

*I came to see how my writing process wasn't the linear one some writers report. I can't sit down and have the story unfold under my pen. No, my process, I realized, was like that of a sculptor who works first in clay. I start with an incident that has been haunting me, then I spend a lot of time scribbling in notebooks, digging the clay through conversations with my characters, through detailing up scenes, through trial runs with voice and so on... Once I get the rough form, I then get to do what I like most of all – shape the story into its most beautiful form.*
*(6)*

Rae sees herself as a guide. Her way of forming and refining a narrative may work for you and then again may not. Her advice is to let her guide you but, in the end, go with what works for you and develop your own process. She asks you, in some form or other, to leave "tip-of-the-iceberg thinking" and immerse yourself in your characters. Let your characters take over and [then] get out of their way.

What's interesting about Rae's readers of written texts and ours of landscape texts is that what she works to produce in her readers' minds, with a great deal of success, is what we produce in our landscape users-as-readers' minds. Interpret what Rae says here as if she was talking about people experiencing a landscape narrative. "Fiction's power comes from the fact that when our readers are feeling and tasting and touching and seeing and hearing in their mind's eye, they're experiencing emotions in their bodies" (Rae, 1996, 27).

The Analogy's narrative-storyboard-design process brings more enjoyment and fun to the landscape architectural design process than ever seemed present when I was a student learning problem definition, inventory, analysis, ideation, design, and evaluation. Rae highlights the fun aspect of writing, I'd say designing, in a quote from Anthony Storr's *Solitude: a return to self* (1988).

*It seems probable that there is always an element of play in creative living. When this playful element disappears, joy goes with it, and so does any sense of being able to innovate. Creative people not infrequently experience periods of despair in which their ability to create anything new seems to have deserted them. This is often because a particular work has become invested with such overwhelming importance that it is no longer possible to play with it.*

*(Rae, 1996, 9)*

Rae talks about images as fiction's building blocks. Given my process, words require focus and selection even when you start by allowing your characters to take over, arrive at, enter into, and move about your design. Your interpretation of those words into images is a major clarification phase of your design process, a process later greatly refined by reflecting your written story off your visual story: reflecting your written narrative off your storyboard.

Rae's advice applies to our drafting our narrative text. "Be willing to write without censoring *anything*. Just let it come off your pen. You'll be surprised by what surfaces. But don't let the surprises scare you" (Rae, 1996, 67). There are two reasons for not being scared. First, any abnormalities will become apparent as you discuss your first- or second-person narrative with others – which you will do. Second, those, what you see as anomalies in your narrative's progression, will be further illuminated when the narrative text emerges as your storyboard and is then discussed with others for its relevance to your written text.

*And, most important, always let your character tell you the story. One of the biggest mistakes we writers make is thinking we have to come up with a story. In fact, all we have to do is listen to our characters. In digging the clay, I've frequently discovered that they often knew better than I the best direction for the story to go.*

*(69)*

One distinction between us as designer-writers and Rae's writers-as-writers is that most writers' approaches can be "slow unforced development." Many writers can allow their unconscious time to roam while designer-authors have clients with schedules and budgets. Don't let this pressure dissuade you from allowing your unconscious to take 24–36 hours to roam, graze, play, or dream during which time is all the more reason to use Rae's exercises, or do what the firm of Zion Breen Richardson Associates used to do. When there was down-time in the office they would prepare designs for settings they had not yet been asked to do. Such was the origin of Paley Park in New York City. As Bob Zion told my students and I, when William Paley asked for a design to memorialize his father Samuel Paley, they went to a collection of already-completed designs and pulled out what became Paley Park, located between Madison and Fifth Avenue in New York City.

In her response to a question about her "dreaming" about her characters while awake, Rae responded,

Figure 8.1 Paley Park, New York City. Photo by author.

*...if there is any place you fall out of the dream, read your story* **aloud.** *You'll hear the places where [the unity of effect is violated] much better than you can ever* see *them.... reading aloud puts a story to even more of a test because it makes the analytical process subordinate to that activist part of us that knows a good story when it hears one.*

*(Rae, 1996, 118)*

The second writer I'd like to introduce is poet Ted Kooser. His book, *The Poetry Home Repair Manual: practical advice for beginning poets* (2005), like Rae's is equally informative for designers whether they use the narrative-storyboard-design approach, or not.

In response to a note that he got from someone who read one of his poems and said that from that day on she'd see a part of her world in a wonderfully different way, Kooser responded, "I'd given her something that changed the way she saw the world, and she was thankful for that" (7). Imagine, one day someone saying that about one of your designs.

*When something happens to catch your attention, and you feel like making a note of it, you can usually trust your impulse. There may well be something there worth writing about. Chances are good that you've noticed whatever it is because somewhere within you, you have a glimmer of feeling about it. At first you may not recognize or be able to elucidate that emotion, but as you work with your words the feelings can be expected to come forward.*

*(14)*

This is one way to feed your intuition. Record your experiences, good and bad. Consider the setting, what senses were reaching out, and what they were bringing back to you? Experience it all.

Think about the following in terms of developing a design.

*After you've written your first drafts, you begin to revise, making critical judgments about what you've written, decisions made in part upon which word goes best in which position, about the appropriateness of a figure of speech, about the use of grammar and punctuation, and about how your writing may affect an imagined reader. Revision, and I mean* extensive *revision, is the key to transforming a mediocre poem into a work that can touch and even alter a reader's heart.*

*(16)*

With a bit of humor, Kooser notes the commitment to revising one's writing with, "Much has been written about revision, but I always liked this passage from Edward Week's book *This Trade of Writing*."

> *A. Edward Newton tells a story of Oscar Wilde at an English house party. Pleading the necessity of working while the humor was on, he begged to be excused from joining the other guests. In the evening his hostess asked him what he'd accomplished, 'This morning,' he said, 'I put a comma in one of my poems.' Surprised, the lady inquired whether the afternoon's work had been equally exhausting. 'Yes,' said Wilde, passing his hand wearily over his brow, 'this afternoon I took it out again'*
>
> *(16)*

> *Though [Edward] Weeks says revision is drudgery, I don't agree. You can learn to love tinkering with drafts of poems till a warm hand from somewhere above you reaches down, unscrews the top of your head, and drops in a solution that blows your ears off.*
>
> *(17)*

In a slightly varied way, Kooser recommends

> *that when you sit down to write you have in mind an imaginary reader, some person you'd like to reach with your words... The more real your imaginary reader seems to you, the easier it becomes to shape a poem that might reach through to that person.*
>
> *(20)*

In your case as a designer-author your "reader" is a little more real.

Two exercises my students actually came to enjoy involved their writing poems and haikus as a means of getting "inside" the experiences a person may have in their design – or, as in some instances, their design experiencing a person. I like J. J. Gibson's strategy "to regard the senses as aggressive, seeking mechanisms and not merely passive receivers" (Bloomer and Moore, 1977, 33) as we move about a landscape. When we think about how we approach and arrive at an anticipated destination, it really isn't a stretch to consider our senses as aggressively seeking out particular stimuli, anticipated stimuli. Remember we are pattern seekers. We look for our expectations to be fulfilled and at times even surprised.

> *I've always liked this very useful passage from John Fowles... It's from a review of William Trevor's book The News from Ireland: 'I remember years ago watching the commercial folk-tale tellers in a Cairo bazaar. All writers ought to have observed this ancient practice of oral narrative – all critics likewise. Getting the audience, I remarked, depended not at all on*

*preaching and philosophizing but very much on baser tricks of the trade: in short, on pleasing, wooing, luring the listeners into the palm of one's hand'.*

*(Kooser, 2005, 23)*

To paraphrase Kooser, the title and first few lines of your narrative "represent the hand you extend in friendship toward your reader…you want to make a good impression. You also want to swiftly and gracefully draw your reader in" (Kooser, 2005, 25). Relate this understanding to what your design is to convey as its first impression. That is your design's hand extended in friendship. How will it lure and woo a person into your design? Another way to consider first impressions is Kooser's thought that "Titles are very important tools for delivering information and setting expectations" (27).

Edgar Allen Poe was a master at not only providing first impressions but reinforcing them multiple times in a story's first page. An excellent example is Poe's three-word title to his short story *The Landscape Garden* (Poe, 1904). If you analyze the three words, keeping in mind the story was originally written in 1848, you come to find out that Poe's title gives you a summary of the story to follow. Even before the actual story's text starts, he further expands upon it by introducing Giles Fletcher's poem just below the title. As for the title, "Landscape" is wild and on its own. "Garden" is the human control or cultivation of land and vegetation. As you work to unpack those two words in the title, ask, why did he say "The Landscape Garden" and not "A Landscape Garden"?

Kooser reinforces Rae's advice to not censor your early drafts when he speaks to rules of writing.

*Well, there are no should or should nots in writing poetry. You can do whatever you feel like doing, pants on or pants off. Part of the joy of writing, or of practicing any art, comes from the freedom to choose.*

*(Kooser, 2005, 35)*

If we consider poets as designers, then Kooser is talking to us when he says:

*Thousands of readers love Robert Frost's 'Stopping by Woods on a Snowy Evening,' but very few of us care that it was written in a complicated form with a demanding rhyme scheme. The beauty of Frost's poetry catches the reader, not the fact that he was able to shoehorn that beauty into a difficult literary structure. Of course the form lends rhythm and music, but we don't start out by remarking upon form. Surely Frost must've been pleased with himself for mastering that structure, for meeting its difficult challenges, but just as surely he would have known that the great majority of his readers would probably appreciate his poem without noticing this technical accomplishment, being, as could be expected, far from engaged by the poem's atmospheric effects: the little horse with its harness bells, the mysterious wood, and the soft and silent snow. The complicated literary form of 'Stopping by Woods' is inseparable from the poem, of course, but it is all but invisible to the average reader. One of that poem's most important achievements is to have kept its structure from calling attention to itself.*

*(Kooser, 2005, 39)*

Whenever I (Bob Scarfo) think, "the best design is one no one knows was designed" I think of Olmsted's forested approach to the Biltmore Estate near Ashville, North Carolina.

Rae, Kooser, and you will read Lamott all believe that a good writer, and I'll add, a good designer, can by carefully selecting their design's details guide their readers or design participants' responses to a great degree. Given the Imperial Tea Garden example (Benzel, 1998, 175–176) you read about back in Chapter 5, I have to say that designers can also guide people's behavioral responses, and their accompanying experiences.

Again, I find the similarity between poetry and design to be striking. In the following quote, replace "poem" with "design" and "poet" with "designer."

*Reading, you turn a page to a poem you haven't encountered before...As you begin to read, to follow the poem word by word, the poet sets out one thing after another [before you]. With every noun the poet sets out, there is something for you to envision, first a chicken, then a washing machine, then a half-melted candle in the neck of a jug. Verbs move these things around, defining their actions.*

*(Kooser, 2005, 111)*

Actually, Hughes, Kooser, and Rae say much the same thing from slightly different perspectives. Bob Hughes' thoughts on computer-generated narrative paths (Hughes, 1997) find that

*Most approaches to computer-generated narrative focus entirely on the creation of paths. But there is much more to narrative than this... Each path is chiefly a route through a particular terrain – and the terrain is the main thing. Each is popular because it is a very good route that begins promisingly, develops interestingly, and concludes satisfyingly. However, the Pyg*

Figure 8.2 Pyg Track, Mount Snowdon, Wales; Shutterstock ID: 1153739200.

*track [up Mount Snowdon] is no use without its surroundings and if you were to divorce it from the landscape, you would have nothing very much.*
*(Hughes, 1997)*

Hughes does make a point worth considering. But, given the control the changing path's surface has over one's experiences in the Imperial Tea Garden, I feel there is an interdependency between path and surrounding terrain. As Hughes notes, "…Pyg track [up Mount Snowdon] is no use without its surroundings…" but then would Mount Snowdon exist in so many people's memories, letters to friends and family, and photo albums without Pyg Track? Take a moment to challenge your thinking. Put Hughes' thinking of path as "chiefly a route through a particular terrain [where] the terrain is the main thing" together with Kandinshy's concept of a line as a "track made by the moving point" (1979, 57). For me, this is what a written narrative and a landscape-as-narrative are all about the interplay between the path and its surroundings. Only you are designing the surroundings and your path brings people to them.

Hughes plays with a number of variations on this idea. If I understand him, a narrative's sentences are segments of a path. As they are read, a landscape forms around them – in the reader's mind. In a playful way this echoes Simonds' idea of designing for experiences and people's expectations and Rae's recommendation to give your character(s) the freedom to take you on their path through your emerging design; or we could say "their emerging design."

Hughes (1997) provides two examples of what he means by different routes having the same goal. Advertising copywriters "In a really good campaign (such as David Abbott's classic press campaigns for Volvo in the 1980s) the start-point of each ad is different, so is the way it develops, but it is still somehow the same story, every time, however you tell it. You are taking the reader "over the same ground" again and again. Good salesmen do this to an even greater extent: "their 'terrain' is the product, and they take every single customer through it by a different route, tailored exclusively for them, on the fly, in real time" yet with the same goal – the sale."

*Four of Abbott's ads, in bold type, along with images of various models of Volvo, say:*

*"TO A RADAR GUN THEY LOOK EXACTLY ALIKE"*

*"How to survive a mid-life crisis with dignity"*

*"Invest in durable goods"*

*(and with a shot of the car's stop lights on, "100 TO 0 IN 3.2 SECONDS"*

Hughes' "second insight came when [he] read Frances M. Yates's 'The Art of Memory' (Hughes, 1997, 4). Yates explains the classical training system for orators, whereby they built up imaginary places for themselves – memory palaces, cities and theatres – in which to store the components of a train of an argument. When preparing an argument, they would plan a route through the palace and place mnemonic objects along it. When the time came to speak, they would simply walk over their ready-planned route, presenting each item to the audience."

*Most poets wish to achieve some definite effect upon their readers, and a reader's collective associations with words have a lot to do that effect.…There are many ways…but one important way is to try to limit the variety of*

*associations a reader may have. You can do this with modifiers, especially adjectives.*
*(Kooser, 2005, 112)*

To say "path" is one thing; "well-trodden path" is another; and "well-trodden muddy path" another still. Take a second and think. Nouns, like objects in a design, bring with them numerous possible associations in people's minds. Modified with the appropriate adjectives you can reduce the number of associations that come to people's minds. In Kooser's words, you can "use adjectives sparingly and with precision to exclude the associations you don't want, and to make the remaining things on the table work toward the effect you want to achieve" (Kooser, 2005, 113).

Consider Kooser's thoughts on nouns and adjectives. Nouns, the things that come to make up your design and delineate activity settings, generate numerous associations. Adjectives modify and refine what the things (nouns) display. In doing so they provide associations related to place and history. Even with little knowledge of a locale's history one still senses something is new or rustic from the materials, the way they are applied and the way they are finished.

*When writing even a brief poem, you have hundreds of decisions to make – choices of words, of syntax, of punctuation, of rhythm, and so on...Every word must be selected for its appropriateness to the task at hand, just as each part of a machine must contribute to its effectiveness. Each choice the poet makes must bring the poem a little closer to its potential. It is impossible to achieve perfection, but any poem will be more effective if it falls just a little short of perfection rather than a long way short.*
*(Kooser, 2005, 120)*

Again, here, the comparison of poet to designer and poem to design is instructional.

All of the above is pertinent to design and the designer-as-author, or poet, or choreographer. Here again, there is a creative progression that goes from general to increasingly specific considerations. As Kooser says, "every word must be selected for its appropriateness to the task at hand" (120). So,

Figure 8.3 Wall is a noun. The modifying adjectives are the wall's treatment or character: stone, poured concrete, vegetated. Photos by author.

identify the task at hand and as Rae says, let your characters run free. Don't force them; go with them. Let them introduce you to the magnitude and nuance of what appears along and gives form and character to their paths. "Let the stories play themselves out on the screen in your mind with a volition all their own" (Rae, 1996, 5).

## GETTING READY TO WRITE

In the same way so many people see what you do as a designer as magic I am sure a lot of you consider what writers do as magic, too. In an effort to dispel that silliness, I couldn't help myself, I had to take some space here to share a bit of Anne Lamott's "Shitty First Drafts." In that chapter, in *Bird by Bird* (1994, 21–27), she offers so much through her down and dirty, candid introduction to writing. What's here is only the briefest of summaries of her message in that chapter. You would do well to find a library copy or purchase a used copy and read the whole chapter. Actually, *Bird by Bird* is not only a wonderful introduction to writing but also living life. Its subtitle is "some instructions on writing and life."

> *Now, practically even better news than that of short assignments is the idea of shitty first drafts. All good writers write them. This is how they end up with good second drafts and terrific third drafts...I know some very great writers...not one of them sits down routinely feeling wildly enthusiastic and confident. Not one of them writes elegant first drafts...The first draft is the child's draft, where you let it all pour out and then let it romp all over the place, knowing that no one is going to see it and that you can shape it later. You just let this childlike part of you channel whatever voices and visions come through and onto the page. If one of the characters wants to say, 'Well, so what, Mr. Poopy Pants?' you let her. No one is going to see it....Almost all good writing begins with terrible first efforts...Start by getting something – anything – down on paper. A friend of mine says that the first draft is the down draft – you just get it down. The second draft is the up draft – you fix it up. You try to say what you have to say more accurately. And the third draft is the dental draft, where you check every tooth, to see if it's loose or cramped or decayed, or even, God help us, healthy.*
> *(Lamott, 1994, 21–26)*

The following is important enough to share at this time. The reason understanding your potential users' emotional state of mind is so critical comes from what I have seen and heard from my landscape narrative workshop participants over the years. Considering their character(s) likely states of mind more so than gender, age, physical abilities, and other usual user descriptors allowed workshop participants to expand their understanding of the kinds of people coming to their site and the array of experiences they might anticipate encountering.

In so doing, the workshop designer-authors were then able develop narratives for quite a diverse collection of characters and their stories. Most of the experiences you'll design for are common to people around the world. For example, everyone wants to experience happiness no matter age, sex, height, weight. Understanding happiness at a conceptual level, or whatever other experience you seek, helps identify generalities that when included in your narrative contribute to a design that works with a larger user cohort. Happiness emerges from what? Less than happy. It has to be compared to something. Seeking a sense of well-being or comfort usually contrasts with darkness, confusion, and limited understanding. Better yet, people will come to your site from some place, an emotional place. Knowing, or at least having a sense of where they are coming from emotionally gives your narrative a starting point. A starting point calls for what follows: sadness to happiness, confusion to comfort, darkness to light. What is it they hope to expand upon, or diminish? This is where you begin to "see" and "describe" what I call environmental metaphors.

Simonds recognized the visual line character of simplistic landscape settings capable of triggering particular emotions. How might your character environmentally experience their state of mind? How might it transition as he/she approaches your site, arrives at its entry, and enters it, progresses through it in your narrative? As an example, imagine you've been asked to design a small urban park, a vest pocket park. You might start by considering your potential users as living with the pressures of business, or chaos of the city. Why would they step off the treadmill and come into your park? What is the array of experiences they would be seeking? How far down the sidewalk leading to your park do they see signals that "say" park? What do the signals indicate? What are the pedestrians expecting to experience as they approach the park entrance? Use Rae's advice and let them help you "see" how their experience unfolds?

Figure 8.4 Lines and objects indicative of emotions. Simonds (2006, 160). Courtesy McGraw Hill.

## SAMPLE NARRATIVES

Below are two examples of CalPoly Pomona landscape architecture students' narratives developed for their Healing Garden workshop. The first narrative's main character may surprise you. A third example was written by a Washington State University landscape architecture student for the Morning Star Boys Ranch in Spokane, Washington. The Boys' Ranch works with troubled youth. Sometimes the young men do not want to have to stay at the ranch, which in many respects, set in the countryside, is a ranch. We were asked to develop a master plan for the facility and did so with the help of the boys, staff, and spiritual leader, Father Joe.

## HEALING GARDEN NARRATIVE #1: THE GARDEN'S DUTY

This time of year is a season of changes. I will soon have a fluffy white blanket adorning me. But for now I am satisfied to show my coat of many colors. Do you hear that? It sounds as though someone is coming. This one shows signs of worry. What can I offer her? I hope she'll find comfort sitting under my friend's gentle branches, rustling in the wind. If she sat there she would surely hear the soft chirping of the robins and the delicate dribble of my fountain.

She wanders by this place I picked for her. But, what else can I offer her? She is starting to cry now, kicking some of my coat's patches as she goes. I hope she will meander a little further so my evergreen friends can help her. Could the pleasing smell of pine have caught her attention? She seems to have found a little cove to nestle in while she cries. This continues for a while, but then she stops. A squirrel darting up and down the tree caught her attention. Is that an attempt at a half smile as my robin friend quickly snatches the squirrel's dropped treasure? I can only hope I have helped ease her pain.

## HEALING GARDEN NARRATIVE #2: THE GARDEN AND ME

I wake up and feel I need to get some fresh air. When I approach the window to open it I notice the garden below. I wonder if the beautiful bouquet of flowers someone placed on my bedside table came from the garden. Intrigued, I am drawn outside to discover the source of the flowers. As I exit the double doors, which face the entrance to the garden, I am stunned by the bright colors of the wildflowers. My eyes are adjusting to the sunlight. The gentle swaying of the foliage beckons me to enter. As I enter I notice these words that provide me with a sense of comfort and safety. (The words were not provided.)

As the gate opens I see that the path diverges and I have a choice to make. I hear wind chimes faintly coming from the direction of one of the paths. As I proceed toward the sound I am mesmerized and enchanted. I just realized that there is a light breeze and the air feels cool and clean. Now I am surrounded and sheltered by trees. I can hardly feel any wind here, but the enchanting sounds of the delicate chimes resonate in this space.

I pause and listen. I notice the sounds of the birds above and the rustle of the leaves. I have been so focused on the wind chimes that I have just now noticed the sound of water, although I cannot tell where it is coming from. I follow its lead. I know it is nearby but the plants around me conceal it from view. The view opens and I feel satisfied, before me are ducks and I can make out the outlines of fish in the pool the water has collected in.

My face detects the moisture and coolness in the air here. I breathe the moisture in and catch the faint fragrance of herbs and orange. I follow my nose to discover a wide-open space, sunny and warm. The sight of growing plants that bear fruit excites me. I think of what we ate for dinner last night and I realize that this is the very place that some of our seasonings came from. The home-grown plants give me life and I begin to feel at home. Someone left a bucket of ripe apples out. A shady picnic table entices me to sit down and I enjoy a fruit. I would like to give back to this garden, perhaps I will be able to tend a patch. I should try to make this place more a home for me.

## MORNING STAR BOYS' RANCH NARRATIVE

### The Beginning

I can't believe they are sending me here. I'd rather be anywhere; even the streets are better than a stupid ranch. They're not making me into a cowboy.

### The Approach

Is that it? There aren't any fences or bars. How do they expect to keep me here? Hey, they've got a baseball field, animals, horse trails, open fields… What is this Club Med? This is going to be too easy. If they're dumb enough not to have bars, they ought to be easy enough to manipulate….Well at least I'll have a roof over my head and food. If I don't like it I can always run away. After all, no bars.

### The Recluse

They can lead a horse to water, but they can't make it drink. I don't care what they do, I won't leave my room. I won't join any of their idiotic activities. I'll show them! See a psychiatrist my ass! I won't say a word to him. I'll sit there in silence the whole time. We'll see how well this "Therapy" BS works! I have to at least go to the activities, huh? Well, you can't force me to join in. You can't make me do anything I don't want to.

## The New Kid

I'm bored. Maybe I'll join in after all. This game sucks, but at least it will give me something to do. What are they staring at? I'll give them something to stare at. Hey you, give me that ball! What? Do you want to make me? Come on you****! I'll show you!

**Judgment**

They aren't even going to punish me? "Go to your room." What am I, a five-year-old? I don't care if I have to stay in here forever. It's better than joining their stupid activities. Look at those guys over there. Doin' whatever they're told, giving into their tactics. Well, I won't give in. Not me. I'll just sit here 'til hell freezes over... They do look like they're having fun though....I wonder what they're doing over there in that field...?

## The Introduction

I hate school. I always feel like such a moron when the teacher calls on me. I hate homework even more. I just won't do it. Instead, I'll....What? You expect me to plant a garden? I don't think so man gardening is for sissies....Look at that punk, weeding like there's no tomorrow. Since you're so good at it, you might as well weed my plot while you're at it. Oh yeah? How would you like to eat this nice juicy slug? Well, you just wait. I'll get you.....I hate school, but gardening is even worse.

## Guilty by Association

I hate the other guys in my living group. They're such a goody-goodies. I wonder why they won't stand up to me. What cowards. And they do everything they're told, without question. I wonder what they know that I don't. Hey, how come he gets to go horseback riding and I have to stay in my room? It's not fair.

**Ownership**

OK, which one of you ran through my gardenias? I spent an hour weeding them yesterday, and when I find out which one of you did it...I get recreation time today? Right on! Maybe this place isn't so bad after all. Don't tell anyone but I'm beginning to have a new respect for my roomie. Did I say respect? Hey Weed Wacker, race you to the basketball court! We'll see who cleans up around here!

## Getting Along

This "family" stuff isn't so bad after all. I wish my real family was more like this. I still hate group therapy, but at least I get a little more respect now. As for the gardening, it's not so bad after all. I never knew there were so many kinds of vegetables! Next week my group gets to go camping. I can't wait maybe if we behave ourselves, we can stay an extra night. I'm beginning to sound like Father Joe.

## Backsliding

Everything was going so well until today. I had everyone right where I wanted 'em. But then I got that question wrong in class, and it all went downhill from there. Who cares what Thomas Edison did? It didn't affect me any. I wish everyone would just go away and leave me alone. No, I don't want to join in your stupid game. Go away before I kick your ass!

## Moving Forward

That's strange. They act like nothing even happened. Not only that, but they're letting me have my own booth at the Harvest Festival. I thought

I'd be in trouble for what happened last week…. They say if I improve my History and English, I can go to the city high school next semester. If I study extra hard, maybe I won't have to retake my junior year….I hope my booth does well at the Festival so I can make some extra money for Christmas.

## Rewarding Actions

I can't believe I won second place in the largest pumpkin contest! I've never won anything before. Did you see the look on their faces when I stepped up on the platform to claim my prize? I've never been this important before. And next year, I'm going to take first. With a little work, I bet I can concoct a better fertilizer than any of these guys.

## A Helping Hand

Look at that new kid over there. He acts just like I did when I first came here. I wonder…? Not good at math, huh? That happens to be the one thing I'm good at – besides growing the biggest pumpkins around. Want some help? OK then. Let me know if you change your mind….Is it all right if I work a few extra hours in the greenhouse? If I can just perfect this new type of string bean I've been working on, I'll have it ready to plant in the spring.

## A Gentle Nudge

Living out here in this apartment has helped a lot. I think I might just be able to make it now "outside."

Buck says I have a good chance of landing a job as an agricultural research company with the experience I gained working in the greenhouse. With the money I've saved over the last year and a half plus the money I'll earn at this year's Spring Fest I should be able to buy that car I've had my eyes on. Now my only worry is passing my last history exam. Then graduation, here I come!….I can't believe it's time to leave already. Seems like I just got here. I'll miss this place. And who will take over my zucchini patch? They better do a good job of it, or I might have to come back…

What is interesting with the narratives above is the lack of explicit descriptors of the spaces. Yet, if you think about the characters' behaviors and attitudes as described, you can see how transferring the written narrative into a storyboard brings the spaces into "view," reinforcing Lefebvre's dictum that space is a social product (1994).

## NOW IT IS YOUR TURN – LET'S GET STARTED ON YOUR NARRATIVE

You have a choice. You can use the Healing Garden Project introduced here and write a narrative for it, interpret it into a storyboard, and then transpose the storyboard into a landscape design. There are points in the book where you are not given step-by-step instructions or visual examples. The reason is my belief that being given answers is not learning. Working for answers brings understanding. What I would like you to do is experience writing a narrative, sketching up a storyboard, even a simple one, and developing a design. These are the suggested steps. Apply them to your own project, or the Healing Garden project laid out below. Experience the process and its idiosyncrasies. Work through them and develop a better system than I'm giving you here. The overall narrative-storyboard-design process is a strong, user-centered approach to design, but it can be made stronger. You can help make it stronger by experiencing it, reformulating those experiences, and trying it out.

For those of you who want to work on the Healing Garden project, here it is.

An Introduction to the Providence Sacred Heart Medical Center Garden Project. This project is dedicated to Sister Bernadette and her wish for people to have a place in which they find solace.

**Situation:** Sister Bernadette, a nun with the Sisters of Providence, called you together to explore design options for a special kind of healing garden. She notes,

> *Sacred Heart Hospital is technologically on the cutting edge with regard to healing the physical body. The Sisters want to see a balance brought to patients' healing experience, that of their families and friends, and the greater staff. You have a chance to introduce a healing environment that works in a more holistic way with people's spirits, minds, and bodies.*

Integrated with the diverse, holistic healing-related programs to be offered in the soon-to-be renovated Providence Center for Faith and Healing building along the western boundary of the site, the healing garden will benefit patients and their families, doctors, nurses, staff, and visiting scholars and the students who study with them. The garden's visitors and the Providence Center's primary clientele are the terminally ill and their families and friends. As part of their welcome to the hospital and the Faith and Healing Center, people will be introduced through an orientation program to the healing garden and available counseling services. The Providence Sisters anticipate that hospital staff who are daily surrounded with terminal illnesses will regularly avail themselves of the garden. Garden strollers will enter the garden either from the double doors on the ground floor of the Providence Center for Faith and Healing (located along the western boundary) or from a gateway just off Eighth Avenue (along the garden's northern boundary).

Sister B's two explicit wishes are that the garden be spiritual but not denominational and that it always be open to those who need it. "There will not be any religious iconography. Yet, people of all religions are to feel welcome." As a meditative place, the garden is to help people find their way, find solace.

Ideally the garden is to help visitors replace the all too frequent fear and anger associated with impending death with a growing understanding and acceptance.

> *This two acres of Eden is to be available to people day in and day out, twenty-four*

Figure 8.5 Aerial of Healing Garden Site; Google Maps; Imagery ©2022 CNES/ Airbus, Maxar Technologies. U.S. Geological Survey Map data © 2022.

*hours a day, through every season. They are to be able to physically and spiritually meander about garden and to find solace in seclusion. In their moving about in a subtle beauty, patients, doctors, staff, and families will all come to a realization that all hasn't been lost. As for myself, I want to walk the garden on a winter's evening and hear snow crunching under my feet.*

**Site:** the approximately two-acre garden area is surrounded on the east, south, and west sides by four-story brick buildings and open to the north. The old garden and now dilapidated swimming pool were used by nurse trainees who resided in the building along the garden's eastern boundary. That building is now a residence for the elderly and for dementia patients. While these folks may use the walkway alongside that building, they will not be able to access the garden. Their outdoor needs are fulfilled by facilities on the other side of their building. However, many of them will be able to view the garden from their rooms and two gathering rooms on the second and third floors.

Starting into the site from Eighth Avenue (on the north), the ground rises six or seven feet, crests at the garden's entryway, and then dips down to a concrete-lined pool at the base of a large basalt outcrop at the back of the site. A concrete-lined stream emanating from the heart of the site flows into the pool. The stream originates in the center of the site, runs north, and then curves west into the pool by the basalt outcrop. The stream's curvature runs along the base of a small knoll topped by two trees, a lamppost, and a bench seat. From the western side of the stream the land slopes up about 7 feet to this centralized high point at the base of the knoll. Halfway down the southern side of the knoll are the remains of a large brick barbecue pit. To the south of that are the empty remains of a large oddly shaped swimming pool which was fed by springs coming down off the hill to the south of the south-bounding building. The springs are still accessible, so the availability of water is no problem. Due to the low site elevation at the southern end of the site and the higher elevation to the north along Eighth Avenue removing surface drainage is a concern.

Figure 8.6 Composite of original healing garden site. Looking south into the back half of the original site. Photos by author.

Looking into the site from Eighth Avenue, the first third of the site is heavily overgrown and will need thinning and to be replaced. All site features are in disrepair and cannot be salvaged.

The four-story brick structure along the south end of the site is the Providence Auditorium. There are windows along the length of the enclosing wall. The doors at both ends of the building wall are emergency exits and must remain unobstructed.

The southern portion of the western enclosing structures houses the soon-to-be torn down hospital laundry. The structure itself is in excellent condition and able to support just about any amount of weight brought to it. While the Sisters of Providence have only talked about the structure as being torn down, their way of thinking leaves the door open for ideas regarding renovations associated with the Providence Center and garden. From inside the garden the four-story brick structure has industrial-sized windows and a four-story wrought iron fire escape. Although the two western buildings' walls abut each other and appear continuous, they are separated on the street side of the buildings by an employee entryway. Also, the laundry facility's street side façade is one story tall. The street side façade is made up of 8′ tall arched windows that look into both the street-level and lower floor of the laundry.

The Providence Center for Faith and Healing, along the western boundary, is to be renovated. The Center will house programs related to various forms of therapy: dance, aroma, herbal, music, sandplay, and the like. It will also, at the south end, house a library and study area for up to five resident scholars here on sabbatical to research aspects of healing. The building is all brick. The four floors are reinforced concrete. There are windows running the length of each of the four floors of the building. The building used to be a convent, so most of the windows open into small rooms. The bank of larger windows along the garden's ground floor opens into what was a cafeteria. You will be working with an architect and an interior designer to recommend changes to both the interior spaces and the exterior façade of the structure.

Directly across Eighth Avenue, to the north, is the hospital's emergency entrance. The arrival and departure of ambulances are a regular occurrence. SPOILER ALERT: the arrival and departure of emergency vehicles and helicopters landing on the roof atop the emergency building are surprisingly not distracting while in the garden area.

## GETTING STARTED

While you got a lot of information on the garden, most of what you read about the site really doesn't matter at this point. It will apply some during your storyboarding and then definitely more when you turn your storyboard into your final design. Consider the following, then maybe take a day off and let your story emerge from your thinking. Then start writing as you allow your character or characters to lead you toward, into, and around your garden.

## CONSIDERATIONS ON WRITING A NARRATIVE

- Prepare yourself. Maybe reading Elisabeth Kübler-Ross's "On Death and Dying" will help you understand people's state of mind. Or, web search reviews of her book for a quicker introduction.
- Write/think in terms of first- or second-person experiences. Do not write about "a person." Write through your character's thoughts and emotions.
- Where are they coming from emotionally? What is their first impression? What expresses it? Where is it, where does their narrative start? What do they experience between their first impressions and arrival at the garden entry?
- What is the theme, its introduction with the first impression, how does it develop, to what kind of conclusion? In what state of mind do you wish the garden participants to leave?
- Make the narrative experiential, feel what your character's senses are drawing in.
- Write with sensitivity; without evaluation (that will come later).
- Let your character(s) take over, don't make them do things. They will inform you as to what they are feeling.

- Watch their actions; consider what's around them that they are reacting to?

## COMPOSITION

- Developing your character's state of mind, personal perspective, transitioning locations, social/cultural values and idiosyncrasies.
- Stage set: foreground, middle ground, background; stage right; stage left. Progression.
- Setting (place/location) reflective of someone's mental state.
    - What is the mental state with which people arrive at the site?
    - How will the garden deal with it and, with luck, transform it? Through what (thematic) progression?
    - Experiential setting; where do they arrive in the theme's progression; in what surroundings do they enter and progress? Always remembering their state of mind.

## BENEFITS OF YOUR NARRATIVE

- Insider's view/experiences versus impersonal plan view.
- Focusing on the design's theme as seen/experienced by your character removes you from the equation.
- When you present your narrative-storyboard through your character(s) clients will ask questions about them and not "why did *you* do this?" or "why did *you* do that?" The first time that happens it is really exciting.
- Build a continuity across the space and time. Once you start a narrative you've focused on a theme to be developed. How does the theme develop? What are the stages? How might they be environmentally expressed?
- What is the language associated with the theme your character is experiencing? What is the language of solace?
- This approach builds on the opportunities and constraints of the setting and place.

Okay, with all this thinking on why narrative brings such a valuable contribution to landscape architectural design, you've succeeded in putting off writing a narrative. Now is the time to turn on Rae's movie in your mind, or what the old Radio Days referred to as The Theater of the Mind. You gotta do it eventually. As Rae says, it is actually quite easy once you experience it. Do it now. Don't worry about how it reads. It is a first draft. A first draft of your narrative, maybe two pages. That is a good start, but only that, a good start. You want to use your understanding of why your characters would likely come to your site. It'll get clearer as you write, and then as you review what you wrote on your own and then with others. This is creativity at work. A somewhat different kind of creativity than you're used to – a kind of collective creativity. Here is where you link meaning and imagination guided by coherence and continuity provided through your narrative's character(s) acting out in your mind's eye.

Take a break, gestate, jot down notes as they come to you, and come back writing.

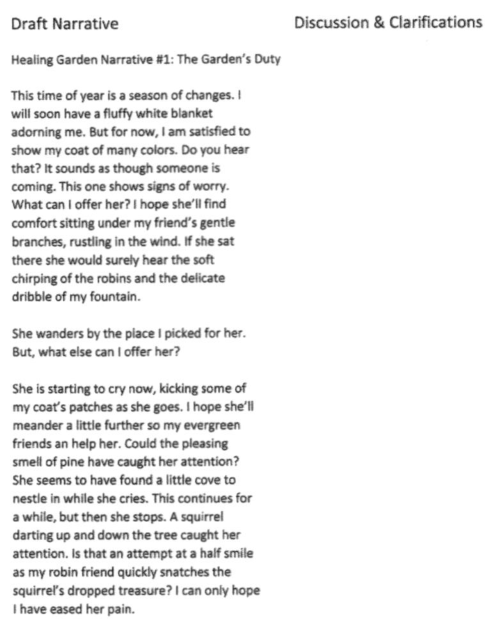
Draft Narrative | Discussion & Clarifications

Healing Garden Narrative #1: The Garden's Duty

This time of year is a season of changes. I will soon have a fluffy white blanket adorning me. But for now, I am satisfied to show my coat of many colors. Do you hear that? It sounds as though someone is coming. This one shows signs of worry. What can I offer her? I hope she'll find comfort sitting under my friend's gentle branches, rustling in the wind. If she sat there she would surely hear the soft chirping of the robins and the delicate dribble of my fountain.

She wanders by the place I picked for her. But, what else can I offer her?

She is starting to cry now, kicking some of my coat's patches as she goes. I hope she'll meander a little further so my evergreen friends an help her. Could the pleasing smell of pine have caught her attention? She seems to have found a little cove to nestle in while she cries. This continues for a while, but then she stops. A squirrel darting up and down the tree caught her attention. Is that an attempt at a half smile as my robin friend quickly snatches the squirrel's dropped treasure? I can only hope I have eased her pain.

Figure 8.7 Note taking worksheet for narrative reviewers. Prepared by author. Credit: Kyaw Thi

## NARRATIVE IS IN THE CAN (WELL, ALMOST)

Once you've got a draft of your narrative, remember Lamott's "shitty first drafts." Don't go overboard stressing out about how complete it is (or isn't), gather a couple, three friends together. Give each a copy of your narrative to follow along as you read it out loud. In fact, test it before you gather your friends or work cohorts. Find a quiet place and read it out loud. I guarantee you will hear gaps, misdirections, incomplete actions or thoughts. That exercise will be your second step to clarification following your writing. Remember shitty first drafts.

Using the Healing Garden Narrative #1 from above, here is what my team's note-taking sheet looked like.

Now get your people together. Tell them what you are doing and what you expect of them. Ask them to listen for and comment on points in the narrative that are vague, that have gaps between moments or people's experiences, or need an additional head turn, glance up or down, sound, or aroma.

Ask your friends to use their copy of your narrative to jot down notes, questions, thoughts on additions, and more and to hold their comments till you're done. Once done, everyone discusses their thoughts. I like to record the discussion.

Work to make changes to your narrative as the discussion goes along. While you will end up with a tightened draft of your narrative, if you have time, put it aside for a day. When you are fresh and snappy bright reread what you produced and make what needed edits pop out at you. Now, on to storyboarding.

# 9

# Storyboarding: Turning Words into Images

The cinematic nature of storyboarding provides a natural progression for your narrative, or scenario, to become a landscape architectural design. Landscape isn't a thing; it is a collection of meaningful and sequential experiences. For those for whom you design, the landscape is the stage on which they live life, raise families, play, learn, earn a living, and grow old. Storyboarding is used by major film directors and logistic planners. Storyboards save movie makers' millions and millions of dollars in production costs as kind of a visual test before the cameras start rolling. Storyboarding, as a serial or sequential visual planning tool, along with your narrative, captures your clients' minds and imaginations by allowing them to participate in your designs. This is where you cannot escape the fact you design living environments for others. This is also where you begin to realize the power you have to influence other people's lives and well-being. As such, this is where you begin to realize the responsibilities that come with designing environments and the lifeworlds for others. Keep in mind, those "others" may be animals, birds, or aquatic or marine life.

The quote below is so rich in the benefits storyboarding brings to landscape architectural, architectural, and interior design I had to include the entire paragraph. As you read it, consider how what you read would appear in an environmental design process or product.

*The storyboard, adapted from the technique of animation cinema, proved, nevertheless, so useful a device...For the director, the storyboard offered a way of planning on several levels: the composition of shots, their size, and their interrelation with the flow of the narrative. It allowed one to determine needed equipment and location sites. It represented an intermediate stage between the script and the editing process: it could, in fact, be seen as a form of pre-editing. It facilitated communication between the director and producer, on the one hand, and between the director and technical crew, on the other. The pictorial schema was often, but not always, supplemented by written instructions as to direction of movement within the shot, duration, camera movement. And since it can, in addition, specify lens changes, dissolves, and the projected use of other optical effects, the range of detail and accuracy of the storyboarding process is, therefore, very wide.*

*(Pace Gallery, 1993, 8)*

DOI: 10.4324/9781003286981-10

Figure 9.1 The Beekeeper Storyboards. Courtesy, Lynne G. Hood, illustrator, animator, author.

By now I'm guessing you are used to substituting landscape architectural terminology for its counterparts in quotes from or about authors and writing, poets and poetry, painters and paintings, and the like. With cinematography brought into consideration, the leap from film making to environmental design should be a small one. If, in that regard, your creative juices really start flowing and you enjoy testing ways to bring cinematic techniques into landscape, architectural, or interior design, get a copy of *Dictionary of Film Terms* (or similar book) (Beaver, 2009) and really challenge your creativity. Fill your journal or notebook with graphic examples of things like a jump cut, flashback, flash-forward, fade in, or fade out as they could appear in an environmental design.

## STORYBOARDING

For our purposes as environmental designers, storyboarding is the molding of sequential messages triggering a landscape participant's experiences into visual images accompanied with notes on direction of movement, progression of view(s), accompanying sounds, and more. Given that film budgets these days average $100 million, and set and on-location complexities, storyboarding allows the pre-visualization of a film and even the logistical program development that will be needed behind the scenes.

A real eye-opener for me regarding storyboarding was displayed at Seattle's Pacific Science Center as part of a program on *Jurassic Park* (1993). Among the displays was a storyboard, not of scenes culminating in the movie, but of heavy cranes, camera booms, semi-trailer trucks, water trucks, electrical sub-stations, crew housing, and more that was needed to produce the scenes seen in the movie storyboard and eventually in the movie. As Tumminello states, "…whereas comic strips entertain, the storyboard is a tool that facilitates production" (2005, 11). In the case of the behind-the-scenes production of *Jurassic Park* we could call it a Logistics Storyboard.

Early storyboards looked more like comic books, three to six drawings to a panel or page. "Because the drawings were confined within the panels, presenting a fresh idea or making a change was not easily done" (Canemaker, 1999, 8).

Storyboarding, as it exists, is a much more flexible tool than its earlier versions. It didn't come into its own until Walt Disney and the Disney Company production staff refined almost two decades of previous illustrators' work. As the number of sketches grew along with the complexity of production, understanding a story's logical flow and content became an increasingly difficult task to which story man Webb Smith had a solution. Around 1930 the confusion of so many sketches and notes spread across floors and desks led Webb Smith to, one day, pick up the sketches and pin them all "up on walls and large sheets of cork or celotexc boards" (Canemaker, 1999, 24).

*He pinned all the sketches and dialog to the wall in a sequence from beginning to end…then reworked the sequence…until he felt that story continuity was achieved…[and] the first cartoon using Smith's method…was the 1933 cartoon The Three Little Pigs.*

*(20)*

As Walt Disney stated, "the first real storyboard was born" (24). With it came a high level of flexibility. Thus illustrators, and later film directors, were using storyboarding to produce one image per page or sheet of paper. The value of storyboarding was well recognized by directors like Hitchcock (*Psycho*, 1960), Martin Scorcese (*Raging Bull*, 1980), and Tim Burton (*Edward Scissorhands*, 1990) who did many of their own illustrations as an aid in preplanning film production (Pace Gallery, 1993). If you are able to locate the book, *Drawing into Film: Directors' Drawings*, that accompanied the 1993 Pace Gallery show, you'll enjoy seeing the simplicity of the various directors' storyboard drawings.

"In a storyboard's sequential presentation of narrative and characters, weak spots are easily seen and changed, and the all-important storyline developed and strengthened" (Canemaker, 1999, ix). The same benefits are realized when storyboarding is applied to the development of a landscape design. Please note, a storyboard's benefits are further enhanced when preceded by a written first- or second-person narrative or what can be considered a script or scenario (Aminzadeh et al., 2016).

It is important that landscape architecture students recognize that

*Storyboard art represents hours of human thought, discussion, and arguments, along with moments of inspiration. It is a [graphic] manifestation of the ideas, emotions, commitment, and physical effort of individual artists and writers who must give birth to story elements daily. Yet, also be willing to 'kill their darlings' Media-like without hesitation and start over again.*

*(Canemaker, 1999, xi)*

As I read about Disney studios being the first to create a Story Department, separate from the illustrators, just for the creation of stories, I had to wonder, what if environmental design firms began to adopt the landscape narrative-storyboard approach to site design and started using story development and storyboards? What if students of animation for film, graphic novels, comics, and video games came together and hired out their services as storyboard illustrators to those environmental design firms?

In *Drawing into Film: Directors' Drawings*, The Pace Gallery (1993) displays storyboard drawings done by 13 movie directors. Among them are Alfred Hitchcock, John Huston, Akira Kurosawa, Federico Fellini, Terry Gilliam, Martin Scorsese, and Tim Burton. In looking at samples of their work, it is obvious they were not the best illustrators in the world. Even so, their simple sketches saved hundreds of hours, millions of dollars, and contributed to numerous classic films. It is important to see the simplicity in what some of the world's best movie directors used as a means to help them quickly and inexpensively gain "maximum control of the factors involved in the composition, filming, and editing of the live-action film in both its narrative and documentary mode" (7).

## STUDENT'S NARRATIVE AND STORYBOARD

The beginning of this student's narrative was captured in three simple sketches. More was eventually expressed (contained) in her narrative than these initial sketches. But, even with this simple set of text and sketches she was off to a good start. When sketches and written narrative were shared

Figure 9.2 Student's simple storyboard panel. Courtesy Jena Jauchius.

with her classmates, the resulting clarifications eventually became a complete and much more comprehensive narrative and storyboard.

To satisfy your curiosity, here is what is written to the left of the three sketches (Courtesy, Jena Jauchius):

Top sketch: Across the wetland, I see a landscape I've never seen before….trees, like tired old men arch over and seem to be entwined with one another. I cross the bridge that crosses the wetland to see if I can't discover what this landscape is. The only thing I really see is the entrance. But I can see that there is much more to it because a snake of trees extends into the field beyond.

Figure 9.3 Student's simple storyboard panel: top sketch; Courtesy Jena Jauchius.

Figure 9.4 Student's simple storyboard panel: middle sketch; Courtesy Jena Jauchius.

Middle sketch: I arrive at the entrance. There is an odd sign suspended in among 2 trees by a tangle of twigs and branches. The sign reads that this landscape is Earth Art….representing the Spokane Aquifer….I look down the path, but can't see much around the bend. That intrigues me and is what draws me in.

Under my feet, stones of different shapes and sizes are fitted together to form a smooth trail. I look up at the living ceiling and notice at how some of the branches have actually grown together. Some low hanging leaves tickle my forehead.

Bottom sketch: As the trail rolls along. Every so often a break in the trees reveals a piece of what I am assuming is earth art and a suspended sign tells me more about what it represents. I also encounter little areas where

Figure 9.5 Student's simple storyboard panel: bottom sketch; Courtesy Jena Jauchius.

exercises and activities take place. In one, I see a teacher and a group of students learn what they can do to protect the aquifer from pollution. Even the tree species, size and color change as I walk inside this art piece. Some trees allow a decent view into the fields beyond while others form a solid wall to divert my attention to the other spaces within the design. The trail takes me over, under, and through the landscape showing me different vistas, revealing more interesting ecological sculptures, and teaching me more about the aquifer. I round a corner and once again catch a glimpse of civilization. I know my exploration is nearing closure. I never knew this landscape could teach me so much about itself.

Take a few moments and reread the above narrative with the aim of finding moments in the narrative that would benefit from further clarification. What do you think when you read "I see a teacher and a group of students learn what they can do to protect the aquifer from pollution." What does the storyboard character see, hear, smell, and possibly touch that tells her that the students are learning "what they can do to protect the aquifer from pollution"? Answering that will call for a few more cards with sketches, edits and additions to the written narrative to be inserted into the storyboard, and in doing so bring greater clarity to the final design.

## STORYBOARDING – FREEHAND OR COMPUTER?

Yes, there is free storyboarding software out there but given what others have sketched, some starting with just stick figures, I can't help but believe you can produce storyboard sketches (the flow of your characters' wishes, movements, and experiences), more quickly than you can search a storyboarding app's "gallery" or "library." Given the speed with which you can produce simple sketches, with comments, that capture exactly what your narrative calls for, I find this comment regarding one storyboarding software a bit off the mark: "You can also choose from a stock library of photographs, graphics and illustrations if you don't have time to create your own." Software's benefits are in how it allows you, once you have your sketches scanned and uploaded to the software, to work with others outside the office, at a distance. Others can look over your boards and help with editing.

How you approach this next step, turning your written narrative into a visual narrative, is up to you. Necessities include your favorite marker or pencil, index cards or plain sheets of paper, pins, and wall space. Get a package of 3″×5″ or 5″×8″ index cards. These sizes pretty much approximate the aspect ratio of human eyesight. Two approaches to interpreting the imagery contained in your written text to your storyboard cards are given as follows:

1. Cut and paste the old way: using scissors cut up your narrative in a way that highlights each scene, then develop a sketch for each scene.
2. Sketch as you read your narrative: the character(s) acting out in your mind's eye knows what is and is not "seen" in each sentence of your narrative.

As you start sketching your visuals don't hesitate to include notes on the appropriate cards that record the particulars for the scene/location: sounds, movement, aromas or odors, plants with special qualities, canopy height, leaf density, and so on that would make your character's experiences as complete as possible. Remember, this process is one of give and take, expansion of scenes, modifications to scenes, and elimination of some scenes. Once you feel comfortable with the scenes you generated, pin them up and have a look. Talk your way through them – out loud. Does the story flow smoothly; does the flow make sense; do you need to add a scene/card here or there; more notes as to supplemental views or sensations? If you can, get away from your narrative/storyboard for a day. Do something else, then come back, go through it again, and then share it with cohorts or friends. Ask them to again review your written narrative as you go through your storyboard. Discuss where there's a need for supplemental information (cards and notes). How's it feeling?

## IDENTIFYING DESIGN CRITERIA FOR PARTICULAR EMOTIONS

I want to step back here and deal with an issue that students brought up on numerous occasions and that gave rise to the following exercise. This exercise, the Emotion Matrix, will help you with your storyboard, and prepare you to move your storyboard to final design and the increasingly detailed questions you'll need to answer. The following exercise grew out of students questioning, "How do I know what I'm suggesting for objects, spaces, materials, and views in the various locations along my character's route in fact contribute to the experiences being called for?"

At some point in authoring your narrative or developing your storyboard or interpreting your storyboard into your final design you will wonder, "How can I be sure what design features, materials, spatial character, color pallets and more contribute to a particular emotional experience?" Over the years, first unconsciously and later consciously, working to understand and apply John Simonds' epiphany found me developing the following exercise.

Looking back to my early days teaching landscape architecture, I obviously took Simonds' epiphany to heart. I didn't realize until years later the extent to which the idea of designing experiences appeared in my teaching and personal design projects. Simonds' observation that we design experiences, not places, spaces, or things, dovetailed with my belief that even a first-semester design student from a small rural community could become a great designer if willing to question certain of their personal experiences and chase down the answers. In the following Emotion Matrix exercise, you challenge your ability to question personal experiences you've taken for granted and, as it were, dig out the answers. The Emotion Matrix is a step-by-step exercise. The worksheet looks like this.

| Things / Senses | 1. Identify and list the things around you. | 2. Identify and list each thing's dominant characteristics. | 3. Describe in as much detail as possible what each of your senses was experiencing for each adjective listed under #2. | 4. Identify objects or landscape features that embody your cumulative adjectives |
|---|---|---|---|---|
| | Nouns | Adjectives | Cumulative Adjectives | Design Criteria |
| Sights | Slowly look about your place and list the things you saw | | | |
| Sounds | List sounds you were hearing, near & further away | | | |
| Touches | List what you were touching and what was touching you | | | |
| Smells | List the smells, aromas, odors | | | |
| Tastes | If you were tasting something, list the tastes. | | | |

Figure 9.6 Emotion Matrix.

Steps to filling out the worksheet:

1. Identify the emotional theme running through your design. Or, if there is a developmental progression to the users' experiences, pick a point in your narrative that calls for your landscape users to have a particular emotional experience. In either case identify that emotional experience: wonder, contemplation, playfulness, curiosity, contemplation. Note it on the line at the top of the worksheet.
2. Find a quiet place. I typically bring a pad of paper, pencil, and recorder. I jot down main ideas and talk out details. Don't worry about taking out loud. Put an ear bud in your ear and people will assume you're on the phone.
3. Take a deep breath, close your eyes, and relax for a moment. When you are calm, think back to a moment when you were totally immersed in the emotion you identified. Take your time. Eyes still closed, slowly look around your mental image. Picture what you were doing; where you were? Talk through this so your recorder gets the things you saw and with which your senses interacted.
4. Work down column #1 entitled "Identify and list the things around you." This is the easiest of the columns to fill in. As your mind's eye slowly looks around the place you related to your selected emotion, identify (in bullet form) those things to which each of your senses reacted. What things do you see: sand, waves, high arching pine trees? Hear: waves hitting the beach; wave sounds punctuated by gull squawks, sounds of approaching footsteps on sand? Touch: bare foot or sandals; breeze on face etc.…you get the idea. Again, really take your time and work to see, feel, smell, touch, and taste it all.
5. Work down column #2 entitled "Identify and list their dominant characteristics." For each bulleted item identified in the previous column, work to identify what it is "saying" to each of your senses. There are numerous websites that list the words we have for variations in things we see, touch, taste, smell, and hear. One such site prepared by Jacqui Murry lists words related to sounds. See those at https://worddreams.wordpress.com/2015/04/06/118-ways-to-describe-sound/

Hers is an interesting website for authors searching to include sounds in their writing. Murry provides numerous ways for authors to describe sounds. Her list is extensive and very helpful in imagining particular kinds of sounds in particular settings. Qualities of sounds not explicitly considered as you go down her list include "direction," "proximity," and "coverage," whether the sound is uni- or omnidirectional (focused or all encompassing). As landscape architectural designers, working in the four-dimensional world, these are critical considerations.

Look carefully at what your mind's eye saw. You'll be surprised at the detail that surfaces. Consider waves. Waves are not all the same. Some are lapping, some are crashing, and some roll up onto a beach and never break. Waves' shapes are different; speed and rhythm are different; coloration from crest to trough; highlights; and so on. As you likely already realize, filling in this column takes some time. It is worth the time. With practice the act of mining your images for more and more detail will become second nature as you become accustomed to seeing beyond an object and into the sensory details that make it what it is.

In fact before you start filling out The Emotion Matrix, it is worth taking time to mine your elementary school memories regarding each of your five senses. Okay, make it six, include the haptic sense. Touch or tactile is a great example of what you likely took for granted when you were immersed in your

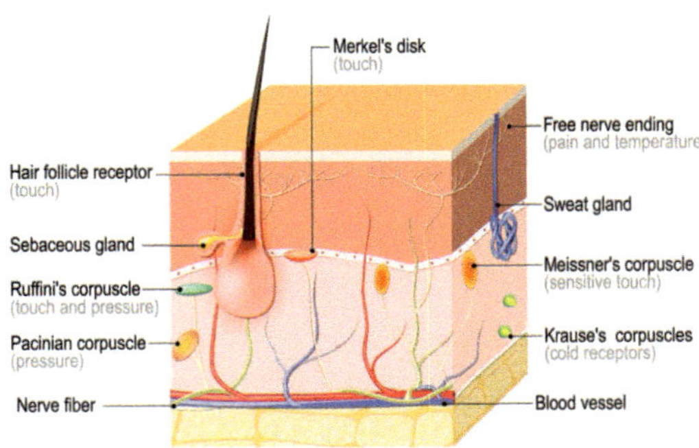


Figure 9.7 Sensory receptors in the skin. Shutterstock Contributor: Designua.

emotional-experience place. Tactile sensations include weight/pressure (light, heavy), texture (smooth, gritty), vibration/rhythm (free, longitudinal, transverse, dampening), temperature (cold, cool, warm, hot), and more. As you bring these sensations to mind you'll likely identify more of what made up your selected emotional experience. We rarely consider the sensations across our seated butt but we still frequently change our seated position and posture to improve our sense of comfort.

6. Now start down column #3. In as much detail as possible, describe what each of your senses experienced. This is the most difficult of the columns to fill out. Doing it causes you to truly understand what registers with each of your senses in regard to a specific emotion. With that recognition you are learning to identify what turns a good design into a great design. Your experiences are a composition of most or all of your senses.

   As we noted, lapping waves are distinct from crashing waves. Rate of speed, color gradients from bottom to top of the curl of the wave. Shape and density of highlights along curve from trough to crest. Sound, can you comfortably talk over it, or do you have to yell? What is the quality of the actual sound – flute like, bass drum, tambourine, or xylophone? Identify what you see, I mean really see, in a wave rising up, cresting and breaking. What are the colors, their shapes, rate of change, motion? Then consider the balloons at a poker party, the squeaking, and kinds of motion, effervescence of a soda or a beer being opened. The detailed descriptors that we associate with waves, balloons, and effervescence are surprisingly similar.

   An excellent example of "seeing" detail in water and waves is found in Frederic Edwin Church's oil painting of *Niagara* (1857). If you get a chance to see his painting, up close, look at the detail along the bottom of the painting. You will see extremely realistic rocks below the surface of the water. You will see the result of the kind of understanding of detail that makes a good painting a masterpiece.

   Once you have a list in column #3, look it over. Surprise! Surprise! What you actually have are the sensual criteria with which to guide the design of your selected experiential setting. The array of sensual characteristics are actually the criteria found in many situations that contribute to your selected emotional experience.

7. In the last column, 4, "Identify objects or landscape features that embody your cumulative adjectives." Consider each grouping of criteria you identified in the previous column. Now identify the objects in which they would be found or from which they would emanate. Work one sensual group at a time. How many landscape features and qualities does each

Figure 9.8 Frederic Edwin Church's 1857 painting *Niagara* measures 3.3′ tall by 7.5′ wide. Courtesy National Gallery of Art, Washington, DC.

grouping bring to mind? The effervescence heard when slowly opening a can of pop comes close to the rustling sound of quaking aspen in a light breeze. The sparkling highlights of lapping waves could be found in slow moving mobile, shimmering silver maple, or banners hung from light posts. Each contributes similar sensations at different costs. If you'd like a somewhat different explanation for this approach, see "Fluency and Flexibility in Thinking" in Adams' *Conceptual Blockbusting* (1976, 79–82).

At this point you have a palette of landscape items and the necessary characteristics and qualities to consider inserting, to some extent, into your text and definitely into your storyboard sketches, accompanying descriptors, and then your final design. Another way to think of this is that you now have a variety of landscape features from which to choose objects, features, spaces and spatial sequences, materials, and more that can make up your proposed design. This is a lot stronger reasoning for design decisions than the usual egocentric "I wanted," "I liked," "I decided."

Out of numerous workshops aimed at identifying the criteria for "Gaiety" my favorite response came from a rather outspoken imposing undergraduate who announced his recalled place experience was the Friday-night fraternity basement poker party. He was the kind of person who liked to stir things up. Everyone else had settled on places like early morning foggy beach, most favorite birthday party, mountain top vista looking over snow-capped range, sitting under a favorite tree or next to a clear stream. I surprised him when I said, "Okay, let's go with it and see what happens." What was most interesting, believe it or not, was that most of the students' final lists of criteria for sensual characteristics drawn from each of their identified memory places were very similar to his Friday night poker party. We learned that "Gaiety" found in seemingly diverse places in fact was often generated by very similar environmental sensations. Is this only indicative of Anglo-American and European cultures? In some instances, probably. In most instances they are probably cross-cultural. Either way, working in a foreign culture you could interview people or ask them to bring pictures that express the emotions you're working to unpack and work through The Emotion Matrix above.

Okay, time to take a break. Asking you to identify and describe things you've long taken for granted can be an exhausting exercise. Some of you, I'll bet, are even frustrated. Hell, you've likely never been this meticulous in considering your surroundings. That is the downside of becoming a designer. You can no longer take your surroundings for granted. Please stick with this, even the simplest list for each sense and each subsequent column is a start. It will pay huge dividends to the designer you are becoming. Believe me you will get better each time and eventually do this automatically.

## HOW DOES A LANDSCAPE NARRATIVE UNFOLD?

Here is another taken-for-granted moment. Consider how landscape narratives unfold as you move through your day. Remember earlier, there are times when messages are delivered opposite to what would be expected. Coming over the top of a hill or knoll, what's in the distance appears before what is closer to you. Walking toward a third-floor railing on a deck the tops of trees appear first, then the body of the trees, then trunk, and then how the tree is set into the ground.

Gibson's diagram, *Ecological Approach to Visual Perception* (1979, 199), provides an excellent example of how, in one environmental configuration, a narrative unfolds for the reader. Given movement through spaces, from one to another, rising or dropping with changes in elevation, the narrative sequence is not always all that straightforward.

In the sequential volumetric spaces one encounters when moving toward, into, across, and out of an environment there is the potential of transitioning in one or more of six directions: up/down, forward/backward, left/right. With these transitioning directional motions a narrative can be choreographed and become evident in a number of interesting ways. Unimpeded movement straight ahead can have you encounter larger to smaller, or smaller to larger spaces. The spaces may get wider or narrower and in doing

**The opening up of a vista at an occluding edge, as seen from above.**
This is a plan view of a passageway that opens on a courtyard from which another passageway leads. As an observer moves along the corridor, the surfaces behind his head progressively go out of sight, and surfaces in front progessively come into sight at one occluding edge and then the other. The hidden portions of the ground are indicated by hatching. The hidden portions of the walls are indicated by dashed lines. The position of the observer is indicated by a black dot.

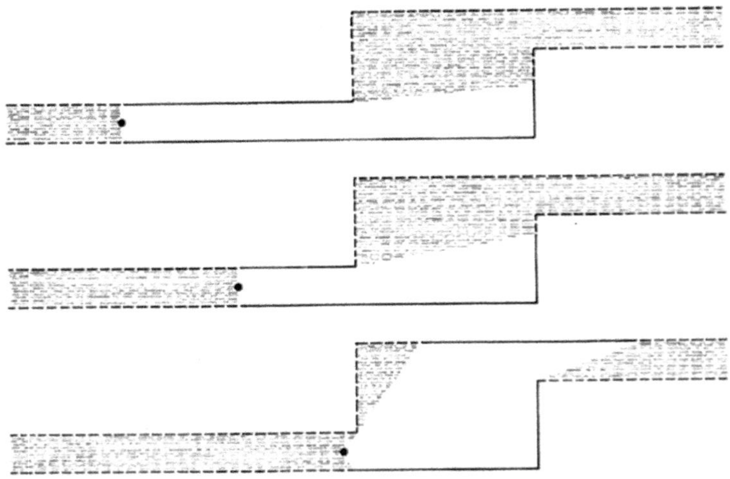

**Gibson, J. 1979. The Ecological Approach to Visual Perception. Boston: Houghton Mifflin, 199.**

Figure 9.9 The ecological approach to visual perception. Gibson (1979, 199). Courtesy of Taylor & Francis Publishers.

so go from brighter to darker due to shadows along with accompanying acoustical and temperature changes. Objects and spaces come into view and, in a sense, as we move forward toward them. They get larger. These are the qualities that can comprise your landscape narrative and be refined frame by frame in your storyboard.

In moving to the right more of what is ahead of me comes into my cone of vision from the right while what is there moves out of my sight to the left. If I move to the left, the appearance and disappearance of what, in a manner of speaking, is coming toward me is reversed.

I find moving right or left around objects at or higher than eye level or up and over crests of hills or down into channels or valleys to provide for more interesting narratives. The sequencing of what comes into view will challenge your creativity but the landscape's end users' experience will be added to by their choregraphed movements.

The sequential delivery of narrative statements as you move toward the crest of a hill or top

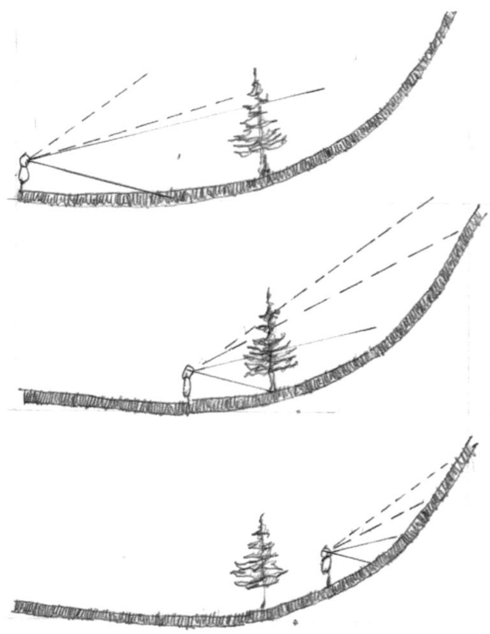

Figure 9.10 Message delivery when moving toward a hillside. Drawn by author.

landing of a stairway as you read in Chapter 5 regarding the stairway at the design firm of Sasaki, Dawson, DeMay moves from the top down.

Years ago I worked to provide a narrative that visitors would experience as they turned the corner from a hallway into a dining room, then moved across the dining room to a porch two stories above the condo's backyard. As they moved across the dining room to and then through a sliding glass door their view was out over a porch rail to the yard below. The goal was to subtly connect the interior and exterior spaces. As you approached the porch's railing two stories above the backyard the sequence of what came into view was the tree tops at the far end of the backyard, then the body of the trees, then the trunks, and then the ground. A few steps closer to the porch rail and the back of the yard came into view first and then the middle of the yard came into view. Reach the rail and lean on it and then almost all of what is below came into view. There is, in this case, your narrative sequence: top to bottom, distant to in close to closest. Not the usual, or expected sequence of messages but one to be taken advantage of.

Now take a first run at translating your written narrative into your storyboard. Take the same approach you did with your written narrative. Remember Anne Lamott's approach to "shitty first drafts": of just getting it down (no one is going to see it), fixing it up, then "the dental draft, where you check every tooth" (Lamott, 1994, 26). Even the dental draft doesn't have to be all that pain staking, you are still going to share your storyboard with peers or friends who will help you to tighten it up considerably.

## YOU'RE ON FINAL APPROACH

When you have what you feel is a good draft of your storyboard get your friends or work cohorts back together. See that they each have a draft of your newest narrative. With all your sketches pinned up on a wall or cardboard panels, or PowerPoint, walk everyone, sketch by sketch, through your storyboard. Ask them to help as they did with your narrative: pointing out what needs clarification, relocation, removal, addition, and filling in with more or less detail. Let the ensuing discussion determine how the storyboard sketches are to be rearranged, tossed out, added to with visuals and written directions: a distant view here, an aroma there, a change in the sound underfoot. As a test of completeness, this would be an excellent point to make a presentation to all concerned: clients and a sample of likely people who would visit the design. Given your previous presentations, before the narrative-storyboard-design approach, you may be surprised by people's, especially your clients', reactions to this presentation.

Remember, your presentation of your design, in storyboard form, will be through the eyes and thoughts of someone participating in the design. It will be an eye-level, personal experience. What you present is much more easily identified with than if it was described using spatial and geometric nomenclature. When sculptor-painter Ken Spiering and I demonstrated our combined narrative-storyboard presentation of the healing garden to the Sisters of Providence and hospital administrators and staff their reaction took us totally by surprise. A number of them started crying. When I realized people were crying I got scared. I'd never experienced anything like that before. Slowly, I began to realize what was happening. The presentation was not the usual description of what people could see in plan view. It was not the language of geometry delivered with the assumption that everyone would be on the same page regarding emotions and experiences derived from what was seen in the geometry in plan view. Our presentation was on the ground and personal and the folks around the table felt it. This was when I learned the power of a first- or second-person narrative approach to design. Our presentation was at eye level, easily identified with, and delivered through the client's intended users' experiences. What we orally and graphically presented was what they hoped their eventual garden would be to those who visited it, what it would bring into people's lives, and as Sister Bernadette had asked, to provide, a balance to the technological side of health care.

To this point in *Landscape Architecture as Storytelling* my focus has been on the value of a narrative approach to design that benefits the designer, that makes learning design easier, and that results in a design process more readily understood by your clients. What's ironic is that prior to developing The Analogy I'd been doing what I'd been trained to do, focusing on things. In working on The Analogy I realized that interacting with the client through narrative was a sensitive and very personal variation of Simonds' goal to design for experiences first and let the places, spaces, and things grow out of the called for experiences.

As said above, "you may be surprised by people's reactions" the first time you make a narrative-storyboard presentation. Now, let's start interpreting your storyboard into a design. Whether you are developing a project of your choosing or the Healing Garden project, it is time to take a run at it.

## TURNING YOUR NARRATIVE-STORYBOARD INTO A DESIGN

We all know there are various ways to approach a design's development. The narrative-storyboard-design approach is one more which, hopefully, you'll find value in and come to adopt. The narrative-storyboard-design approach to site design moves you from general issues, questions, concerns to increasingly specific questions, answers, and design decisions. Given the general-to-specific flow of the narrative-storyboard-design process, I see three possible approaches to interpreting your storyboard into your final design: first, move directly from storyboard to design in your existing site; second, turn your storyboard into a concept diagram without a strict reference (an overlay) to your site; or third, lay trace paper over your site's base plan and massage your storyboard into a concept diagram. Personally, I go with the second approach and develop a concept diagram first.

Understanding a general-to-specific approach to design decision-making highlights two reasons to move from storyboard to concept diagram before detailed design. First, you avoid the confusion that comes with mixing general questions and answers with more specific ones. Second, concept diagrams deal with relationships and not things. This fact relates back to working from general, fundamental, to specific, detailed. It is not unusual, especially for newcomers to design, to find themselves struggling with questions requiring detailed answers while they still haven't answered the broader general ones. Trying to determine the detailed design of a bench seat when you haven't an idea what the seat, given the design's theme, might look like, or maybe even exactly where it is to be placed contributes to a lot of unnecessary confusion. This isn't all that ridiculous. Many times, I've sat with a student or team of students only to hear them simultaneously struggling to answer both general and detailed questions. Adopting a narrative-storyboard-design sequential approach helps alleviate that problem, or at least minimize it. Its very nature is that of a decision-making process that moves from broad to increasingly more detailed decisions.

In an effort to reduce struggling with a variety of questions/answers along the general to specific spectrum, I like to go from storyboard to fairly detailed concept diagram, to refined concept diagram, to a schematic as applied to the site and its surroundings. Some of you will likely work to interpret your storyboard directly into your site. And, some of you may be able to do it, most likely those of you who have been doing design for a few years and are now testing the narrative-storyboard-design approach. The temptation here is to assume you have a handle on your storyboard and to focus on doing the design. If that is the case, I ask that as you develop your design, most likely working to organize activity areas and circulation patterns, frequently looping back to your storyboard. Trust in it. You've done a lot of valuable self-reflection to identify and express your design's users' experiences, emotions, and expectations in the making of your storyboard. Don't ignore all that now. You'll find that this work will also pay off when you present your narrative-storyboard to your client(s).

A last thought on the usefulness of concept diagrams is that when done, with or without the site plan, you have the activities, circulation, enclosure and openness, and spatial character's interrelationships set. The beauty of a concept diagram is that it expresses relationships more than things. It can be massaged, squeezed here, elongated there, and continue to maintain its original relations between areas varying activities.

In referring to your storyboard, many of you will likely realize your narrative's character(s) encountered vertical features that directed their movement and vision and overhead features that provided stronger or lesser enclosure. While plan view concept diagrams are common to design, vertical concept diagrams are less so. However, your storyboard sketches have given you a head start in visualizing vertical and overhead planes. Doing both helps you develop your design. Done well, the result can literally be considered an experiential choreography from which, as you add more and more detail from your storyboard, becomes a human choreography. After all, life is a dance and if choreographed well, your site design will, along with your users' experiences, whatever the theme, flow well, flow smoothly.

If you haven't already, now is the time to have your base plan ready to help you understand the site's ecological, social, and legal opportunities and constraints. "A site design deals with three elements: the pattern of activity, the pattern of circulation, and the pattern of sensible form that supports them" (Lynch and Hack, 1989, 127). Don't be constrained, think beyond the property boundaries. Hopefully your narrative started with your site visitors' first impression somewhere at a distance from your site. Don't hesitate, even though it is not within your site's boundaries, to include this environment in your design. There will be opportunities to work with the client, surrounding landowners, and local authorities to include your "first-impression" zone in your design proposal. Nothing lost by trying.

As you move back and forth between storyboard and concept diagram, you bring greater detail to your diagraming. In addition to greater detail, as derived from your storyboard, a schematic brings scale into play. As you refine your horizontal, vertical, and sequential relationships, you close in on the need to determine the scale of spaces, their dimensions, and the materials and the finishes making them up. Your schematic begins to capture both the explicit and the more subtle site qualities contributing to your narrative character's experiences.

Sitte (1945), Lynch (1960; 1971), Cullen (1961), Ashihara (1970), Bacon (1974), Arnold (1991), and more deal with scale, particularly what is referred to as human scale, in terms of both relative sizes and ratios of height of objects or plant masses to distance from viewer. "Scale is not size, it is the inherent claim to size that the construction makes to the eye" (Cullen, 1961, 79).

The concept of scale, captured in Ashihara's use of the word "feel," is related to human scale or relative size of an object or a space to people, typically an average-sized adult. In the following, "D" refers to distance; "H" refers to height.

*According to my observation D/H=1 is the critical point at which the quality of exterior space radically changes. In other words, if D/H becomes larger than 1, we feel that the distance between buildings*

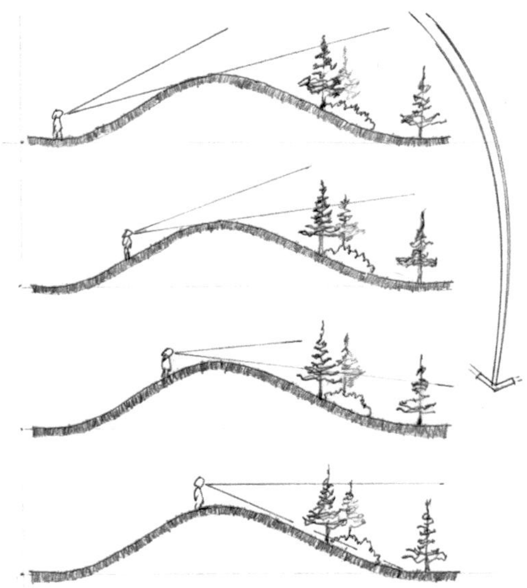

Figure 9.11 Message delivery when moving toward a hill's crest. Drawn by author.

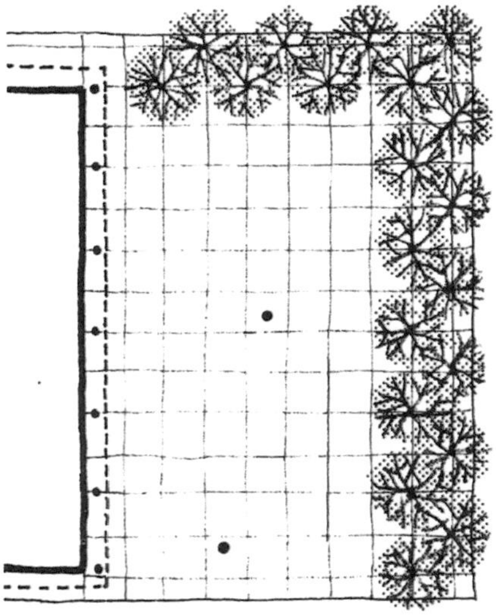

4-27 Plan: Horizontal definition of space. The observer at "A" has sense of enclosure on three sides since the distance from observer to wall (trees) is less than 3 times the tree height. The observer at "B" begins to lose sense of spatial definition at far end of plaza because distance is greater than 3 times tree height.

Figure 9.12 Defining urban open space with trees as edge (Arnold, 1991, 59). Courtesy Wiley Publishers.

*become rather great, while if D/H becomes smaller than 1, then we feel that the distances become rather small. When D/H becomes equal to 1, then, we feel a balance between building height and the space between buildings.*

*(Ashihara, 1970, 43)*

*Creating a horizontal enclosure with trees is different from using architectural building materials. Unlike the solid enclosure of a building… when viewing a space from the ground, there is a feeling of enclosure when the surroundings are tall enough to intercept our entire cone of vision. A convenient rule of thumb is that the observer should not be further away from the defining edge than three times the enclosure height. The vertical definition of space with trees is important… the ever-changing translucent canopy of leaves and branches allows for a simultaneous experience of the smaller space within the larger volume. In this sense, trees are more valuable than a building arcade for reducing the overwhelming scale of large structures.*

*(Arnold, 1991, 59)*

"The practice of using trees to make urban spaces is, in its most advanced form, inspiring. Urban trees are powerful symbols. Like words, they can be organized to create confusion or poetry" (Arnold, 1991, 47). "Trees organize spaces in two distinctly different ways – horizontally and vertically. Horizontally, they do it by visually enclosing, completing, or defining an area of open space. Vertically they define space by creating an airy ceiling of branches" (Arnold, 1991, 59).

Continue to refer back and forth between your narrative and storyboard, drawing by drawing, locate your design's participant in your site, think through what would ideally make up each space, each transition between activity areas: underfoot, overhead, up ahead, peeking around a corner;

sight, sound, touch, smell, and taste. Bring in what you determined best contributes to the site visitor's experiences. This again is developing the choreography mentioned earlier. What I find interesting is that the storyboard deals more with your site visitors' choreography while the design captures more the opportunity for orchestration: rhythm, tempo, harmony in overlapping sounds, and more.

In your experiential exercise, think about the qualities of the sounds your narrative called for then think about the things that provided those sounds. Consider where and when those sounds can be reflected or absorbed. Gravel underfoot, gravel mixed with moss. Overhead a dense foliage for maximum sound absorption (but also heavy shade) or a less dense foliage with reduced sound absorption and lighter shade. Give consideration to what those selections mean for fall and winter. Think back to Arnold's observation regarding honey locust. Its sparse branching structure provides light summer shade and after losing its leaves the tree nearly disappears in winter. The idea that the honey locust almost disappears because of its sparse branching structure means more sunlight during the colder months. These are the kinds of natural changes that come with landscape architecture's primary medium, plants that benefit from having been considered for how they change from season to season.

## VARIOUS DESIGN PROCESSES

As we noted back in Chapter 7 along with Figure 7.15, Design Process Sampler, there are all sorts of variations on what makes for a design process. Even so, they do have some commonalities. There is always a problem to be solved or an issue to be dealt with. There are various forms of inventory, analysis, idea generation, and evaluation of those ideas. However, as Motloch notes, "In rare instances, the characteristics occur consciously as discrete entities. More often they occur intuitively and without apparent organization" (Motloch, 1991, 239).

*Design processes are not linear in character nor definite in length. Rather, they are cyclical and ongoing. They have no predetermined beginning; the designer may enter by observing a problem, by having an idea, by physically creating something, or by evaluating some given situation.*
*(Motloch, 1991, 239)*

However executed the process remains goal oriented. I like Motloch's summation that "Herein lie the frustration and excitement of design" (239). But the goal is usually modified again and again right to the end. Design is a learning process all the way up to and through construction. That is why after all the modifications are made and noted on the respective drawings and the last shovel and frontend loader has left the site all the drawings are stamped AS BUILT.

There is really no reason to try to walk you through turning your storyboard into a design. At this point, given a fairly tight storyboard, I'm confident you will soon want more wall space or more corkboard or foamcore.

## CLOSING THOUGHTS ON THE BENEFITS OF A NARRATIVE-STORYBOARD-DESIGN APPROACH

Although I've mentioned this a few times it is worth restating the magnitude of John Simonds' aha moment when he discovered within his background as a landscape architect, that what counts as design is that "one plans not places, spaces or things – one plans experiences" (1961, 225). Simonds' epiphany is the foundation of a narrative approach. And, you know well by now, by "narrative approach" I do not mean a description of the physical environment, that would still bring into play assumptions of what's good, what works, and what fits. By "narrative approach" I mean through the mind's eye first- or second-person experiential expression of what will come to make up your design. So, what comes with believing in and practicing, an experience-based narrative approach to developing a design?

*Simonds' homage to Corbusier: "For he perhaps of all contemporary planners can, in the simplest diagram, most powerfully grip the mind and kindle the imagination. He thinks, I could now see, not primarily in terms of lines and masses, but rather in terms of impacts. He designs and shapes and arranges human experiences."*

*(1961, 225)*

There are benefits that come with each phase of a more traditional design process. Just the same, what The Analogy's narrative-storyboard-design process has you do is reorder the sequence to allow you to take better advantage of those benefits. Plan view is a critical tool of our profession, an important part of your design process, but toward the end of the design process, not the beginning. What I mean here is that a design seen in plan view allows spatial, structural, and horizontal relationships to be seen and discussed. Elevations and cross sections provide for horizontal and vertical relations to be considered. An overall plan view aids in understanding how an existing and eventually proposed site fits into the fabric of its surroundings. The end product says your client cares about where the site is located with regard to its surroundings, who her neighbors are, and how they live, work, and play around the site. All these plan-view benefits are best realized toward the end of the design process, having already determined shape, form, color palette, character, and more from your empathically arrived at users' experiential expectations. In other words, the narrative-storyboard-design process starts with why a design, as it will be designed, exists, from a user's perspective, not the designer's perspective. The narrative process functions to provide a developmental progression of particular experiences, in particular locations, and in particular sequences. Those experiences as drawn in your storyboard will help determine the length, width, and height dimensions of the volumetric spaces and their defining objects.

*After you've written your first drafts, you begin to revise, making critical judgments about what you've written, decisions made in part upon which word goes best in which position, about the appropriateness of a figure of speech, about the use of grammar and punctuation, and about how your writing may affect an imagined reader. Revision, and I mean extensive revision, is the key to transforming a mediocre poem into a work that can touch and even alter a reader's heart.*

*(Kooser, 2005, 16)*

Starting your design process with a narrative brings forth four benefits. First, the act of writing brings with it the need to selectively start your narrative, to thoughtfully pick that first word. It could be your character's state of mind, or her first impression of the site. Either way you are taking time to put thought into a process that has you considering what words to use that best capture your character's attitude, expectations, place, and progression. In that you make conscious decisions to use one word over another, you initiate a thoughtful, experience-based process regarding approach, arrival at, entry into, and movement around and through the site design. Even when you start by allowing your characters to determine how your written narrative starts, that requires a focused selection of words and sentences. That selection process is further refined when it comes to reflecting your narrative off the visuals making up your storyboard. One side note in regard to clarity in writing is, to the best of your ability, never start a sentence with a pronoun, especially "it." You will be amazed as to how much clearer your writing will be and as a result your visuals.

Second is the fact that your design is based on and grows out of people's experiences. To a great extent they are not assumptions on your part. You've encountered many situations throughout your life that are similar to those you will design. If not, if you find yourself dealing with a situation with which you are not familiar, then it is time for research or bringing in a paid specialist. With a first- or second-person narrative as the starting point of your design, what makes up your emerging landscape, architectural, or interior design is prompted not through assumptions but through empathy; your ability to recall similar experiences, and the character of those places giving rise to those experiences. As a result, your skill at sharing the feelings and behaviors of your narrative's characters greatly helps determine the spaces, places, and things that come to be your design. You are building an experiential foundation on which the sustainability of your design is likely perpetuaterd through people's everyday routines.

Third, accepting various authors and poets' call to allow your narrative's characters to show you the design, their actions and reactions illuminate those sensory triggers to be included in your design.

Allowed to enjoy your design, your characters' participation helps you identify the, all too often taken-for-granted, sensory triggers that make a place memorable. Almost like reverse engineering, your sensitivity to your characters allows you to see, smell, touch, hear, and maybe even taste the things and spaces that make up your design's physical qualities. In this way the increasingly more detailed design decisions are put off till later, when they are more easily addressed. Many of your characters' sensations are of things not usually seen in a plan view.

A fourth benefit comes with your narrative-based discussions with your client that revolve around your characters and the spaces, places and things they experience. You no longer state design decisions as having come from you. You may go through an entire presentation of your narrative and never be asked, "Why did you do...?" or "Why did you decide (or worse, feel) this belonged here?" Working through your narrative removes those so often heard statements on your part: "I wanted," "I decided," "I liked," and "I felt." Instead, the discussion revolves around your characters or, more typically, your characters' experiences. What you are doing is stimulating your client's ability to empathize with someone participating in what will be their design. Questions directed at why particular objects, spaces, materials, etc. are in the design remove you from the hot seat. Your answers can be drawn from The Emotion Matrix, relating experiences, emotions, and expectations to past similar encounters with which your client probably identifies. You could reference related studies. You may even involve the client by asking them, "Do you remember the last time you felt _____? I'd bet this or that was reaching out to you and interacting with this or that sense."

The four benefits give you a strong head start on what makes up your design, and what makes it what it will become.

I haven't tried this, but I see no reason why you shouldn't make a presentation to your client of just your narrative, or narratives, without the storyboard. Yes, you may author more than one narrative of people experiencing your emerging design in different seasons, early morning and later in the evening, as youngsters and as adults. Your character gets to comment to him or herself, "I have to come back here tonight and see if...." Making a presentation earlier in the process without the storyboard will help with making corrections earlier rather than later, and better guarantee your narrative's characters (yes, I know it is really you) correctly grasp the client's wishes.

While it may seem I've spent more time on writing, on word selection, than storyboarding, I may have. I've long felt that the better a designer can write, can narrate her design as experienced, then the better that designer can see her design. As a result, moving from written narrative to visual storyboard is made all the easier. This is an arduous task but so was doing design in your first studio. Remember Kooser's thought on the creativity in writing poetry: "It is impossible to achieve perfection, but any poem will be more effective if it falls just a little short of perfection rather than a long way short" (Kooser, 2005, 120).

# 10

# Ethics in Landscape Architectural Design Practice

Ethics in the practice of landscape architecture concerned me for 50 years. There was always a niggling feeling that in our concern for the environment, although we talked about people, about those who would use our designs, those who got left with our designs, what we did wasn't enough. My feelings finally got put into words as I struggled to nail down what it meant for me as a landscape architect to claim to be a steward of the land (Scarfo, 1988), to teach landscape architecture (Scarfo, 1987; 1989), and to clarify if what I say as an environmental designer working in the service of others actually shows up in what I do (Scarfo, 1992).

> *The received wisdom of today belongs more and more to experts who hold their licenses without ever having asked themselves or having been asked whether immediate experience with the subject at hand might in any particular way be relevant to their qualifications.*
> *Murry Kempton, (Gratz, 1989, 15)*

> *The ethics of the individual architect were replaced by the ethics of the architectural office, and the more the architectural office resembled businesses in general, the more did its ethics resemble those of the business world.*
> *(Boyle, 1977, 342)*

Over the years, as I worked on *Landscape Architecture as Storytelling*, I thought I had put aside ethics and the practice of landscape architecture; I thought I had replaced it with my desire to better understand how to develop, consistently develop, really good landscape designs. As luck had it, the more I explored John Simonds' belief that we design experiences and not places, spaces, and things I realized that people came first. As landscape architects, people were our main concern. Plants, ecology, topography, soil mechanics, laws and regulations, and business practices were all important areas of knowledge, practiced through their respective skill sets in support of those for whom we practice our profession. But, when we put people first, when we approach design as a stage on which people experience life as a wonder, we more readily realize the other areas of study are employed in support of people's experiences, in support of satisfying their expectations as they move about the outdoors. The more I thought of this the more I remembered the limited attention my profession gave people as an ethical concern and wondered if it had changed.

## BACKGROUND: HOW WE GOT HERE

> *If we are to design psychologically healthy urban settings rather than confusing alienating*

DOI: 10.4324/9781003286981-11

*ones, we must understand the individual as an instrument for decoding information, imparting meaning, and emotionally responding to settings. We must also understand settings as physical entities with encoded meanings that are read phenomenally as wholes, rather than as aggregates of pieces.*
*(Motloch, 1991, 278)*

Walking the boardwalk along the Chesapeake Bay at the Aspen Wye Institute one morning I met and struck up a conversation with a woman also enjoying the early morning quiet along the edge of the bay. When she found out I was a designer, she shared a personal story. She and her husband had recently contracted an architect to remodel their kitchen. Three times they shared their ideas with the designer only to be disappointed each time with what they were given. "He wasn't listening to us. What we were seeing wasn't what we were asking for." As she described it, the third presentation/discussion was terminated with her looking the designer in the eye, pointing to her lips, and saying "Read my lips. You're fired."

Was the designer exercising his ethical responsibilities in not providing the clients with what they asked for? I would say "no." Yet, at the same time, I'm sure the designer was not being malicious. I'm sure the designer was doing basically what he had been taught to do. Just as I am sure the designer was being sincere in his efforts to provide the client the best design possible. Still, the bottom line remains. In ignoring his clients' experiential goals he was not exercising an ethical approach to their wishes. For these reasons I want to speak to an area of ethical practice that does concern the public, not just its health, safety, and welfare, but the public's daily, experiential lives and emotional well-being.

Being recognized as a steward of the land used to rest on an individual's ability to work on the land, know its needs and those of its animals and plants. That was a necessity in order to provide sustenance for one's family. Over the years, from 700 AD to about the 1800s, the stewards, or later yeoman farmers, became increasingly recognized for the academic degrees they earned more than their personal experience with flora and fauna, soil and weather (Scarfo, 1988; 1989; 1989a). Today we graduate well-educated landscape architectural designers, some of whom have never planted a tree.

My sense is that two social forces, both from within the still emerging profession of landscape architecture, contributed to the self-avowed stewards of the land being separated from the very thing they claimed to serve. One driving force was the effort to be recognized as a profession (Larson, 1977). This is what social scientists today see as one of the steps in the professionalization process (Vernon, 1987, 13). That effort, becoming a recognized professional within a government-accepted profession, requires the popular and legislative support of the public. With legislative approval a professional association may exercise the powers of self-evaluation, such as exercised by the Council of Landscape Architectural Registration Boards (CLARB) and delivered through the Landscape Architect Registration Examination (LARE). The parent association, the American Society of Landscape Architects (ASLA), may also exercise self-control and "become almost immune to external regulation. The fact remains, however, that their privileges can be lost" (Larson, 1977, xii). Most ASLA state chapter associations have confronted sunsetting: the potential withdrawal of public support resulting in the loss of professional licensure.

The other social force separating landscape architects from the world they design is the profession's desire to be accepted more for the sciences and technologies it employs than a social-psychological awareness of their design participants' experiences. From the very beginning, the founders of the ASLA noted two purposes for the professional society.

## *The constitutional statement of the goals of the Society*

*[were]... 1) to promote fellowship among the members, and 2) to increase the 'efficiency' (i.e., the technical expertise) of the profession. The members made no mention in the goals of service to the public.*

*(Vernon, 1987, 14)*

Reinforcing the lack of attention given those they serve, even before focusing on the science and technology applied through professional practitioners' skills, the Society's concern for honorable business practices saw the development of ethical codes "ostensibly for the benefit of the client... their primary purpose is the protection of the practitioner" (Vernon, 1987, 14).

Over the years my approach to landscape architectural design shifted from thinking mostly in plan view to seeing designs through the user's experiential views at eye level. The shift was from objects and spaces as objects along with the assumption I knew what those things meant to a design's participants to letting the users' expected experiences guide my design's development. Unexpectedly, for me, that shift brought with it a growing ethical awakening, not one found in the American Society of Landscape Architect's (ASLA) Code of Professional Ethics. To this end I take this closing chapter to discuss designers' ethical responsibilities not as businessmen and women, not as ecologically responsible designers and planners, but as the designers of people's lifeworlds, of those places that exist in support of people's daily routines, and, even more so, as supported by those daily routines.

## BACKGROUND: THE EMERGING PROFESSIONAL SOCIETY

*The first 'Official Statement of Professional Practice: The Practice of Landscape Architecture,' as published in 1911, consisted of three sections: Scope and Values, Methods of Practice, and Official Statement. Under 'Scope and Value,' the authors described the work of the profession in three ways. First, landscape architecture was 'the art of fitting land for human use and enjoyment'*

*(Vernon, 1987, 16)*

rather than the other way around – the art of fitting human use and enjoyment to the land. Today, our ethical responsibilities are the culmination of what started in the 8th century with the introduction of the concept of stewardship and grew into a professional educational process in the early 20th century that separates landscape architects from the land and the people they claim to serve (Scarfo, 1988; 1992). In becoming recognized as a profession those same people took on the responsibility to determine what constitutes human use and enjoyment and then design landscapes that provided for them. In so doing, ever since then, the professional association has worked to formalize and sustain its decision-making authority.

While the 1911 "Statement" makes professional landscape architects responsible for their actions and interactions with their profession, their peers, clients, contractors, and the plants used in their designs their understanding of the recipient-users of their designs, those left to live with their designs day in and day out, remain "assumptions" behind their design and regulatory decisions. Section Two, "Methods and Practice," outlined design process and generally understood products, e.g., park and playground design. The Third section dealt with the professional's expected business behaviors. There was no mention of the reception of their designs on the part of the users. I have to imagine that what could be deemed an inward look on the part of a profession so concerned with financial and business ethics played a role in the fact that it still "did not garner the public visibility which landscape architects sought" (Vernon, 1987, 16).

As Vernon notes, the "1917 report of the Committee on Professional Practice and Ethics... did not regulate the behavior of professionals toward their clients [or landscape users] so much as regulated the behavior of professionals toward other professionals" (17). This ethical responsibility to each other, and their clients was further elaborated upon and adopted in the 1927 Code of Professional Ethics. They did come to refer to "societal goals" which today are "public health, safety, and welfare" but still did not focus on designing for people's particular experiences in particular places. Professional responsibilities of that nature are generalities left first to be interpreted by designers, then typically diminished by public regulations, and then even further diminished in public hearings.

Pretty safe to say there is little attention in the American Society of Landscape Architecture Code of Professional Ethics, Constitution, or Bylaws given those who live with the design and planning we do. Professional Ethics are all about business and public relations. My guess is that the assumption, of the profession's leadership, although not stated, is that the Landscape Architectural Accreditation process sees to it that new designers' education instills in them an ethical concern for those who live with and use their designs. But nowhere is it mentioned or are there expressed standards to be met.

As you'll remember me saying earlier, when I was introduced to landscape architecture, I wondered what the formula was for good design. Today, the question would probably be, "Is our understanding of what constitutes the basis of good design evidence based and how do we arrive at it?" As we've found there are numerous case studies, a growing body of research, and publicly approved projects used as precedents of good design. Yet, other than the science and technology of site engineering and horticulture, researchers' findings, and the legal regulations surrounding building codes and zoning regulations, the recognition of people and what constitutes good design is to a very great extent left to individual designers or teams of designers. Given the level at which design is explored, the findings and recommendations are given to rather than derived from those we as designers, not businesspeople, serve. A good example can be drawn from the Americans with Disabilities Act (ADA). Design guidelines drawn from the ADA assume to understand the needs of a particular social cohort. That understanding relates to function, physical capabilities, and access to and use of various environments but omits understanding individuals' emotional states, their concerns, fears, and wishes (Rowles, 1978; Rowe and Kahn, 1998) and how we, as landscape architects, can help improve the lives of the elderly, the infirm, the young, those with intellectual and developmental differences, and those considered active and healthy.

Such design guidelines as derived from the ADA and more would benefit greatly from the work of gerontologists and geographers, public health professionals, and epidemiologists. How would a landscape narrative differ for an active, middle-aged individual versus an elder relying on a walker and looking at the landscape ahead of them, while wondering if they did, in fact, take their morning medications or how far away a bathroom is? What kinds of concerns would each narrative express at the functional level (getting from here to there), at the social-cultural level (why doesn't my community care about me?), and from the deeper, collective unconscious level (why am I being made to feel like an outsider?).

The world of designers' assumptions can, to a great extent, be replaced with an approach based on first- and second-person narratives and a storyboarding of those narratives that give rise to designs. What is most interesting is that it wouldn't be difficult to build a professional program around such an approach that meets the Landscape Architectural Accreditation Board (LAAB) accreditation process and in so doing turn what are now designs based primarily on assumptions and opinions into designs that grow out of identifiable and defensible facts.

Think back to your introduction to the healing garden in Chapter 8 and let's consider a non-narrative approach to the healing garden's design. Clients have been listened to. Objects have been, as they say, moved about on the surface of the earth until their arrangement looks good. In plan view the garden respects Sister Bernadette's wishes. In elevations and perspective drawings the views are inviting, green and colorful. When all is said and done a garden has been designed. It is a visually pleasing environment. Your clients are happy with it, as it is seen in the drawings. It has plenty of seasonal color. Lots of greenery in contrast to its surrounding urban, concrete, and brick setting. There are benches to sit on. It has at least one if not two water features emanating from existing basalt outcrops. Jets of water shoot into the air. Night lighting, paths, benches, and slopes respect the Americans with Disabilities Act.

Now, step back for a moment and consider where you are in your narrative-storyboard-design approach to designing the healing garden or your own design. You've maybe not designed it yet but at least you wrote a narrative and played with turning it into a storyboard. By writing a first-person narrative your approach is sensitive to the functional, social, psychological, and cultural contexts as dealt with and as experienced at eye level, by individuals looking for more than a garden – looking for a place of respite, rejuvenation, or, ideally, emotional comfort and calm.

Given our understanding of the majority of healing garden users' feelings, their confronting loss, their feeling lost, being confused with many more questions than answers, in the final analysis, is the garden developed in the plan-view approach described above as much of a healing place as yours is becoming using the narrative approach? In its layout and contents, the typically developed garden realizes people want a place in which to escape but it likely doesn't realize (or express) the particular healing needed to address an individual's emotional trauma. It may provide respite but we can only assume it helps visitors come to terms with the turmoil in their lives and lives of their loved ones, know they are not alone, know there is light (relief) beyond this otherwise suddenly dark and confusing mindset that comes with being told you or a loved one is dying. The plan-view designer, like most designers, is sincere, well-meaning, and concerned. Anyone would look and say that as a professional the designer met the standards outlined in the ASLA Code of Professional Ethics....but, knowing now what

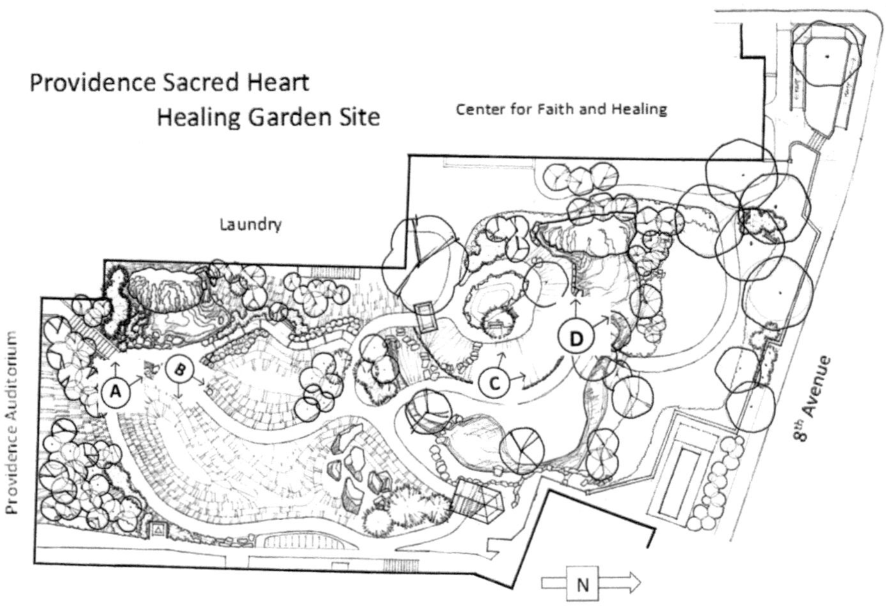


Figure 10.1 Healing garden design keyed to Ken Spiering's Sketches.

Figure 10.2 A – Garden viewpoint: looking north along west side of garden. Courtesy Ken Spiering.

Figure 10.3 B – Garden viewpoint: looking north along east side of garden. Courtesy Ken Spiering.

you know of design from an experiential-narrative perspective is designing in plan view early on in the process the way to go? More to the point, is it ethical?

The fountains may spew water 4′ to 6′ in the air, not trickle lightly down the outcrops' surfaces. The sound of the jetting water falling to the basalt echoes off the three buildings, as you get closer to a fountain the noise likely distracts you from what you are dealing with, from thoughts of a loved one or you facing an uncertain future. The colors may be too much, and

Figure 10.4 C – Garden viewpoint: looking south from north side of garden. Courtesy Ken Spiering.

Figure 10.5 D – Garden viewpoint: looking west to basalt outcrop and waterfall.

if so, distracting. There isn't a meeting between what the user is experiencing inwardly and what is around her. The garden is a well-designed garden, don't get me wrong. The designer was sincere in what he or she did. But, considering the garden space through the minds and eyes of the anticipated users can bring nuances into the design that are so much more appropriate to their states of mind and outlooks on life. All these concerns are much more easily identified and expressed in an experiential narrative combined with accompanying eye-level storyboard sketches than a plan view. Plan views are products, end points, to be arrived at later in the narrative-storyboard-design process and not developed at the beginning. The beginning is what you allowed your first- or second-person narrative characters to bring to you as per Rae (1996), Kooser (2005), and Lamott's (1994) recommendations, the manifestation of their behaviors as an outgrowth of their experiences with your emerging design.

After working through the pros and cons of plan view versus eye-level perspectives derived from a first-person narrative I realized the simple question is, which tells the more personal, the more sensual story, plan view or narrative come storyboard and eye-level perspectives?

## A DESIGNER'S POWER

By now you may be wondering, "What's driving Scarfo's concern? It can't just be his wish to find a formula for good design." Well, it is and it isn't. In my effort to understand how to do consistently good, no, great designs, meet people's expectations, and be able to help others learn what I've learned, I realized the magnitude of our actions as the designers of other people's worlds.

Designers wield a lot of power, yet few realize it. Thirty years ago I did a small survey of about 30 landscape architects in order to understand the profession's and professionals' impact on land and life. While I'm sure the numbers are different today I am equally sure the message provided by the findings remains. My survey's question to practitioners focused on their previous year's involvement in single-family residential design. Among the respondents, one had not worked on any single-family residential design during the previous year. One had done one 1/8th acre plot. Still the estimates came to a ratio of each designer having influenced 353 acres in the previous year and an estimated 37.5 people's lives each day.

Extrapolated over the approximate number of 10,000 registered LA's back then, we get, per landscape architect, 3,530,000 acres in single-family residences influenced per year, and an estimated 375,000 users' lives per day. Keep in mind that is just in single-family residential design. Today there are 267,000 people employed as landscape architects in the United States and 2,340 in Canada. You do the math.

## ACCEPTING THE POWER

As we become aware of, and admit to, the power we wield, then the question becomes, to what extent are we willing to accept the responsibilities that come with designing the environments in which people will live, work, play, and grow up? Over the years I asked hundreds of senior landscape architecture students, within a few weeks of graduation, "If I randomly selected one of your classmates to design the environment you would live in for the next five years, would you accept that?" Over 95% responded "No," some quite emphatically.

To my asking "Why?" the dominant response, by far, was "They don't know me." "But," I'd respond,

> *most of you have been together for four (and in some schools, five) years. You've gone as a group to classes. You've pulled all-nighters together. You've dated, broken up, and travelled together and more. How can you say you don't know each other?*

The discussion would continue. Usually about an hour into the discussion someone would blurt out something to the effect

> *We don't want someone we know and who knows us to design our living environment, and yet we're about to graduate and start doing that to people we don't know. From there on the discussion would turn to "What can we do to close that gap?*

## WE'VE COME A LONG WAY (OR HAVE WE?)

In *The Reflective Practitioner* (1983), Don Schon provides an excellent example of an environmental design professor and student's interaction (79–102). This may be dated, but given my design studio experiences with architecture, interior design, and landscape architecture students up to just recently not much has changed. Schon sets the context. The students received "...a set of design specifications...for [the design of] an elementary school...Each student is to develop his own version of the design, recording his results in preliminary sketches, working drawings, and models" (88). During one review the student describes how she is stuck with "a set of problems she cannot solve" (80). Laying trace paper over the student's work the instructor says, "The kindergarten might go over here…then you might carry the gallery level through – and look down into here" (80). Schon does an excellent job of describing a typical design instructor, student interaction; or what we know as the socialization of a future designer.

Whether the teacher or student "speak in words or drawings, their utterances refer to spatial images which they try to make congruent with one another" (80). In the socialization of soon-to-be designers to an abstract design process and language Schon sees "the language of designing is a language game which [the instructor models for the student] displaying for her the competencies he would like her to acquire" (81).

Schon sees the design language game as a conversation between words and drawings, speaking while sketching. "Drawing and talking are parallel ways of designing, and together make up what I [Schon] call the *language of designing*" (80). Schon's recognition of a "language of design" supports the benefits gained when you reflect your narrative text off your storyboard. The main difference between Schon's instructor-student dialog and your narrative-storyboard dialog is that yours deals with people's expectations and experiences while the other deals with geometry of spaces and objects.

Within the transcript Schon provides, the language-drawing dialectic is a back-and-forth reflection of geometric figures expressed in the integrated language of geometry and basic design. As presented in the teacher-student interaction, they are intertwined and gain direction from the instructor and student's likes, wants, and wishes, not an expressed understanding of the elementary school students, teachers, or school administrators. Note the "I" viewpoint in the following. "I chose the site because it would relate to the field there but the approach is here. So I decided the gym must be here…I have the layout like this" (82).

*I had six of these classroom units, but they were too small in scale to do much with. So I changed them to this much more significant layout (the L shapes). It relates one to two, three to four, and five to six grades, which is more what I wanted to do educationally anyway. What I have here is a space in here which is more of a home base. I'll have an outside/outside which can be used and an outside/inside which can be used -- then that opens into your resource library/language thing.*

*(83)*

The student "has also experimented with the size and arrangement of her classroom units, she has found that the classrooms must reach a threshold of scale in order to be 'significant' enough for design" (84). What that threshold is and how it was determined is never discussed or questioned let alone in terms of teaching and learning.

The professor's suggestion to the student, "now this would allow you one private orientation from here and it would generate geometry in this direction" (85), is further elaborated upon by Schon as needing "coherence [that] must be given to the site in the form of geometry – 'a discipline' – which can be imposed upon it" (85).

The only actual reference in 23 pages (Schon, 1983, 79–102) to any defined living, working, or learning activities in any part of the elementary school being designed is "we could have as much as 5-foot intervals, which for a kid is maximum height, right?" (88). And even that is stretching it.

Throughout the transcript decisions are expressed in terms of geometry as a "discipline," and "I" and "We" as the bases of design decisions. Never are the design decisions expressed in terms of teachers or students. References to geometry are numerous by the architecture professor and Schon. The assumption, all along, is that a particular geometry, seemingly picked out of thin air, would provide for the best learning environment: "...because he finds that [playing with geometric figures] has produced something that he likes" (100). Schon goes further in referring to the professor as "a chess master who develops a feeling for constraints and potential of certain configurations of pieces on the board" (104). Schon gives him too much credit as a designer of spaces and places in which people gather to teach and to learn. With the human dynamics of teaching and learning never discussed as an influence for how the geometry is determined to be good or bad, the question again has to be asked, "Is this an ethical approach to designing an environment that will, for at least 40 to 50 years, support the maturation and socialization of individuals who come to make up society?"

I'm sure you can already answer that question with some authority. If, though, you'd like some help, three great sources are urban anthropologist Tim Sieber's "Schoolrooms, Pupils, and Rules: The Role of Informality in Bureaucratic Socialization" (1979), anthropologist Norris Brock Johnson's "Education as Environmental Socialization: classroom spatial patterns and the transmission of social norms" (1982), and social scientist Paul Willis' *Learning to Labor: how working class kids get working class jobs* (1977). The three authors focus on the fact that what is taught and what is learned are often two very different things. Being instructed in Read'n, Write'n, and Rithmatic is what Sieber refers to as formal education. But as Sieber (1979) and Johnson (1982) show us "a school's dual function as an organization that at once educates children formally through instruction in 'the three R's,' [also] socializes them informally in the basic modes of behavior appropriate to bureaucratic settings" (Sieber, 1979, 274). Norris Brock Johnson (1982) complements Sieber's looking into the socialization of youth through their formal and informal interactions with authority figures by drawing our attention to not what is said and carried out but where it is said and done.

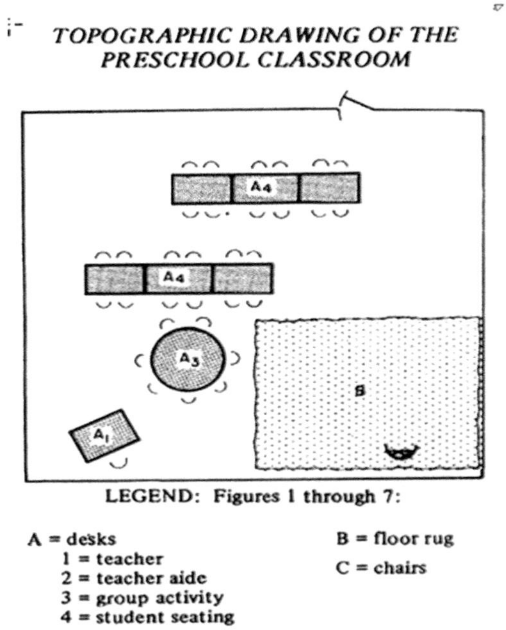


Figure 10.6.1 Preschool floor plan (Johnson, 1982, 34). Courtesy of the *Anthropological Quarterly*.

Johnson shows us how, through our formative years, we are socialized through the configuration, content, and character of our material culture. "As well as interacting with other human beings, people invariably interact socially and culturally with the materially defined forms and spaces they inhabit" (Johnson, 1982, 31). This not only makes sense given the recognition that space is a social product, but also adds to Churchill's dictum that "We shape our buildings and afterwards our buildings shape us." Consider how, over your first few years in elementary school, your indoor preschool play area disappeared and was replaced by desks all facing forward aimed at the teacher's much larger desk and blackboard. What were you learning from the evolving spatial makeup of your classrooms?

Johnson's study (1982) of a midwestern elementary school shows the transition from, in a sense, a "free-range" educational environment to institutionally structured Cartesian environment. While his findings spanned preschool through

TOPOGRAPHIC DRAWING OF A THIRD GRADE CLASSROOM

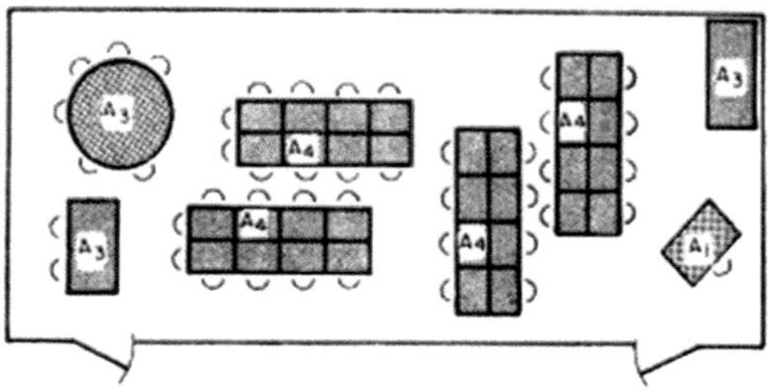

Figure 10.6.2 Third grade floor plan (Johnson, 1982, 38). Courtesy of the *Anthropological Quarterly*.

TOPOGRAPHIC DRAWING OF FIFTH/SIXTH GRADE CLASSROOM

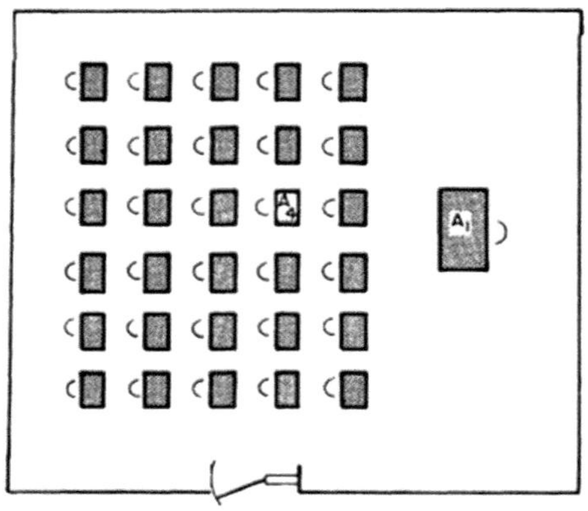

Figure 10.6.3 Fifth- and sixth-grade floor plan (Johnson, 1982, 39). Courtesy of the *Anthropological Quarterly*.

sixth grade, many of us likely experienced the same transition in a shorter time span, e.g., preschool to first or second grade.

Johnson refers to Edward Hall (1969) when he says "School settings themselves, the contexts of everyday classroom life, are a "hidden dimension." Then Willis (1977), while taking a different tact entirely, shows how socialization occurs among youth and between youth. "He shows how kids, through their own activity and ideological development, reproduce themselves as a working class" (1977, xi). With these individuals' insights into learning, what we must admit to is that space is not only a social product but also a socializing product involving the material environment and much of what transpires in it.

FIGURE 3.2
*Normative Design Domains*

| Domains | Definitions | Examples |
|---|---|---|
| Program/Use | Functions of buildings or building components; uses of buildings or site; specifications for use | "gym," "auditorium," "classroom"; "5', which is maximum height for a kid"; "no city will plow a road that steep" |
| Siting | Features elements, relations of the building site | "land contour," "slope," "hill," "gully" |
| Building Elements | Buildings or components of buildings | "gym," "kindergarten," "ramp," "wall," "roof," "steps" |
| Organization of Space | Kinds of spaces and relations of spaces to one another | "a general pass-through," "outside/outside," "layout" |
| Form | 1) Shape of building or component 2) Geometry 3) Markings of organization of space 4) Experienced felt-path of movement through spaces | "hard-edged block," "a geometry of parallels," "marks a level difference from here to here," "carry the gallery through and look down into here, which is nice" |
| Structure/Technology | Structures, technologies, and processes used in building | "a construction module for these (classrooms)" |
| Scale | Magnitudes of building and elements in relation to one another | "a 20' parallel," "too small in scale to do much with," "just the right scale" |
| Cost | Dollar cost of construction | (none in this protocol) |
| Building Character | Kind of building, as sign of style or mode of building | ("warehouse," "hangar," "beach cottage"—but not in this protocol) |
| Precedent | Reference to other kinds of buildings, styles, or architectural modes | "an artifice . . . the sort of thing Aalto would invent" |
| Representation | Languages and notations by which elements of other domains are represented | "look at it in section," "1/16 scale model" |
| Explanation | Context of interaction between designer and others | "the sort of verbal order you could explain to someone" |

Figure 10.7 Schon's normative design domains (1983, 96). Courtesy Hachette Books Group.

Granted, in Schon's transcript, there are accompanying graphics the professor and student produced as they talked, but what is interesting is that in designer speak versus first-person narrative speak of someone experiencing the spaces you often don't even know what they are talking about. A teacher's story, or a student's story, would have made this entire dialog clearer from the beginning.

Schon, as a language analyst, presents us with Figure 3.2 (on page 96 of *The Reflective Practitioner*), Normative Design Domains. It lists 12 normative design domains in all. No Design Domain identifies any group of people who would be left to live with the design if constructed…The "Domain" of Program/Use comes closest to recognizing the users with "such terms as 'classroom,' 'administration,' and 'kindergarten' buildings labelled according to their uses" (97).

Keep two things in mind. Knowing university professors and instructors as I do, I feel safe saying that neither the professor nor the student in Schon's transcript had any formal teaching/learning background. Yet, they both discuss the spatial configurations that would foster and support teaching and learning without ever mentioning those activities while they are both situated in a teaching and learning environment: their studio and building are dedicated to education. In addition, I'm sure both had attended kindergarten through sixth grade. The form of "reflection" they could have but didn't carry out could have been their own learning-experience stories which they then could have brought to their design process/discussion to help determine the "geometry." Remember, "...the places, spaces, and things take their form from the planned experiences" (Simonds, 1961, 225). But in Schon's instructor-student interaction the places, spaces, and things would have given rise to unexpressed and unexpected experiences.

While I don't believe it, I couldn't keep the thought out of my mind that just maybe the landscape students were correct in not accepting the random selection of one of their classmates to design the environment they would live in for five years.

## ON TRACK ACADEMY

Years later, in 2018, one of Spokane, Washington's alternative high schools, On Track Academy (OTA), received sufficient state funding to contract the design and construction of a flexible Makerspace as part of a larger middle-school high-school campus. In April 2019 Integrus Architecture, in Spokane, produced the Educational Specifications document (the cover of which is shown here) for On Track Academy and Spokane Public Schools. The 134-page document followed the interaction between the architects, OTA administration, faculty, staff, students, and community representatives between September 2018 and April 2019 and reported the results in three sections: Vision Development, Process, and Program. Discussions leading up to the "geometry" that would make up On Track's

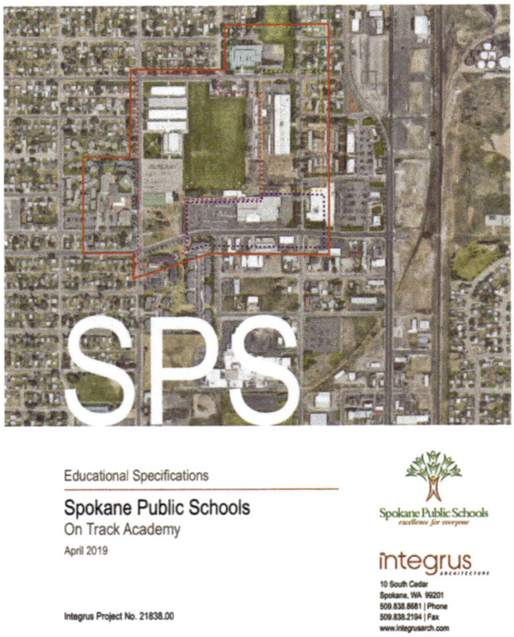


Figure 10.8 Integrus architecture. Report cover page, *Educational Specifications for On Track Academy*. Courtesy Integrus Architecture & Spokane Public Schools.

makerspace and new building were ripe with references to and feedback from teachers about teaching, students about learning, and the greater community and its involvement in the eventual school programs. In fact, in the architects' document, the words student and students appear 203 times, teach, teachers, and teaching 59 times, and learn or learning 136 times. On February 11, 2020 Integrus architects made a public presentation that was almost entirely about people, their asked for activities, the kinds of interactions that would occur between the kids, the teachers, the kids and their teachers, and the activities they would carry out. All this as the delineated physical structure and visuals of the various activity spaces were presented but not overtly described for their geometry.

If there is a spectrum along which we can see how designers are taught to interact with clients and user groups, Schon's architecture professor and student interactions are at one end and the Integrus architects are at the other. For me, I couldn't help but be curious as to how the Integrus architects came to practice such a people-first, user-centered approach to design. Did they learn it in school or did they develop it over the years after they graduated? So, I asked. Five of the designers' responses indicted they developed their sensitivities to users after graduation. The feeling was that their schooling dealt more with the "theory and philosophy" of design and spaces and that interacting with user groups "was completely learned after graduation." One of the architects, previously a student of mine at the Interdisciplinary Design Institute in Spokane, WA, shared the following:

*Personally, school taught me skills for presentation, vocabulary, how to think outside the box, look critically at things etc., and understand design. Studio focused on coming up with an idea and translating that into space, it was far more theoretical. From my sampling of a few other architects, consensus was it seems very little of our training was actually about the practice of architecture. It did not focus as much on interactions with clients, user groups, or putting together a set of construction documents.*

*I was asked once during an interview to show the process we would use to start the layout of a building on a site devoid of any input from the interview group (they wanted to see how we worked together in the approach of a design). I stopped half way through and said "THIS IS SOOOO TOUGH....because this isn't how we design!" A good design is absolutely based on input.... What are issues with drop off and pick up? What type of experience do you want students to have as they enter? Are they interested in outdoor learning? Where are the prevailing breezes? What are some of the security concerns? The list goes on and on!! The students, teachers, and public who use the building are the ones who know what works and what doesn't work and have the hopes and dreams for what it could become. We ended up getting the job... which was awesome.... and it validated for me how critical the time and energy spent meeting with the users was to the success of the overall design.*

## WHAT COMES WITH THE NARRATIVE-STORYBOARD-DESIGN APPROACH?

*And there is no doubt that equal access has improved society immensely, creating a society that is more tolerant, respectful, and inclusive. But equal access suffers from one major shortcoming: it aims to maximize individual opportunity on average by ensuring that everyone has access to the same standardized system, whether or not the system actually fits.*

*(Rose, 2016, 186)*

***Side note: if you apply this thinking to education you begin to understand why outstanding alternative high schools such as On Track Academy in Spokane, WA exist and come to outshine traditional high schools with a 94% graduation rate while the State average is 81%.***

***(see https://www.greatschools.org/washington/spokane/5759-On-Track-Academy/)***

I'm sure something similar to The Analogy is out there, but 35 years of working with a narrative approach to design, an approach that starts with a narrative of people's experiences and expectations, I have yet to find it. Designers, researchers, and authors of design studies are all sincere in their efforts to expand our understanding of how best to approach projects sympathetic to land and life. Their efforts focus on people, society, and culture, health and well-being, economic and resource benefits, deep ecology, anthropocentricism, and so much more. But their identified benefits regarding the recipients of each particular approach often remain, at best generalized across society, and at worse assumed. The benefits noted remain expressed as demographics, or a geographical area or region. Assumptions are averages and according to Todd Rose, there is "no such thing as average" (Rose, 2016, 4). What the narrative-storyboard-design approach does is flip the outside-looking-in to an inside-looking-out; the benefits of an ecological responsibility, health and well-being, social justice are still there but as most likely experienced through the eyes, ears, noses, skin, and taste buds of those who will be a part of the final design, the real subjects as living, breathing gotta-get-through-the-day people. As my previous student said, above, they "are the ones who know what works and what doesn't."

So, now, let's ask the question again. Given your understanding of the sequential, human experiences that can be identified and designed for in a narrative cum storyboard, is it ethical to move designer-selected objects about in plan view until the objects' layout "looks good" without reflecting on and specifically expressing the users' experiences and expectations?

## ETHICAL RESPONSIBILITIES

One major difference exists between design decisions associated with planning a trip and those associated with developing a design. That difference, and it is one I have never heard faculty, students, or practitioners discuss, is that the trip's decisions are made by the person taking the trip. Many of the decisions associated with a landscape, architectural, or interior design are made by a designer who is working in the service of someone else who, in a sense, will be taking the trip. Although licensed to work in the service of others, designers tend to work within their personal realm, or worldview. As a result, they tend to argue for designs as though the designs were theirs. Designers often talk as though they own the product, forgetting that the product is a set of drawings, and that they were paid to do them and therefore relinquish ownership. Insinuated throughout most arguments for the goodness of a design is the unspoken belief that "I will live in my product."

As designers we rarely live with our products, even while they are under construction. Still we argue (I wanted; I needed; I liked) for the goodness of our work as though we will be the residents, the problem is that in our presentations we are actually promoting the goodness of a commodity for sale. As a professional, whose reputation rests on the quality of my products, this is a vastly different view of my work from the false belief that "I will have to live in my product." Our professional codes of conduct clearly state that we work in the service of others. Others, not us, are left to inhabit and sustain our designed products. Therefore, there is a need for designers to recognize the social structures inherent in the systems that will come to sustain each design. With that recognition of the other's (clients and users) structuring systems we are able to unfold the social rationale in which our design decisions are legitimized and the physical palette from which the design gains its component parts. This approach complements Spirn's thinking throughout *Language of Landscape* (1998) and particularly her thought that

> *Architects and landscape architects tend to focus not on process, but on form and material; when they fail to be sustained in the real world, it is often because designers ignore the processes that shape them during and after construction.* (96)

There are a couple ways to consider our responsibilities to those we serve, those who live with our products as part of their daily routines. First, we are responsible for what we do for (and to) others, and second, done well, we are to an extent responsible for what they do for us, themselves and the earth.

The ethics I'm talking about here are those that guide one group of people who have the socially conveyed authority to modify landscapes, structures, and interiors in which others live, work, play, learn, raise families, grow old, and to an extent direct society. There are other forms of ethics dealing with ecology, natural resources, social justice, feminism, and more. But here, I want to shine a particular light on the ethical responsibilities designers accept with every project they take on. The responsibility to understand and anticipate not their wishes, wants, and likes but the wishes, wants, and likes of the people for whom they design. Given the interaction between teacher and student in *The Reflective Practitioner*, that cannot happen.

Given the rapid, interconnected pace of life today, designers can't, on their own, take full responsibility for the stewardship of the earth and the resources it provides. But, by developing designed environments with which the users identify, find personally meaningful, and sustain a sense of personal, and existential orientation in their individual and collective lives they are as likely as designers to carry out the responsibilities of stewards of the land. Remember earlier, the examples of people maintaining and sustaining those environments in which they "see themselves; with which they identify and in which they find meaning"? This approach and outcome, by its very nature, embodies the public acting as stewards of their environments and includes a high degree of social justice at a more detailed and personal level than related to natural resources, ecology, and land planning.

Designers can do this by adopting an empathetic outlook to those for whom they design – both the clients who pay to maintain the designs and those who use them. When the users sense they have been recognized and respected by the clients (usually not considering the designers' role), they are less likely to vandalize and more likely to care for a site and its features. As I noted earlier, I believe that most of what is considered vandalism is actually an effort on the part of the users to reformulate a place to that with which they identify and what identifies them to others. Again, go move some pencils, a cup of coffee, a notepad, or a laptop around on someone's desk while they are away. Call that your design and see what the person does when she or he returns to their desk. You can almost hear them saying, "I have to regain a sense of comfort with my space so it reflects who I am." As for me, Bob Scarfo, I still can't understand how anyone can leave a perfectly ordered desk at the end of each workday. That just isn't me.

Users who feel respected, who identify with and find meaning in their designed spaces, care for those spaces, steward those places and in so doing reduce the client's cost to maintain them. Merged with a design that similarly respects the local ecology, and to an extent maintains itself, maintenance costs are even lower. This is the kind of stewardship that comes with understanding users' values, expectations, and emotional attachments. This comes not with the language of geometry and basic design, but with the stories or narratives that make up the expected experiences of those for whom a design is being produced.

"Finally, consequentialist ethics often drives our thinking about future landscapes, how we should judge our current behavior in terms of its expected consequences" (Fisher 2013, 44). I always enjoyed Tony Bauer's (of Bauer Ford Reclamation) presentations of his approach to mine and quarry reclamation. I was most impressed with how he would, with the client's help, study the mining or quarrying processes and come to know the consequences of those processes. With an understanding of the years-long site operation he would then develop a program whereby waste, as it was being produced, was to be moved to those locations where it'd be most easily, with minimal additional movement, introduced into the reclamation design once the operation closed down. The reclamation design became a planned act of stewardship (Fisher would say, "Consequences") on the part of the mining or quarrying company's actions. By developing designs through the intended users' expectations we are, in fact, acting on the consequences of our actions as empathetic designers.

Our ethical responsibilities extend beyond what the American Society of Landscape Architects proclamations state. They shouldn't but they do. As professional landscape architects, yes we are a business with legal responsibilities, yes we are professionals with financial responsibilities to our peers, outside contractors, and clients. But we are also designers of the day-to-day worlds in which thousands upon thousands of people

Figure 10.9 Friends of the garden fund raising event, June 28, 2017. Photo by author.

live, work, play, raise families, go to church, and learn. This last area of responsibility, to those who use, live in, and grow old in our designs, is to a great extent neglected. Yet annually landscape architects influence hundreds to thousands of acres of land and even more people's lives in all kinds of land uses.

An outgrowth of authoring landscape, architectural, and interior narratives sensitive to and anticipatory of people's experiences is that we become much more aware of and sensitive to how well we can serve people of all ages and backgrounds. This is the area of ethics to which we need to be better attuned.

How we accomplish that needed attention is built into The Analogy. This area of professional design ethics that comes as a natural outgrowth of a designer-as-author, landscape-as-text, and participant-as-reader is a consequence of the narrative-storyboard-design approach to what we do as landscape architects.

In this day and age of tightening budgets (which seems like all the time) the Healing Garden received the love and attention of hospital and healing center administrators and staff in the form of a fund raiser to pay for needed maintenance and upkeep. The value to those who found the garden to be a refuge over the last 20 years speaks to the garden design's recognition of people's wishes and hopes. An even greater symbol of the garden's success in meeting people's expectations appeared in an odd occurrence. From the beginning Sister Bernadette said there was to be no food in the garden. Yet, somewhere along the line, people did

Figure 10.10 Friends of the garden fund raising event, June 28, 2017. Photo by author.

start to bring their lunches. Contrary to Sister B's concerns, and although there were no trash receptacles in the garden, there was also no trash left in the garden. The bringing of personal lunches into the serenity of the healing garden has expanded to summer events where there is music and food.

## CLOSE

I am sure that at some point you have thought "What's the likelihood of two designers employing the narrative-storyboard-design approach to producing designs with mutually agreed upon design outcomes?" The usual approach to design would not likely produce similar, mutually agreed upon design outcomes. Yet, in a CalPoly workshop, in which 27 students from freshman to third-year graduates, using the experiential narrative-storyboard-design approach, did in fact, and to everyone's surprise especially after hearing all 27 narratives, unexpectedly find a great deal of similarity across the designs. If there was a phase of mutual understanding and agreement in the narrative-storyboard-design process, it was with our defining the potential users, in terms of their states of mind, their reasons for visiting the garden, and their likely expectations. Understanding the likely users was the key to success. Remember the differences in the narratives. You read a tree's concern with a visitor to the garden. Another was a gentleman wondering if the flowers by his bed came from the garden. And one you didn't read was a dishwasher upset with the way his day was going, kicking trash cans out of his way, and stomping across the street to escape into the healing garden. With those 27 different narratives and more you can understand our surprise at the fact the resulting designs were so similar. I'm sorry I haven't copies of any of the designs to share here.

To date, the American Society of Landscape Architects guiding principles remain primarily aimed at business relations, working relations with peers, and financial practices with clients. What mentions there are regarding servicing the public's health, safety, and welfare, the substantive nature of design services, remains vague. Yet, somewhat more defined guides to professional practice are found in the ASLA Code of Environmental Ethics. The opening umbrella statement – "Members of the American Society of Landscape Architects (ASLA) should make every effort within our sphere of influence to enhance, respect, and restore the life-sustaining integrity of the landscape for all living things" (https://www.asla.org/ContentDetail.aspx?id=4308) – is followed by an extensive list of ethical standards expressed as professional practices (land planning and design behaviors) and anticipated environmental outcomes. This addendum to the ASLA Code of Professional Ethics and Standards of Professional Practice opens the possibility for a second addendum to the Standards of Professional practice that focus on design practices and outcomes based upon landscape users' anticipated experiences, their cultural and social backgrounds, and their diversity. It might be called ASLA Code of Site Design and Planning Ethics. As landscape architects continue to strive to be publicly recognized on par with architecture and engineering, a publicly expressed step toward design outcomes in terms of users' anticipated experiences would likely be favorably received by clients and the general public. I have to imagine such added clarity to our design and planning responsibilities, practices, and outcomes would garner greater public support for the Society and its members.

If a form of Site Design and Planning Ethics became a reality, then I could see comparable modifications being brought into Landscape Architecture Accreditation Board (LAAB) standards and associated assessment criteria. If ASLA Professional Site Design and Planning Ethics were introduced and related modifications made to the LAAB assessment criteria then it isn't difficult to imagine related modifications being made across Professional University Programs, the Council of Landscape Architectural Registration Boards (CLARB), and Landscape Architect Registration Exam (LARE).

Would all this remove the part played by assumptions in the work of landscape architects? I wish I could say "yes"; maybe over a number of years. For now, it would constitute a major step toward ensuring that landscape architects' design products contribute to the social and emotional makeup of the public's health, safety, welfare, and individuals' lifeworlds.

# Epilogue

I am still learning about my profession, my potentials as a landscape architect, and my responsibilities. Working my way through The Analogy's experiential narrative approach to design has helped a lot. The designer-as-author, landscape-as-text (or narrative), and user, participant-as-reader approach to design still has so much be illuminated and dealt with explicitly. I ask you to stay open and alert to what you read, see in photographs, watch in films. How many ways can one characterize an emotional experience in a landscape? How many ways can one trigger a planned for emotional experience; a particular emotional experience through vegetation, building materials, water, sculpture, cloth banners, and more? How can these items be used, not as primary expressive features but as a means of attracting your eye or ear to the primary feature, expression, statement; to the user-participant's sequential experiences?

Figure 11.1 Exley Falls, Horseshoe Lake, WA; Courtesy Karen Shields.

Searching for that formula that enables us to produce consistently good landscape architectural designs has been exciting and fun. Like John Simonds, only after much thought, lots of observation, and a number of loops back to my early days as a landscape architecture student, I realize the power in Simonds' epiphany. We design experiences. Not sure I've discovered the formula for good design yet but designing for people's experiences is certainly a key factor. My hope is that current landscape architects (or architects or interior designers) or some of you among those soon-to-be designers will pick up my formula-hunting challenge, dig deeper than I've gone, and help make learning and doing design less ambiguous, more focused in its approach, and more consistently appropriate to the people and places we influence.

In reading this far, you have uncovered so much regarding designing landscape experiences. My hope is that even if you do not adopt the narrative-storyboard-design approach, you will employ what you gained from *Landscape Architecture as Storytelling*. I hope you have become sufficiently sensitized that you are open and alert to what you read in your surroundings, in photographs, in films and apply that to designing a better world for others.

## LOOKING BACK TO POSSIBILITIES

As designers we can learn a lot from fiction and non-fiction writers and those who walk the blurry line between the two: e.g., Henry David Thoreau (1973), Wendell Berry (1968; 1969; 2013), Barry Lopez (1987; 2004; 2019), and Loren Eiseley

(1957; 1978). All along we've talked about the movement of the user as a reader of landscapes and a little about sun and shadow, changing seasons, and rustling leaves in trees. Before leaving narrative, we should think about the other forms of motion and other forms of seeing that can be used in a landscape text to supplement a narrative and its associated meanings.

First to mind, for me, is our senses' natural attraction to water. As for meanings associated with water, there is so much to be considered: birth, rebirth, renewal, baptism, and cleansing. Most readily though is the passage of time. It could be slow and subtle or raging and loud or our passing by it or sitting and enjoying the visuals and wind chime-like sounds. When considering water, as with many other landscape features, the changing seasons bring a lot to be considered: liquid, vapor, ice in its various forms and beyond the water's edge; movement of stars, planets, the agrarian rhythms of the four seasons, or maybe your tomato garden and the salads and salsas it provides come fall.

For an exciting journey into what nature has to offer you as a designer-author through the four seasons and day and night times, read Thoreau's *Walden* (1973) and learn to see and read landscapes as Thoreau did. As a sample:

*The phenomena of the year take place every day in a pond on a small scale. Every morning... the shallow water is being warmed more rapidly than the deep... and every evening it is being cooled more rapidly until the morning. The day is an epitome of the year. The night is the winter, the morning and evening are spring and fall, and the noon is the summer. The cracking and booming of ice indicate a change of temperature... the pond began to boom about an hour after sunrise, when it felt the influence of the sun's rays... it stretched itself and yawned like a waking man...*

*(301)*

*The [pond's] first ice is especially interesting and perfect, being hard, dark, and transparent, and affords the best opportunity that ever offers for examining the bottom where it is shallow... only two or three inches distant, like a picture behind a glass, and the water is necessarily always smooth then... but the ice itself is the object of most interest... the morning after it freezes, you find that the greater part of the bubbles, which at first appeared to be within it, are against its under surface, and that more are continually rising from the bottom... these bubbles are ... very clear and beautiful, and you see your face reflected in them through the ice.*

*(246–247)*

In *Walden*, Thoreau provides so much more, basically a daily, if not in some instances hourly, diary of the pond and its surrounds. So many examples of what can be seen, heard, felt, smelled, and tasted when people slow down and look to their feet, above their heads, up ahead, and back to where they've been versus where they're going.

If you'd like to be fascinated and enhance your understanding of nature as a medium of art and aesthetics, you have to explore Andy Goldsworthy's ephemeral earth art (1990; 2015). A sample is found here: https://kellyeimwrapped.weebly.com/andy-goldsworthy.html

## READING AND WRITING

I had to close *The Analogy* with examples of a few authors', to use Kevin Lynch's term in a different context, sensuous writing. Fiction and non-fiction writers help us read environments so much more clearly than we usually do.

Lawrence Durrell, "Landscape and Character" in *Spirit of Place*:

*'You write' says a friendly critic in Ohio, 'as if the landscape were more important than the characters.' If not exactly true, this is near enough the mark, for I have evolved a private notion about the importance of landscape, and I willingly admit to seeing 'characters' almost as functions of landscape. This has only come about in recent years after a good deal of travel…[I'm] a 'residence writer'. My books are always about living in places, not just rushing through them.* (1969, 156)

Figure 11.2 Spangle, WA. Photo by author.

Can you imagine people, the landscape readers of your narratives, wanting to reside in your designs rather than, without thought, just rushing through them?

D. H. Lawrence, *Sketches of Etruscan Places:*

*And at the top [of the hill] is a fragment of public garden, and a look-out…We walk to the parapet, and suddenly are looking into one of the most delightful landscapes I have ever seen: as it were, into the very virginity of hilly green country. It is all wheat green and soft and swooping, swooping down and up, and glowing with green newness, and no houses. Down goes the declivity below us, then swerving the curve and up again, to the neighboring hill that faces in all its greenness and long-running immaculateness. Beyond, the hills ripple away to the mountains, and far in the distance stands a round peak, that seems to have an enchanted city on the summit.*

*Such a pure, uprising, unsullied country, in the greenness of wheat on an April morning!*

*The hill opposite is like a distinct companion. The near end is quite steep and wild, with evergreen-oaks and scrub, and specks of black-and-white cattle on the slopes of common. But the long crest is green again with wheat, running and drooping to the south. And immediately one feels: that hill has a soul, it has a meaning. (2007, 353)*

Wendell Berry, (2000) *Life Is a Miracle: An Essay Against Modern Superstition*:

*No human being has ever known, let alone imagined, the entire planet. And even in an age of 'world travel,' none of us lives on the entire planet; in fact, owing to so much mobility, a lot of people…don't live anywhere. But if we are to know any part of the planet intimately, particularly, precisely, and with affection, then we must live somewhere in particular for a long time. We must be able to call up to the mind's eye by name a lot of local places, people, creatures, and things.*

Figure 11.3 Sunrise on the Palouse, Eastern, WA. Courtesy Evelyn Georgeson, Spokane, WA.

*One of the most significant costs of the economic destruction of farm populations is the loss of local memory, local history, and local names. Field names for instance, even such colorless names as 'the front field' and 'the back field' are vital signs of a culture. If the arts and sciences ever waken from their rapture of academic specialization, they will make themselves at home in places they have helped to spoil, and set about reconstructing histories and remembering names.*

*(2000, 138)*

Berry (1981) continues in "A Country of Edges":

Figure 11.4 Red River Gorge, KY.
Photo by author.

*It is a country of overwhelming edges. Again, and again, walking down from the wooded ridge tops above the Red River Gorge one comes into the sound of water falling – the steady pouring and splattering of a tiny stream that has reached its grand occasion... One looks up twenty or thirty or fifty or more feet to where the water leaps off the rock lip, catching the sunlight as it falls. Maybe there will be a rainbow in the spray. The trail may have passed through the little shelves or terraces covered with wild iris in bloom. Or along the stream sides below the falls there may be pink lady's slippers. The slopes will be thickly scrubbed with rhododendron, darkened by the heavy green shade or hemlocks. And always on the wet faces of the rock there will be liverwort and meadow rue and mosses and ferns.*

*The critical fact about water, wherever you find it in the Red River Gorge, is motion. Moving, it is gathering. All the little seeps and trickles of the slopes, the tiny streams heading up near the ridge tops and*

Figure 11.5 Providence sacred heart healing garden, Spokane, WA. Photo by author.

*leaping and tumbling down the steep ravines – all are moving toward their union with the river…*

*And in the movement of its waters the place also is in motion; not to the human eye, nor to the collective vision of human history, but within the long gaze of geologic time the Gorge is moving within itself, deepening, changing the outline of its slopes; the river is growing into it like a great tree, steadily incising its branches into the land.*

*(1981, 224)*

Barry Lopez, *Arctic Dreams*; and *Vintage Lopez*

Lopez has a thought quite similar to one I've had regarding landscape narratives. Over the years, authors of all kinds have opened my mind to what's around me. They have an eye for so much that many of us don't. My reaction when reading Lopez, or Durrell, or Lawrence, or Jackson is so often "Oh my God, I've looked at that 100 times and never seen it like that before." What I do so enjoy about what I find in this and that author's work and then being able to bring an aspect of it into a landscape narrative is how they reach in and touch our subconscious, only to have us wonder hours or days later, "Why did I feel that way?" Given so many of their observations and ways of expressing those observations, I can't help but believe that those who participate in our landscape narratives – those who walk past, touch unknowingly – see so casually that they too register in their subconscious the subtle beauty and symbolic meanings only to, hours or days later, wonder "Why did I feel that way?" That is why reading such authors as mentioned throughout *Landscape Architecture as Storytelling*, remaining open and sensitive to such settings, provides us with an ever-increasing design palette – wordy, expressive, meaning-full design palette.

*Lying there in the tent, I knew, as does everyone I think who spends some time*

Figure 11.6 El Potrero Chico's limestone cliffs and spires up to 2,000′ in the Mexican state of Nuevo León. Photo by author.

*hunting with Eskimos, that they are… a people, some of them, still close to the earth, maintaining the rudiments of an ancient philosophy of accommodation with it that we have abandoned. Our first wisdom as a species, that unique metaphorical knowledge that distinguishes us, grew out of such an intimacy with the earth; and, however far we may have come since that time, it did not seem impossible to me that night to go back and find it.*

*(Lopez, 1987, 35)*

*It is through the power of observation, the gifts of eye and ear, of tongue and nose and finger, that a place first rises up in our mind; afterward it is a memory that carries the place that allows it to grow in depth and complexity. For as long as our records go back, we have held these two things dear, landscape and memory.*

*(Lopez, 2004, 96)*

Imagine capturing Wendell Berry's waterfall imagery, Lopez's contemplative moment, or Durrell's panoramic orchestration of a landscape in even one design in a lifetime.

J. B. Jackson, *A Sense of Place, a Sense of Time*:

*Like millions of other Americans I have no great liking for wilderness and forest, but like the majority of Americans I am fond of trees: individual trees, trees in rows along the street or in orchards, trees in parks. I continue to plant them when and where I can – to such an extent that when their leaves start to fall I look forward to many months of raking and transplanting in preparation for the spring.*

*(Jackson, 1994, 95)*

Many of my students wondered in a "Really? Are you kidding me?" kind of way, why I had them do so much reading and writing. Easy. The better

Figure 11.7 Manito Blvd, Spokane, WA. Photo by author.

reader you are, the better you can express yourself; the better your insight and eye-sight, the better designer you are. People who see the world in ways that we so rarely do have so much to bring to our design palettes. As Italo Calvino shows us, what you read does not have to be from nature writers, agrarian philosophers, or travelers:

Italo Calvino, *Invisible Cities:*

> *In Eudoxia, which spreads both up and down, with winding alleys, steps, dead ends, hovels, a carpet is preserved . . . But if you pause and examine it closely, you become convinced that each place in the carpet corresponds to a place in the city and all the things contained in the city are included in the design, arranged according to their true relationship, which escapes your eye distracted by the bustle, the throngs, the shoving. All of Eudoxia's confusion, the mules' braying, the lampblack stains, the fish smell is what is evident in the incomplete perspective you grasp; but the carpet proves that there is a point from which the city shows its true proportions, the geometrical scheme implicit in its every, tiniest detail*
>
> *(1974, 96)*

Read, read, and read some more and keep notes on the ways you think to transfer what you read, what's behind what you read, or what is between the lines, into design features. Who knows,

Figure 11.8 Salem's Gateway to the Far East Fountain, Salem, MA. Shutterstock ID: 1804746124. Contributor: Wangkun Jia.

Calvino's carpet, may just have given rise to the Gateway fountain in Salem, Massachusetts.

The Fountain at East India Square is a stone fountain, designed by John Collins, showing the changes in Salem's topography as the waterfront area was dredged and filled. The fountain's top layer shows the shoreline of Salem before many areas were filled in. The bottom layer is Salem's shoreline today. Called "Salem's Gateway to the Far East" the fountain was completed in 1976 and acts not only as a fountain but also as a small amphitheater. See

http://www.noblenet.org/salem/wiki/index.php/East_India_Square_Fountain

## I HAVE TWO HOPES

I hope my Dad has followed my personal and professional journey since he passed away in 1991 and come to agree with me that landscape architecture is so much more than planting bushes. I wish Mom and Dad were alive today to see how the hell I put them through in making me read out loud for an hour a night paid off in so many ways. Thanks you two.

And finally, if we started *Landscape Architecture as Storytelling*, as just that, a comparison between landscape and an authored narrative, given all that is written, and all we've discussed about text, narrative, oral history, semiotics (sign, symbols, signified, signifier), it is safe to say landscape is more than an analogy. Landscape is a text in its own right, not of individual statements but of narratives capable of providing us with experiences that at least meet, if not exceed our expectations.

# Bibliography


Abbott, E. A. 1991. *Flatland: A Romance of Many Dimensions*. Princeton: Princeton University Press.

Adams, J. 1976. *Conceptual Blockbusting: A Pleasurable Guide to Better Problem Solving*. San Francisco: San Francisco Book Company.

Al-Husain, L., Kanjo, E. and Chamberlain, A. 2013. "Sense of Space: Mapping Physiological Emotional Response in Urban Space." *UniComp'13* (September): 1321–1324.

Alexander, C. 1977. *A Pattern Language: Towns, Building, and Construction*. New York: Oxford University Press.

Alexander, C. 1979. *The Timeless Way of Building*. New York: Oxford University Press.

Alon-Mozes, T. 2006. "From 'Reading' the Landscape to 'Writing' a Garden: The Narrative Approach in the Design Studio." *Journal of Landscape Architecture* (Wageningen, Netherlands) 1, no. 1: 30–37.

Amick, B., Levine, S., Tarlov, A. and Walsh, D. 1995. *Society and Health*. New York: Oxford University Press.

Aminzadeh, B., Motevali, M., and Nikooparast, S. 2016. "A Proposal for Landscape Design Process Based on Scenario Writing Phases in Cinema and its Application in the Darabad Route, Tehran, Iran." *Urban Design International* 21, no. 2: 175–189.

Anderson, S. (ed.) 1986. *On Streets*. Cambridge, MA: MIT Press.

Ardalan, N. and Bakhtiar, L. 1973. *The Sense of Unity: The Sufi Tradition in Persian Architecture*. Chicago: University of Chicago Press.

Arnheim, R. 1969. *Visual Thinking*. Berkeley: University of California Press.

Arnold, H. 1991. *Trees in Urban Design*. New York: John Wiley & Sons.

Ashihara, Y. 1970: *Exterior Design in Architecture*. New York: Van Nostrand Reinhold.

Auster, P. 1990. *Moon Palace*. New York: Penguin.

Avishay, A. 2018. "Gang Borders Create Invisible Walls in Los Angeles." KCRW Housing & Development section (13 March). Viewed November 15, 2021. https://www.kcrw.com/culture/shows/design-and-architecture/gang-borders-create-invisible-walls-in-los-angeles

Bacon, E. 1974. *Design of Cities*. New York: Viking.

Barnes, T. and Duncan, J. 1992. *Writing Worlds: Discourse, Text, & Metaphor in the Representation of Landscape*. New York: Routledge.

Barthes, R. 1972. "The New Citroen." In *Mythologies*. New York: Hill and Wang: 88–90.

Barthes, R. 1979. *The Eiffel Tower and Other Mythologies*. Trans. by Richard Howard. New York: Hill & Wang: 3–18.

Barthes, R. 1982. "The Eiffel Tower." In Sontag, S. (ed.) *A Barthes Reader*. New York: Hill and Wang: 136–150.

Basinger, J. 2019. *Movie Musical*. New York: Knopf Doubleday.

Baum, H. 1983. *Planners and Public Expectations*. Cambridge, MA: Schenkman.

Beaver, F. 2009. *Dictionary of Film Terms: The Aesthetic Companion to Film Art*. 4th editon. New York: Peter Lang.

Beecher, C. and Stowe, H. B. 1987. *American Woman's Home, or Principles of Domestic Science*. Hartford, CT: Stowe-Day Foundation.

Benzel, K. 1998. *The Room in Context: Design beyond Boundaries*. New York: McGraw-Hill.

Berendt, J. 1994. *Midnight in the Garden of Good and Evil*. New York: Vintage.

Berger, A. 1998. "Archeological Space: Negotiating Narrative Content in the Landscape Continuum." *Landscape Review* (Canterbury: Lincoln University) 4, no. 1: 3–15.

Berger, J. and Sinton, J. 1985. *Water, Earth, and Fire: Land Use and Environmental Planning in the*

*New Jersey Pine Barrens.* Baltimore, MD: Johns Hopkins University Press.
Berger, P. 1963. *Invitation to Sociology: A Humanistic Perspective*. New York: Doubleday.
Berger, P. and Luckmann, T. 1967. *The Social Construction of Reality: A Treatise in the Sociology of Knowledge*. New York: Doubleday.
Berleant, A. 2005. "Ideas for a Social Aesthetic." In Light, A. and Smith, J. (eds.) *The Aesthetics of Everyday Life.* New York: Columbia University Press: 23–38.
Berry, W. 1964. *The Broken Ground.* Harcourt Brace & World.
Berry, W. 1968. *Openings.* New York: Harcourt, Brace & World.
Berry, W. 1969. *The Long-legged House*. New York: Harcourt, Brace & World.
Berry, W. 1981. "A Country of Edges." In Berry, W. (ed.) *Recollected Essays, 1965–1980.* San Francisco, CA: North Point: 223–229.
Berry, W. 2000. *Life is a Miracle.* Washington, D.C.: Counterpoint.
Berry, W. 2013. *The Country of Marriage*. Berkeley, CA: Counterpoint. Reprint edition.
Bloomer, K. and Moore, C. 1977. *Body, Mind, and Architecture.* New Haven, CT: Yale University Press.
Bonta, J. 1979. *Architecture and Its Interpretation: A Study of Expressive Systems in Architecture.* New York: Rizzoli.
Borges, J. L. 1962. *Ficciones.* Trans. by Emece Editores. New York: Grove Press.
Boyer, E. and Mitgang, L. 1996. *Building Community: A New Future for Architectural Education and Practice.* Princeton, NJ: The Carnegie Foundation.
Boyle, B. 1977. "Architectural Practice in America 1986-1965- Ideal and Reality." In *The Architect*, ed. S. Kostof. New York: Oxford University Press: 309–344.
Brady, E. 2005. "Sniffing and Savoring: The Aesthetics of Smells and Tastes." In Light, A. and Smith, J. (eds.) *The Aesthetics of Everyday Life.* New York: Columbia University Press: 177–193.
Bring, M. and Wayembergh, J. 1981. *Japanese Gardens: Design and Meaning.* New York: McGraw-Hill.
Bronowski, J. 1973. *The Ascent of Man.* Boston, MA: Little, Brown & Company.
Burger, D. 1965. *Sphereland: A Fantasy about Curved Spaces and an Expanding Universe.* Trans. Cornelie Rheinbolt. New York: Harper & Rowe.
Buttimer, A. 1980. "Social Space and the Planning of Residential Areas." In Buttimer, A. and Seamon, D. (eds.) *The Human Experience of Space and Place.* London: Croom Helm, 21–54.
Calvino, I. 1968. *Cosmicomics.* New York: Harcourt Brace & Company.
Calvino, I. 1974. *Invisible Cities.* Trans. William Weaver. San Diego, CA: Harcourt Brace Jovanovich.
Campbell, J. 1949. *The Hero with a Thousand Faces.* New York: Pantheon.
Campbell, J. 1969. *The Masks of God: Primitive Mythology*. New York: Penguin.
Campbell, J. 1968. *The Masks of God: Creative Mythology*. New York: Penguin.
Campbell, J. 1974. *The Mythic Image*. Princeton: Princeton University Press.
Canadian Geographic. (2015). "Sabrina Doyle" (December 2). Viewed March 2022. https://www.canadiangeographic.ca/article/geography-word-week-palimpsest#:~:text=%5BPAL%2Dimp%2Dsest%5D,traces%20of%20something%20preceding%20it.
Canemaker, J. 1999. *Paper Dreams: The Art and Artists of Disney Storyboards.* New York: Hyperion.
Capra, F. 1982. *The Turning Point: Science, Society, and the Rising Culture.* New York: Simon and Schuster.
Chatman, S. 1980. *Story and Discourse: Narrative Structure in Fiction and Film.* Ithaca, NY: Cornell.
Chatman, S. 1990. *Coming to Terms: The Rhetoric of Narrative in Fiction and Film.* Ithaca, NY: Cornell University Press.
Ching, F. D. K. 1978. *Architecture: Space, Form, and Order.* New York: John Wiley & Sons.
Church, T. 1983. *Gardens are for People.* New York: Reinhold.
Collins, G. and Collins, C.C. 1986: *Camillo Sitte: The Birth of Modern City Planning.* New York: Rizzoli.

Conron, J. 1974. *The American Landscape: A Critical Anthology of Prose and Poetry*. New York: Oxford University Press.

Conron, J. 1985. Fitz Hugh Lane and the American Picturesque." *Revue Francaise d'Etudes Americines*, (November): 383–401.

Conron, J. 2000. *American Picturesque*. University Park, PA: Pennsylvania State University Press.

Conzen, M. (ed.) 1990. *The Making of the American Landscape*. London: Harper Collins Academic.

Cooper, A. 2020. *A Most Beautiful Thing*. New York: Flatiron Press.

Cooper, J. F. 1980. *The Last of the Mohicans*. New York: Penguin.

Cornish, J. 2019. *Dementia with Dignity: Living Well with Alzheimer's or Dementia using the DAWN Method*. Amazon eBook: rdp.com.

Cosgrove, D. 1984. *Social Formation and Symbolic Landscape*. London: Croom Helm.

Cronon, W. 1983. *Changes in the Land: Indians, Colonists and the Ecology of New England*. New York: Hill and Wang.

Cronon, W. 1991. *Nature's Metropolis: Chicago and the Great West*. New York: W.W. Norton.

Cuff, D, 1991. *Architecture: The Story of Practice*. Cambridge, MA: MIT Press.

Cullen, G. 1961. *Townscape*. New York: Reinhold.

Deetz, J. 1977. *In Small Things Forgotten: The Archeology of Early American Life*. New York: Anchor.

Derrida, J. 1978. *Writing and Difference*. Trans. by Bass, Ala. Chicago: University of Chicago Press.

Didion, J. 1979. "Many Mansions." In Didion, J. (ed.) *The White Album*. New York: Pocket Books: 67–73.

Dillard, A. 1990. *Three by Annie Dillard*. New York: Harper Perennial.

Domosh, M. 1996. *Inventing Cities: The Creation of Landscape in Nineteenth-century New York and Boston*. New Haven, CT: Yale University Press.

Dondis, D. 1973. *A Primer of Visual Literacy*. Cambridge, MA: MIT Press.

Dos Passos, J. 1953. *Manhattan Transfer*. Boston: Mariner Books.

Downing, A. J. 1841. *A Treatise on the Theory and Practice of Landscape Gardening, Adapted to North America*. New York: Wiley and Putnam. Digital Collection, UMass Amherst. Viewed February 2020. https://archive.org/details/treatiseontheory41down/page/n8/mode/2up

Downing, A. J. 1969. *The Architecture of Country Houses*. New York: Dover.

Downs, R. and Stea, D. 1973. *Image and Environment. Cognitive Mapping and Spatial Behavior*. Aldine, Chicago.

Duncan, J. 1990. *The City as Text: The Politics of Landscape Interpretation in the Kandyan Kingdom*. New York: Cambridge University Press.

Duncan, J. and Ley, D. (eds.) 1993. Place/Culture/Representation. New York: Routledge.

Dunlop, D. 2020. "A Brief History of Clouds." Viewed May 2020. https://www.paintingclass.net/a-brief-history-of-clouds-chapter-i/

Durrell, L. 1969. *Spirit of Place: Letters and Essays on Travel*. New Haven, CT: Leete's Island Books.

Eckstein, B. and Throgmorton, J. 2003. *Story and Sustainability: Planning, Practice, and Possibility for American Cities*. Cambridge, MA: MIT Press.

Eco, U. 1994. *Six Walks in the Fictional Woods*. Cambridge: Harvard University Press.

Edwards, B. 1999. *The New Drawing on the Right Side of the Brain: A Course in Enhancing Creativity and Artistic Confidence*. New York: Distributed by St. Martin's Press.

Eiseley, L. 1957: *The Immense Journey*. New York: Vintage.

Eiseley, L. 1978. "The Innocent Fox." In Eiseley, L. *The Star Thrower*. New York: Time Books: 26–36.

Eliot, T. S. 1963. "Little Gidding." In *Collected Poems 1909–1962: T. S. Eliot*. New York: Harcourt Brace Jovanovich: 202–209.

Emerson, R. W. 1971. *Nature, Addresses, and Lectures*. Cambridge, MA: Harvard University Press.

Evernden, N. 1992. *The Social Creation of Nature*. Baltimore, MD: Johns Hopkins University Press.

Ewen, S. 1976: *Captains of Consciousness: Advertising and the Social Roots of the Consumer Culture*. New York: McGraw-Hill.

Ewert, E. 1999. "Setting the Pacific Northwest Stage: The Influence of the Natural Environment." In Goble, D. and Hirt, P. (Eds.),

*Northwest Lands, Northwest Peoples: Readings in Environmental History*. Seattle: University of Washington Press. 1–26.

Faris, W. 1988. *Labyrinths of Language: Symbolic Landscape and Narrative Design in Modern Fiction*. Baltimore: Johns Hopkins University Press.

Farrel, B. and Farrel, P. 2001. *Men are Like Waffles, Women are Like Spaghetti*. Eugene, OR: Harvest House.

Fisher, T. 2013. The Ethics of Landscape. *Landscape Architecture Frontiers*, 1, no. 2: 44–51.

Forester, J. 1985. "Critical Theory and Planning Practice." In Forester, J. (ed.) *Critical Theory and Public Life*. Cambridge, MA: MIT Press: 202–230.

Forester, J. 1989. *Planning in the Face of Power*. Berkeley: University of California Press.

Forman, R. and Godron, M. 1986. *Landscape Ecology*. New York: John Wiley and Sons.

Foster, J. 2009. "Environmental aesthetics, ecological action and social justice." In Bondi, L., Cameron, L. Davidson, J. and Smith, M. (eds.) *Emotion, Place and Culture*. Aldershot, England: Ashgate Press: 97–114.

Fowles, J. 1965. *The Magus*. New York, NY: Dell.

Fowles, J. 1979. *The Tree*. Don Mills, Ontario, CA: Collins.

Gans, H. 1962. *The Urban Villagers: Group and Class in the Life of Italian Americans*. New York: The Free Press.

Garfinkel, H. 1963. "A Conception of and Experiments with 'Trust' as a Condition of Stable Concerted Actions." In Harvey, O. J. (ed.) *Motivation and Social Interaction*. New York: Ronald Press: 187–238.

Genette, G. 1980. *Narrative Discourse: An Essay in Method*. Ithaca: Cornell University Press.

Gibson, J. J. 1979. *The Ecological Approach to Visual Perception*. Boston: Houghton Mifflin.

Giddens, A. 1979. *Central Problems in Social Theory*. Berkeley: University of California Press.

Giddens, A. 1984. *The Constitution of Society*. Berkeley: University of California Press.

Girouard, M. 1978. *Life in the English Country House*. New Haven, CT: Yale University Press.

Goffman, E. 1963. *Behavior in Public Places: Notes on the Social Organization of Gatherings*. New York: Free Press.

Goffman, E. 1959. *The Presentation of Self in Everyday Life*. New York: Doubleday.

Golden/La Jolla, F. 1999: "Lying Faces Unmasked: The Most Gifted Fibbers Can Fool Almost Anybody, But a New Computer Can Look Right Through Them." In *Time Magazine*, Science section. New York: Time Life. April 5, 1999: 153, no.13.

Goldhagen, S. W. 2017. *Welcome to Your World: How the Built Environment Shapes Our Lives*. New York: Harper Collins.

Goldsworthy, A. 1990. *Collaboration with Nature*. New York: Abrams.

Goldsworthy, A. 2015. *Ephemeral Works: 2004–2014*. New York: Abrams.

Golledge, R. 1999: "Human Wayfinding and Cognitive Maps." In Golledge, R. (ed.) *Wayfinding Behavior: Cognitive Mapping and Spatial Processes*. Baltimore, MD: Johns Hopkins University Press: 5–45.

Gombrich, E. H. 1972. *Art and Illusion: A Study in the Psychology of Pictorial Representation*. Princeton: Princeton University Press.

Goode, W. 1957. "Community within a Community: The Professions." *American Sociological Review*, 22, no. 2: 194–200.

Goodman, P. 1954. *The Structure of Literature*. Chicago: University of Chicago Press.

GoodTheraphy, 2019. Collective Unconscious. Viewed December 18, 2020. https://www.goodtherapy.org/blog/psychpedia/collective-unconscious#:~:text=The%20collective%20unconscious%20is%20a, of%20universal%20figures%20and%20relationships.

Gordon, W. 1961. *Synectics*. New York: Macmillan.

Gottdiener, M. 1985. *The Social Production of Urban Space*. Austin: University of Texas Press.

Gratz, R. 1989. *The Living City*. New York: Simon & Schuster.

Gray, J. 1992. *Men Are from Mars, Women Are from Venus: A Practical Guide for Improving Communication and Getting What You Want in Your Relationships*. New York: Harper Collins.

Greene, H. 1962. *Mind & Image: An Essay on Art and Architecture*. Lexington: University of Kentucky Press.

Grove, N. 1990. "Greenways, Paths to the Future." *National Geographic.* 177, no. 6: 93–94.
Guilpin, W. 1792. *Three Essays: On Picturesque Beauty*. London: Printed for R. Blamire.
Guilpin, W. 1802. *An Essay on Prints*. London: Straham, Codell, and Davies.
Guth, H. 1961: *Concise English Handbook*. Belmont, CA: Wadsworth.
Haapala, A. 2005. "Ideas for a Social Aesthetic." In Light, A. and Smith, J. (eds.) *The Aesthetics of Everyday Life*. New York: Columbia University Press: 23–38.
Habermas, J. 1970. *Towards a Rational Society*. Boston: Beacon.
Habermas, J. 1979. *Communication and the Evolution of Society.* Trans. McCarthy, T., Boston: Beacon.
Harbermas, J. 1981. *The Theory of Communicative Action: Reason and the Rationalization of Society.* Boston: Beacon.
Hagerstrand, T. 1970. "What about People in Regional Science?" *Papers of the Regional Science Association* 24, no. 1: 621.
Hall, E. 1969. *The Hidden Dimension: An Anthropologist Examines Man's Use of Space in Public and in Private*. New York: Doubleday.
Hall, E. 1981. *The Silent Language*. New York: Anchor.
Hall, E. 1983. *The Dance of Life: The Other Dimension of Time.* Garden City, NJ: Anchor.
Halprin, L. and Burns, J. 1974. *Taking Part: A Workshop Approach to Collective Creativity*. Cambridge, MA: MIT Press.
Harington, D. 1986. *Let Us Build a City*. San Diego, CA: Harcourt Brace Jovanovich.
Hartshorne, R. 1939. *The Nature of Geography: A Critical Survey of Current Thought in Light of the Past.* Lancaster, PA: Association of American Geographers.
Hawken, P. 1993. *The Ecology of Commerce: A Declaration of Sustainability*. New York: Harper Collins.
Hayden, D. 1995. *The Power of Place: Urban Landscapes as Public History*. Cambridge, MA: MIT Press.
Hester, R. 2006. *Design for Ecological Democracy*. Cambridge, MA: MIT Press.
Heuer, K. (ed.) 1987. *The Lost Notebooks of Loren Eiseley.* Boston, MA: Little, Brown and Company.
Horton, T. 1987. *Bay Country: Reflections on the Chesapeake.* New York: Ticknor & Fields.
Horton, T. 2002. *The Great Marsh: An Intimate Journey into a Chesapeake Wetland.* Baltimore, MD: Johns Hopkins University Press.
Hough, M. 1990. *Out of Place: Restoring Identity to the Regional Landscape*. New Haven, CT: Yale University Press.
Howard, B. 2014. "Buried Stream Sees the Light of Day" (November 27, 2014). Viewed May 2020. https://www.nationalgeographic.com/news/2014/11/141125-dc-daylighting-broad-branch-stream-restoration-science/
Hughes, B. 1997. *Narrative as Landscape.* Presented at the Meno Workshop on Hypermedia and Narrative. Brighton UK April 1997. Viewed May 2020 http://www.dustormagic.net/NasL.html.
Integrus Architecture. 2019. *Educational Specifications, Spokane Public Schools, On Track Academy*. Spokane, WA: Integrus Architecture. Project No. 21838.00.
Itoh, T. 1981. *Space & Illusion: In the Japanese Garden*. New York: Weatherhill/Tankosha.
Jackson, J. B. 1970. "The Stranger's Path." In Zube, E. (ed.) *Landscapes: Selected Writings of J. B. Jackson*. Amherst, MA: University of Massachusetts Press: 92–106.
Jackson, J. B. 1977. "Ghosts at the Door." In Zube, E. and Zube, M. (eds.) *Changing of Rural Landscapes*. Amherst, MA: University Massachusetts Press: 41–52.
Jackson, J. B. 1994. *A Sense of Place, a Sense of Time*. New Haven, CT: Yale University Press.
Jacobs, J. 1961. *The Death and Life of Great American Cities*. New York: Vintage.
Jauchius, J. 2017. *Landscape for Autism: How Nature-based Play and Learning Environments Benefit ASD*. Spokane, WA: SPVV.
Jewett, S.O. 1956. *The Country of the Pointed Firs and Other Stories*. Garden City, NY: Doubleday & Company.
Johnson, B. 1982. *Ojibway Ceremonies*. Lincoln, NE: University of Nebraska Press.
Johnson, N. B. 1982. "Education as Environmental Socialization: Classroom Spatial Patterns and

the Transmission of Sociocultural Norms." *Anthropology Quarterly* 55, no. 1: 31–43.
Johnston, R.J., Gregory, D., Haggett, P., Smith, D., and Stoddart, D.R. (eds.) 1981. *The Dictionary of Human Geography*. Oxford: Blackwell.
Jordanova, L. (ed.) 1986. *Languages of Nature: Critical Essays on Science and Literature*. New Brunswick, NJ: Rutgers.
Juster, N. 1963. *The Dot and the Line: A Romance in Lower Mathematics*. New York: Random House.
Kandinsky, W. 1979. *Point and Line to Plane*. New York: Dover.
Kearns, R. and Gesler, W. (eds.) 1998. *Putting Health into Place: Landscape, Identity, And Well-Being*. Syracuse, NY: Syracuse University Press.
Kemmis, D. 1990. Community and the Politics of Place. Norman, OK: University of Oklahoma Press.
Kemmis, D. 1995. The Good City and the Good Life: Renewing the Sense of Community. New York: Houghton Mifflin.
Kendig, L. 1980. *Performance Zoning*. Washington, DC: Planners Press, American Planning Association.
Knapp, B. 1986. *Archetype, Architecture, and the Writer*. Bloomington, IN: Indiana University Press.
Kooser, T. 2005. *The Poetry Home Repair Manual: Practical Advice for Beginning Poets*. Lincoln, NE: University of Nebraska Press.
Kuo, F. and Taylor, A. 2004. "A Potential Natural Treatment for Attention-Deficit/Hyperactivity Disorder." *American Journal of Public Health* 94, no. 9: 1580–1586.
Kryder-Reid, E. 1996. "The Construction of Sanctity: Landscape and Ritual in a Religious Community." In Yamin, R. and Metheny, K. B. (eds.) 1996. *Landscape Archaeology: Reading and Interpreting the American Historical Landscape*. Knoxville: University of Tennessee Press, 228–248.
Kuhn, T. 1970. *The Structure of Scientific Revolutions*. Chicago: University of Chicago Press.
Lakoff, G. 1987. *Women, Fire, and Dangerous Things*. Chicago: University of Chicago Press.
Lamott, A. 1994. *Bird by Bird: Some Instructions on Writing and Life*. New York: Anchor.
Langer, S. 1967. *Mind: An Essay on Human Feeling. Volume 1*. Baltimore: Johns Hopkins University Press.
Langer, S. 1980. *Philosophy in a New Key: A Study in the Symbolism of Reason, Rite, and Art*. Cambridge, MA: Harvard University Press.
Lao Tse, 1948. *The Wisdom of Laotse*. Lin Yutang (ed.). New York: Modern Library pub. By Random House.
Lao Tzu, 1955. *The Way of Life: Tao Te Ching*. Trans. Blakney, R.B. New York: Mentor.
Lao Tzu, 1963. *Tao Te Ching*. Trans. Lau, D. C. New York: Penguin.
Larson, M. S. 1977. *The Rise of Professionalism: A Sociological Analysis*. Berkeley: University of California Press.
Laseau, P. 1980. *Graphic Thinking for Architects and Designers*. New York: Van Nostrand Reinhold.
Lawrence, D. H. 2007. "Sketches from Etruscan Places." In Lawrence, D. H. (ed.) *D. H. Lawrence and Italy*. London, England: Penguin. 327–441.
Lefebvre, H. 1976: "Reflections on the Politics of Space." *Antipode*. 8(2) 30–37.
Lefebvre, H. 1979. "Space: Social Product and Use Value." In Freiberg, J.W. *Critical Social Theory: European Perspectives*. New York: John Wiley and Sons, 285–295.
Lefebvre, H. 1994. *The Production of Space*. Trans. Nicholson-Smith, D. Cambridge, MA: Blackwell.
Le Guin, U. 1974. *The Dispossessed*. New York: Avon.
Leone, M. 2015. *The Archaeology of My Own Ignorance: Excavations in Annapolis*. Berkeley: University of California Press.
Ley, D. 1974. *The Black Inner City as Frontier Outpost: Images and Behavior of a Philadelphia Neighborhood*. Washington, DC: Association of American Geographers.
Ley, D. 1983. *A Social Geography of the City*. New York: Harper & Row.
Light, A. and Smith, J. (eds.) 2005. *The Aesthetics of Everyday Life*. New York: Columbia University Press.
Lincoln University Press. 1996. *Theme: Languages of Landscape Architecture – Theory into Practice* (Canterbury, NZ) 2, no. 3: 1.
Lippard, L. 1983. *Overlay: Contemporary Art and the Art of Prehistory*. New York: Pantheon Books.

Lopez, B. 1987. *Arctic Dreams: Imagination and Desire in a Northern Landscape.* New York: Bantam.

Lopez, B. 2004. *Vintage Lopez.* New York: Vintage.

Lopez, B. 2019. *Horizons.* New York: Alfred A. Knopf.

Lowenthal, D. and Bowden, M. 1976. *Geographies of the Mind.* New York: Oxford University Press.

Lund, K. 2012. "Landscapes and Narratives: Compositions and the Walking Body." *Landscape Research.* 37, no. 2: 225–237.

Lutwack, L. 1984. *The Role of Place in Literature.* Syracuse: Syracuse University Press.

Lynch, K. 1960. *Image of the City.* Cambridge, MA: MIT Press.

Lynch, K. 1971. *Site Planning.* Cambridge, MA: MIT Press.

Lynch, K. 1972. *What Time is this Place?* Cambridge, MA: MIT Press.

Lynch, K. and Hack, G. 1989. *Site Planning.* Third Edition. Cambridge, MA: MIT Press.

Lynes, R. 1954. *The Taste-Makers.* New York: Harper & Brothers.

Mann, W. 1993. *Landscape Architecture: An Illustrated History in Timelessness, Site Plans, and Biography.* New York: John Wiley & Sons.

Marcus, C. C. 1995: *House as a Mirror of Self.* Berkeley, CA: Conari.

Marcus, C. C. 1999. *Healing Gardens: Therapeutic Benefits and Design Recommendations.* New York: John Wiley.

Marx, L. 1981. *The Machine in the Garden: Technology and the Pastoral Ideal of America.* New York: Oxford University Press.

Maslow, A. 1970. *Motivation and Personality.* 3rd edition. New York: Harper and Row.

McKim, R. 1972. *Experiences in Visual Thinking.* Belmont, CA: Wadsworth.

McPhee, J. 1968. The Pine Barrens. New York: Farrar. Strauss and Giroux.

McPhee, J. 2000. *Annals of the Former World.* New York: Macmillan.

Meinig, D. W. 1979. "The Beholding Eye: Ten Versions of the Same Scene. In Meinig, D. W. and Jackson, J. B. (eds.) *The Interpretation of Ordinary Landscapes: Geographical Essays.* New York: Oxford University Press: 33–48.

Mitchell, W. J. T. 1981. *On Narrative.* Chicago: University of Chicago Press.

Michener, J. 1965. *The Source.* New York: Fawcett.

Mollison, B. 1988. *Permaculture: A Designer's Manual.* Tyalgum, Australia: Tagari.

Moore, C., Mitchell, W., and Turnbull, W. 1988. *The Poetics of Gardens.* Cambridge, MA: MIT Press.

Moreno, L. 2012. "A Review of 'Henri Lefebvre on Space: Architecture, Urban Research and the Production of Theory.' " *Annals of the Association of American Geographers* 102, no. 4: 885–890.

Morgan, F. 2006. *The Power of Community: How Cuba Survived Peak Oil.* DVD, 53 minutes. Yellow Springs, Ohio: Community Service Inc.

Morris, W. 1970. *The American Heritage Dictionary of the English Language.* Boston: Houghton Mifflin.

Motloch, J. 1991. *Introduction to Landscape Design.* New York: van Nostrand Reinhold.

Moughtin, C. 2003. *Urban Design: Street and Square.* 3rd edition. New York: Elsevier.

Muir, J. 1997. *Nature Writings.* New York: Literary Classics.

Muir, J. 1988. *Wilderness Essays.* Salt Lake City, UT: Peregrine.

Nabham, G. 1993. In "The Foreword." Thoreau H. D. *Faith in a Seed.* Washington, D.C.: Shearwater. xvi.

Nabokov, P. and Easton, R. 1989. *Native American Architecture.* New York: Oxford University Press.

Nash, R. 1989. *Rights of Nature.* Madison: University of Wisconsin.

Ndubisi, F. 1988. "Community Planning for Amerindian Cultures: A Phenomenological Approach." *Landscape Architecture* 78, no. 7: 88–90.

Nesbitt, P., Pond, A., and Allen, W. 1959. *The Survival Book.* Princeton, NJ: van Nostrand.

Newman, O. 1972. *Defensible Space: Crime Prevention through Urban Design.* New York: The Macmillan Company.

Newton, N. 1971. *Design on the Land.* Cambridge, MA: Harvard University Press.

Norberg-Schulz, C. 1985. *The Concept of Dwelling: On the Way to Figurative Architecture.* New York: Rizzoli.

Norberg-Schulz, C. 1980. *Genius Loci: Towards a Phenomenology of Architecture.* New York: Rizzoli.

Novak, B. 1980. *Nature and Culture: American Landscape and Painting, 1825–1875.* New York: Oxford University Press.

Pace Gallery, 1993. *Drawing into Film: Directors' Drawings*. New York: The Pace Gallery.

Parsons, T. 1964. "Evolutionary Universals in Society." *American Sociological* Review 29, no. 3: 339–357.

Philips, S. 1982. "The Language Socialization of Lawyers: Acquiring the 'Cant'." In Spinder, G. (ed.) *Doing the Ethnography of Schooling: Educational Anthropology in Action*. New York: Holt, Rinehart and Winston: 176–209.

Piaget, J. and Inhelder, B. 1967. *The Child's Conception of Space*. New York: Norton Library.

Pinkham, R. 2000. *Daylighting: New Life for Buried Streams*. Snowmass, CO: Rocky Mountain Institute.

Poe, E. A. 1904. *The Works of Edgar Allen Poe, Volume Six. Tales – Horror and Death*. New York: Funk & Wagnalls.

Pollan, M. 1991. *Second Nature: A Gardener's Education*. New York: Atlantic Monthly Press.

Potteiger, M. and Purinton, J. 1998. *Landscape Narratives: Design Practices for Telling Stories*. New York: John Wiley & Sons.

Prince, G. 1970. *The Practice of Creativity*. New York: Macmillan.

Rae, C. M. 1996. *Movies of the Mind: How to Build a Short Story*. Santa Fe, NM: Sherman Asher.

Rapoport, A. 1969. *House Form and Culture*. Englewood Cliffs, NJ: Prentice-Hall.

Rapoport, A. 1982. *The Meaning of the Built Environment: A Nonverbal Communication Approach*. Beverly Hills: Sage.

Raver, A. 1993. "In Georgia's Swept Yards, a Dying Tradition." *New York Times*, August 8, 1993, Section 1, Page 1.

Reid, G. 1987. *Landscape Graphics*. New York: Whitney Library of Design.

Relph, E. 1987. *The Modern Urban Landscape*. Baltimore, MD: Johns Hopkins University Press.

Reps, J. 1969. *Town Planning in Frontier America*. Princeton: Princeton University Press.

Repton, H. 1794. *Sketches and Hints on Landscape Gardening*. London, Printed by W. Bulmer and Co. http://digital.library.wisc.edu/1711.dl/DLDecArts.ReptonSketches

Repton, H. 1803. "Theory and Practice of Landscape Gardening." In Loudon, J. C. (ed.), *The Landscape Gardening and Landscape Architecture of the late Humphry Repton, Esq.*: London, 1840.

Roberts, A. 2018. *Churchill: Walking with Destiny*. eBook. New York: Viking.

Rose, T. 2016. *The End of Average: How We Succeed in a World that Values Sameness*. New York: Harper One.

Rosenthal, S. and Bourgeois, P. 1990. "Mead and Merleau-Ponty: meaning, perception, and behavior." In Tymieniecka, A. (ed.) *The Moral Sense and Its Foundational Significance: Self, Person, Historicity, Community, Phenomenological Praxeology, and Psychiatry*. Boston: Kluwer Academic: 401–409.

Rowe, J. and Kahn, R. 1998. *Successful Aging*. New York: Random House.

Rowles, G. 1978. *The Prisoners of Space? Exploring the Geographical Experience of Older People*. Boulder CO: Westview.

Rowles, G. and Bernard, M. 2013. "The Meaning and Significance of Place in Old Age." In Rowles, G. and Bernard, M. (eds.) *Environmental Gerontology: Making Meaningful Places in Old Age*. New York: Springer: 3–24.

Rubin, D. 1995. *Memory in Oral Traditions: The Cognitive Psychology of Epic, Ballads, and Counting-out Rhymes*. New York: Oxford University Press.

Rudofsky, B. 1969. *Streets Are for People: a primer for Americans*. Garden City, NY: Double Day.

Ruskin, J. 1859. *The True and the Beautiful in Nature: Art, Morals, and Religion*. New York: Wiley & Halsted.

Rybczynski, W. 1986. *A Short History of the Idea of Home*. New York: Viking.

Samuels, M. 1978. "Existentialism and Human Geography." In Ley, D. and Samuels, M. (eds.) *Humanistic Geography: Prospects and Problems*. Chicago, IL: Maaroufa.

Sartre, J. 1966. *Being and Nothingness*. New York: Washington Square Press.

Sauer, C. 1925. "The Morphology of Landscape." Reprinted in Leighly, J. (ed.) 1974: *Land and Life: A Selection from the Writings of Carl Ortwin*

*Sauer.* Berkeley: University of California Press: 315–350.

Scarfo, R. 1987. "Misstructuring the Structure of Design Education." In *Proceedings, Council of Educators in Landscape Architecture*: 389–396.

Scarfo, R. 1988. "Stewardship and the Profession of Landscape Architecture." *Landscape Journal* 7, no. 1: 60–68.

Scarfo, R. 1989. Transforming the Meaning of Space: The Socialization of Professional Landscape Architects. Ph.D. Dissertation, Geography Department, Clark University, Worcester, MA.

Scarfo, B. 1989a. "Consensus: The Aim of Teaching and Learning." In *Proceedings, Council of Educators in Landscape Architecture*, Gainesville, Florida: 59–64.

Scarfo, B. 1992. "Ethics in Practice: What We Say and What We Do, " *Landscape Architecture Review* 13, no. 3/4: 6–8.

Schama, S. 1995. *Landscape and Memory.* New York: Alfred Knopf.

Schmidt, J. 1975. *Urban Development: a review and Chicago Case Study.* Department of Geography, Research Paper 161. Chicago: University of Chicago Press.

Schon, D. 1983. *The Reflective Practitioner.* New York: Basic Books.

Sharpe, W. and Wallock, L. (eds.) 1987. *Visions of the Modern City: Essays in History, Art, and Literature.* Baltimore, MD: Johns Hopkins University Press.

Sieber, T. 1979. "Schoolrooms, Pupils, and Rules: The Role of Informality in Bureaucratic Socialization." *Human Organization* 38, no. 3: 273–282.

Simonds, J. O. 1961. *Landscape Architecture: A Manual of Site Planning and Design.* New York: McGraw-Hill.

Simonds, J. O. 2006. *Landscape Architecture: A Manual of Site Planning and Design.* 4th edition. New York: McGraw-Hill.

Sitte, C. 1945. *The Art of Building Cities.* Trans. Stewart, C. Westport, CT: Hyperion Press.

Smith, N. 1984. *Uneven Development: Nature, Capital, and the Production of Space.* New York: Basil Blackwell.

Smith, P. 1977. *The Syntax of Cities.* London: Hutchinson.

Smithson, R. 1996. *Robert Smithson: The Collected Writings.* Berkeley: University of California Press.

Smitten, J. and Daghistany, A. (Eds.) 1981. *Spatial Form in Narrative.* Ithaca: Cornell University.

Snyder, G. 1996. *Mountains and Rivers without End.* Washington, D. C.: Counterpoint.

Solnit, R. 2000. *Wanderlust: A History of Walking.* New York: Penguin.

Sommer, R. 1969. *Personal Space: The Behavioral Basis of Design.* Englewood Cliffs, NJ: Prentice-Hall.

Sommer, R. 1974. *Tight Spaces: Hard Architecture and How To Humanize It.* Englewood Cliffs, N.J.: Prentice-Hall.

Sontag, S. (ed.) 1994. *A Barthes Reader.* New York: Hill and Wang.

Sorvig, K. 1996." Linguistics and the 'Language' of Design." *Landscape Review*, Canterbury, NZ: Lincoln University Press, 2(3): 2–12.

Spirn, A. W. 1984. *The Granite Garden: Urban Nature and Human Design.* New York: Basic Books.

Spirn, A. 1998. *The Language of Landscape.* New Haven, CT.: Yale University Press.

Sternberg, E. 2009. *Healing Spaces: The Science of Place and Well-Being.* Cambridge: Harvard.

Stevens, A. 1982. *Archetypes: A Natural History of the Self.* New York: William Morrow and Company.

Storr, A. 1983. *The Essential Jung.* Princeton: Princeton University Press.

Storr, A. 1988. *Solitude: A Return to Self.* New York: The Free Press.

Swaffield, S. 1995. "Theme: Languages of Landscape Architecture." *Landscape Review: Journal of Landscape Architecture – South Pacific.* 2.

Swaffield, S. 1996. Theme: Languages of Landscape Architecture – Theory into Practice. *Landscape Review: Journal of Landscape Architecture – South Pacific.* 2, no. 3.

Tara, A., Lawson, G., and Renata, A. 2021. Measuring Magnitude of Change by High-rise Buildings in Visual Amenity Conflicts in Brisbane. In Elsevier, *Landscape and Urban Planning* 205 (2021) 103930.

Thiel, P. 1961. "A Sequence-Experience Notation: For Architectural and Urban Spaces." *The Town Planning Review* 32, no. 1: 33–52.

Thiis—Evensen, 1987. *Archetypes in Architecture.* Oslo: Norwegian University Press.

Thoreau, H. D. 1973. *Walden.* Princeton: Princeton University Press.

Thoreau, H. 1993. *Faith in a Seed.* Washington, DC: Island Press.

Tippett, K. 2016. *Becoming Wise: An Inquiry into the Mystery and Art of Living.* eVersion. New York: Penguin.

Treib, M. 1991. "Sources of Significance: The Garden in Our Time." In Wrede, S. and Adams, W. A. (eds.) *Denatured Visions: Landscape and Culture in the Twentieth Century.* New York, NY: The Museum of Modern Art: 106–109.

Tuan, Yi-fu, 1974. *Topophilia: A Study of Environmental Perceptions, Attitudes, and Values.* Englewood Cliffs, NJ: Prentice Hall.

Tuan, Yi-fu. 1978. "Space, Time, Place: A Humanistic Frame." In Carlstein, T., Parkes, D., and Thrift, N. J. *Timing Space and Spacing Time.* London: Edward Arnold. Vol. I: 7–16.

Tumminello, W. 2005. *Exploring Storyboarding: An In-Depth Guide to the Art and Techniques of Contemporary Storyboarding.* Clifton Park, NY: Thomson Delmar Learning.

Turner, V. 1980. "Social Dramas and Stories about Them." *Critical Inquiry* 7, no. 1: 141–168.

Turner, V. 1981. Social Dramas and Stories about Them. In Mitchell, W. J. T. (ed.) *On Narrative.* Chicago: Chicago University Press, 137–163.

Twain, M. 1984. *Life on the Mississippi.* New York: Penguin.

Vernon, N. D. 1987. "Toward Defining the Profession: The Development of the Code of Ethics and the Standards of Professional Practice of the American Society of Landscape Architects, 1899–1927." *Landscape Journal* 6, no.1: 13–20.

Vitruvius, 1960. *Vitruvius: The Ten Books on Architecture.* Trans. Morgan, M. H. New York: Dover.

Vogler, C. 1992. *The Writer's Journey: Mythic Structure for Storytellers and Screenwriters.* Studio City, CA: Michael Wiese Productions.

Waltower, S. 2019. "New Stop Sign on Grand Blvd. and High Drive Confuses Some Motorists." *Spokesman Review,* November 6, 2019. Viewed February 2020. https://www.krem.com/article/traffic/new-stop-sign-on-grand-blvd-and-high-dr-confuses-some-drivers/293-3a2cc0f7-f238-43f9-b346-6770536f690d.

Watts, M.T. 1975. *Reading the Landscape of America.* New York: Macmillan.

Way, D. 1973. *Terrain Analysis: A Guide to Site Selection Using Aerial Photographic Interpretation.* Stroudsburg, PA: Dowden, Hutchinson, & Ross.

Wester, L. 1990. *Design Communication for Landscape Architects.* New York: Van Nostrand Reinhold.

Westmacott, R. 1992. *African-American Gardens and Yards in the Rural South.* Knoxville, TN: University of Tennessee Press.

Whyte, W. H. 2001. *The Social Life of Small Urban Spaces.* Ann Arbor, MI: Edward Brothers.

Wildlife Wonder. 2020. *Darwin's Thinking Path.* Viewed July 2020. http://www.wildlife-wonder.co.uk/darwins-thinking-path/#:~:-text=While%20working%20on%20his%20famous, as%20his%20'thinking%20path'.

Williams, W. C. 1965. Rain. In Alan, G. W., Rideout, W. and Robinson, J. *American Poetry.* New York: Harper & Row. 745–748.

Williams, R. 1973. *The Country and the City.* New York: Oxford University Press.

Williams, R. 1983. *Key words: a vocabulary of culture and society.* New York: Oxford.

Willis, P. 1977. *Learning to Labor: How Working-class Kids Get Working Class Jobs.* New York: Columbia University Press.

Wing, R. L. 1979. *The I Ching Workbook.* New York: Doubleday.

Wikipedia, 2016. *Structuralism.* Viewed April 24, 2016. https://en.wikipedia.org/wiki/Structuralism

Wittgenstein, L. 1958. *Philosophical Investigations.* Trans. Anscombe, G.E.M. New York: MacMillan.

Wolff, J. 1981. *The Social Production of Art*. New York: New York University Press.

Wolkomir, R. 1996. "Decibel by Decibel, Reducing the Din to a Very Dull Roar." In *Smithsonian* 26, no. 11: 56.

Wong, W. 1972. *Principles of Two-Dimensional Design*. New York: John Wiley and Sons.

Wong, W. 1977. *Principles of Three—dimensional Design*. New York: Van Nostrand Reinhold.

Wood, D. and Beck, R. 1994. *Home Rules*. Baltimore: Johns Hopkins University Press.

Yamin, R. and Metheny, K. B. (eds.) 1996. *Landscape Archaeology: Reading and Interpreting The American Historical Landscape*. Knoxville: University of Tennessee Press.

Young, D. 2015. "Charles Darwin's Daily Walks." *Psychology Today*. Viewed, July 2020, https://www.psychologytoday.com/us/blog/how-think-about-exercise/201501/charles-darwins-daily-walks

Zube, E. and Zube, M. (eds.) 1977. *Changing Rural Landscapes*. Amherst, MA: University of Massachusetts Press.

Zimring, C., Joseph, A., Nicoll, G., and Tsepas, S. 2005. "Influences of Building Design and Site Design on Physical Activity." *Am. Journal of Preventive Medicine*: Elsevier 28, no. 2S2: 186–193.

Zucker, P. 1959. *Town and Square*. Cambridge, MA: MIT Press.

# Index

Note: *Italic* page numbers refer to figures.